石油教材出版基金资助项目

高等院校特色规划教材

大学物理实验

（英汉双语 · 富媒体）

主　编　李东明
副主编　陈国祥　杨阿平　李娟妮

石油工业出版社

内 容 提 要

本书在参照教育部《理工科类大学物理实验课程教学基本要求》的基础上，充分考虑了复合型、应用型本科专业人才培养模式的特点及课程结构，采用英汉双语形式介绍了25个实验，包括基础性实验、综合性实验、设计性实验、研究性实验4个层次，形成了层次化、模块化的教学体系，让学生掌握基本知识、基本技能，提高学生综合应用能力及科学研究能力，培养学生的创新意识、创新精神和创新能力。本书体系完整、内容新颖，附有富媒体资源，方便学生学习。

本书可作为来华留学生的大学物理实验教材，也可作为国内学生的双语教材。

图书在版编目(CIP)数据

大学物理实验：英汉双语·富媒体/李东明主编.
—北京：石油工业出版社，2022.3
高等院校特色规划教材
ISBN 978-7-5183-5149-7

Ⅰ.①大… Ⅱ.①李… Ⅲ.①物理学—实验—高等学校—教材—汉、英 Ⅳ.①O4-33

中国版本图书馆CIP数据核字(2021)第273170号

出版发行：石油工业出版社
(北京市朝阳区安定门外安华里2区1号楼 100011)
网 址：www.petropub.com
编辑部：(010)64523579 图书营销中心：(010)64523633
经 销：全国新华书店
排 版：北京密东文创科技有限公司
印 刷：北京中石油彩色印刷有限责任公司

2022年3月第1版 2022年3月第1次印刷
787毫米×1092毫米 开本：1/16 印张：18
字数：490千字

定价：44.90元
(如出现印装质量问题，我社图书营销中心负责调换)

前言

“大学物理实验”是高等院校理工科专业的必修基础课程，是本科生接受系统实验方法和实验技能训练的开端。其目的是：培养学生的基本科学实验技能，提高学生的科学实验基本素质，使学生初步掌握实验科学的思想和方法；培养学生理论联系实际和实事求是的科学作风，认真严谨的科学态度，积极主动的探索精神，遵守纪律，团结协作，爱护公共财产的优良品德；培养学生的科学思维和创新意识，使学生掌握实验研究的基本方法，提高学生的分析能力和创新能力。随着国际交流日益频繁，越来越多的留学生到国内高等学校交流学习，但是适合留学生的物理实验的双语教材并不多。根据教育部《理工科类大学物理实验课程教学基本要求》，结合西安石油大学十多年的留学生“大学物理实验”课程的教学经验，笔者在原来英语物理实验讲义的基础上编写了这本教材。

本书采用英汉双语的形式，英文和中文部分均分四章：第一章介绍了物理实验的主要目的和要求；第二章介绍了误差理论、不确定度、有效数字概念、数据处理方法；第三章介绍了常用仪器和使用方法；第四章精选了 25 个实验，包括基础实验、综合实验、设计性实验和研究性实验，内容涉及力学、热学、电磁学、光学和近代物理实验。为了更好的阅读体验，书中的英文部分和中文部分并没有逐字逐句地对应，而是尽量发挥两种语言各自的优势。

本教材凝聚着西安石油大学物理实验中心承担留学生“大学物理实验”课程教师的共同努力和辛勤劳动。具体编写分工如下：(1) 英文部分，李东明编写 Chapter 1、Chapter 2 和 Chapter 4 的 Exp. 1、Exp. 2、Exp. 13、Exp. 16、Exp. 17、Exp. 19；陈国祥编写 Chapter 3；李娟妮编写 Chapter 4 的 Exp. 3；陈海霞编写 Chapter 4 的 Exp. 4；杨阿平编写 Chapter 4 的 Exp. 5、Exp. 14、Exp. 21、Exp. 22、Exp. 23、Exp. 24、Exp. 25；和万霖编写 Chapter 4 的 Exp. 6；刘帅编写 Chapter 4 的 Exp. 7；杨昌旗编写 Chapter 4 的 Exp. 8；禹大宽编写 Chapter 4 的 Exp. 9 和 Exp. 15；马成举编写 Chapter 4 的 Exp. 10、Exp. 11 和 Exp. 12；王向宇编写 Chapter 4 的 Exp. 18 和 Exp. 20。(2) 中文部分，李东明编写第一章、第二章、第三章、第四章的实验 1 ~ 实验 19、参考文献和附录；杨阿平编写第四章的实验 20 ~ 实验 25；实验的富媒体由白亮整理和制作。全书

由李东明统稿和担任主编，陈国祥、杨阿平和李娟妮担任副主编，陈国祥负责审核工作，李娟妮和杨阿平承担英文实验的修改和校稿工作。

本书在编写过程中参考了一些国内外相关高等院校的教材，在此谨致以深切的谢意。

由于编者的水平有限，本书难免有错误和不妥之处，敬请批评指正。

编　者

2021 年 10 月

英文部分(English part)

中文部分(Chinese part)

富媒体资源目录

Contents of media resources

英文部分

(English part)

Chapter 1 Introduction

Since Antiquity and until the Renaissance in order to understand natural phenomena, it was only necessary to develop a model, to make observations, to present a hypothesis and by reason alone to arrive to the model. Galileo is considered the father of modern science due to his insistence in introducing measurements, i. e., data acquisition and its subsequent analysis by mathematical formulation, in order to arrive to a verifiable model. Since then, at the core of all work in science lies a set of procedures that we call " the scientific method. " First there is an observation of natural phenomena, then a hypothesis is developed to explain it. In order to test this hypothesis an experiment is designed with the purpose of making precise and accurate measurements. The data obtained in the experiment is analyzed using the appropriate mathematical formulation.

The main objective of the method is to arrive to a conclusion that is verifiable, a theory in which the observations and the mathematical formulations agree, hopefully a theory that would fit within a larger model. Physics distinguishes itself from the other sciences by its strong emphasis and focus on precise and accurate measurements. Therefore, every experiment should follow this method. The corresponding laboratory report should also reflect the scientific method, as it will be explained in the following section.

The book has 25 experiments involving mechanics, electrics, magnetics, optics and modern physics. Physics experiments aim to provide understanding of various physical principles better, while introducing experimental objectives, apparatus, principle, procedures, data and calculations, precautions.

Experiment primarily consists of careful observations or measurements, recording the data in an orderly form, followed by data analysis, and finally drawing conclusions. Some details as to how to conduct the experiments are discussed in the introduction. The introduction is not trivial and you should not expect full comprehension after reading it once. With the help of your instructor you should strive to master these skills by the end of the fourth or fifth experiment.

1.1 Objectives of physical experiment

Physical experiment is a compulsory course for students to systematically acquire experimental methods and skills training after entering the university, which lays a necessary foundation for students to study experiments in subsequent courses and independently carry out engineering experiments in the future. The main objectives of the physical experiment class are (Video 1 - 1 - 1):

Video 1 - 1 - 1 Objectives of physical experiment

(1) Understand basic physical concepts and theories.

(2) Gain familiarity with a variety of instruments and learn to make reliable measurements.

(3) Learn how to make a measurement precisely with given instruments and the size of the measurement error.

(4) Learn how to do calculations and make the appropriate number of significant figures. The students should know how to process data, analyze experimental results and write experimental reports.

(5) Learn how to analyze data by calculations and plotting graphs that illustrate functional relations.

(6) Learn how to make an accurate and complete laboratory measurement.

1.2 Experimental report

Video 1 - 2 - 1 Experimental report

In general, loose-leaf paper is not appropriate for recording data or doing experimental calculations. It is recommended that you record data and do calculations directly in your specified experiment report (Video 1 - 2 - 1). Each experiment in your report should contain the following items.

(1) Title, time, partner.

(2) Objectives: summarize the purpose of the entire experiment in one or two sentences.

(3) Sketch the principle diagram, free handed, but with parts labeled.

(4) Data: all original data are to be recorded directly in the report where all can see them, not on scratch paper. Be sure to indicate clearly what is being measured and its unit.

(5) Measured quantities should include a figure of uncertainty or error.

(6) Results and conclusion: tell briefly what you did and how it came out. If you measured a physical constant, how does it compare with the accepted value in the light of your estimated errors?

1.3 Physical laboratory rules

(1) Safety is important firstly in the laboratory. You must ensure the safety of personnel and equipment during the experiment.

(2) Read experiment text book before entering the laboratory and take a seat appointed by

instructor, inspect instruments from the appearance, if any shortage, please report to the instructor in time, do not carry other group's instruments.

(3) During the experiment, we should have a serious and scientific attitude, abide by the discipline of the laboratory strictly, and work quietly in laboratory.

(4) Take good care of the instruments and equipment. If there is any damage or loss, please report to the instructor immediately. If the instrument is damaged due to carelessness or violation, compensation must be made as required.

(5) The circuit must be checked and approved by the instructor before the power can be turned on.

(6) After the experiment, the measured data should be checked and signed by the teacher. Before leaving the laboratory, the instruments should be arranged back.

(7) The experimental report should be finished and submitted on time.

Chapter 2 Measurement error and uncertainty analysis

Laboratory work in physics involves the use of apparatus to make measurement which can be used for two purposes, one is to precisely measure a physical quantity, such as the elementary electric charge, the wavelengths of lights, the refractive index of air, etc. and the other is to verify the laws of physics. In both cases, the observations must be faithfully recorded and should be as reliable as possible. We have only to examine anytime measurement carefully. Consider, for example, a carpenter who must measure the height of a doorway before installing a door. As a first rough measurement, he might simply look at the doorway and estimate its height as 210cm. This rough "measurement" is certainly subject to uncertainty. If pressed, the carpenter might express this uncertainty by admitting that the height could be anywhere between 205cm and 215cm. If he wanted a more accurate measurement, he would use a tape measure and might find the height is 211.30cm. This measurement is certainly more precise than his original estimate.

The verification of a physical law or the determination of a physical quantity involves measurements. A reading taken from the scale on a voltmeter, a stop-clock or a meter stick, for example, may be directly related by a chain of analysis to the quantity or law under study. Any uncertainty in these readings would result to an uncertainty in the final result. A measurement by itself, without a quantitative statement as to the uncertainty involved, is of limited usefulness. It is therefore essential that any course in basic laboratory technique include a discussion of the nature of the uncertainty in individual measurements and the manner in which uncertainties in two or more measurements are propagated to determine the uncertainty in the quantity or law being investigated.

In science, the word error does not carry the usual connotations of the terms mistake or blunder. Error in a scientific measurement means the inevitable uncertainty that attends all measurements. As such, errors are not mistakes; you cannot eliminate them by being very careful. What you can do is to ensure that errors are as small as reasonably possible and to have a reliable estimate of how large they are. Most textbooks introduce additional definitions of error, and these are discussed later. It should be noted error and uncertainty are different concepts, and they are both different and related, and cannot be confuses or replace one another.

2.1 Types of experimental errors

According to the measurement laws that errors follow, there are two types of experimental errors in physical experiments, one is systematic error and the other is random error.

Systematic errors are due to identifiable causes and can, in principle, be eliminated. Errors of this type result in measured values which are consistently too high or consistently too low. Four possible sources of systematic error are as follows:

(1) Instrumental errors, e. g. a poorly calibrated instrument such as a thermometer that reads 102℃ when immersed in boiling water and 2℃ when immersed in ice water at atmospheric pressure. Such a thermometer would result in measured values which are consistently too high.

(2) Human errors, e. g. parallax in reading a meter scale.

(3) Environmental errors, e. g. an electrical power "brown out" which causes measured currents to be consistently too low.

(4) Theoretical errors, e. g. due to simplifications of the model system or approximation in the equations describing it, e. g., if a frictional force is acting during the experiment but such a force is not included in the theory, then the theoretical and experimental results will consistently disagree.

The law and source of systematic error may be known, which is called determinable systematic error. It may also be unknown, known as undetermined systematic error. For the former, certain measures can generally be taken to reduce, eliminate, or correct in the measurement results, while for the latter, it is generally difficult to correct, and only the range of its value can be estimated.

Random errors do not have a definite change rule, and are completely random. However, when a measured quantity is measured under the condition of equal precision and the number of measurements is sufficient, the random error generally obeys the statistical law.

(1) Observational errors, e. g. errors in judgment of an observer when reading the scale of a measuring device to the smallest division.

(2) Environmental errors, e. g. unpredictable fluctuations in line voltage, temperature, or mechanical vibrations of equipment.

Random errors, unlike systematic errors, can often be quantified by statistical analysis; therefore, the effects of random errors on the quantity or physical law under investigation can often be determined.

The distinction between random errors and systematic errors can be illustrated with the following example. Suppose the measurement of a physical quantity is repeated five times under the same conditions. If there are only random errors, then the five measured values will be spread about the "true value"; some will be too high and others too low as shown in Fig. 2 - 1 (a). If in addition to the random errors there is also a systematic error, then the five measured values will be spread, not about the true value, but about some displaced value as shown in Fig. 2 - 1 (b).

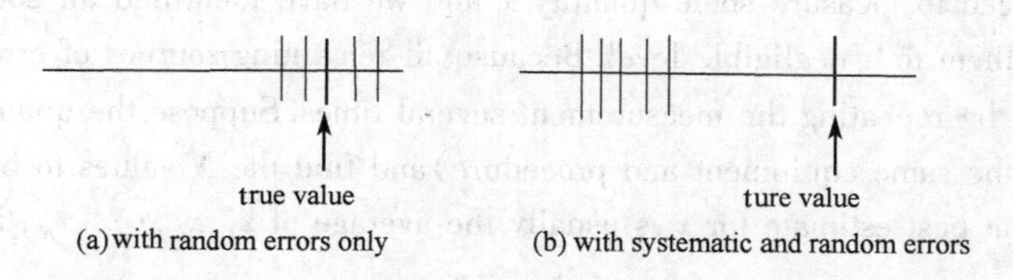

Fig. 2 - 1 - 1 Set of measurements with random and systematic errors

2.2 Error calculation

Error is defined as the result of a measurement minus the true value of the measure. It is sometimes called absolute error.

$$\text{Error} = \text{Result of measurement} - \text{True value} \tag{2-2-1}$$

It should be noted that absolute error can be positive or negative and should not be confused with absolute value of error, which is the modulus of the error.

True value is consistent with the definition of a given particular quantity. It means that it is a value that would be obtained by a perfect measurement and by nature indeterminate.

Relative error is the error of measurement divided by a true value of the measurement. It is commonly expressed as a percentage and is also called percent error of an experimental value.

$$\text{Relative error} = \frac{\text{Absolute error}}{\text{True value}} \times 100\% \tag{2-2-2}$$

In both formula(2 - 2 - 1) and formula(2 - 2 - 2) a conventional true value or accepted value is used since a true value cannot be determined.

For example a cylindrical object is measured to have a diameter d of 20.24mm and a circumference c of 63.85mm. What is the value of π determined from the measurements? What are the absolute error and the relative error, respectively, if the accepted value of π with two decimal places is 3.14?

Solution with $d = 20.24$mm and $c = 63.85$mm

$$c = \pi d \text{ or } \pi = \frac{c}{d} = \frac{63.85}{20.24} = 3.15 \tag{2-2-3}$$

then

$$\text{Absolute error} = \text{Result of measurement} - \text{True value} = 3.15 - 3.14 = 0.01(\text{mm})$$

and

$$\text{Relative error} = \frac{\text{Absolute error}}{\text{True value}} \times 100\% = \frac{0.01}{3.14} \times 100\% = 0.3\% \tag{2-2-4}$$

2.3 Statistical analysis of random errors

2.3.1 The mean and the standard deviation

Suppose we need to measure some quantity x and we have identified all sources of systematic error and reduced them to a negligible level. Because all remaining sources of errors are random, we should detect them by repeating the measurement several times. Suppose the quantity x be measured N times (all using the same equipment and procedure) and find the N values to be, $x_1, x_2, \cdots, x_N$.

Once again, the best estimate for x is usually the average of $x_1, x_2, \cdots, x_N$, that is

$$x_{\text{best}} = \bar{x} \tag{2-3-1}$$

$$\bar{x} = \frac{x_1 + x_2 + \cdots + x_N}{N} = \frac{\sum_{i=1}^{N} x_i}{N} \tag{2-3-2}$$

Given that the mean $\bar{x}$ is our best estimate of the quantity x, it is natural to consider the difference $d_i = x_i - \bar{x}$. This difference, often called the deviation (or residual) of x_i from $\bar{x}$, tells us how much the i th measurement x_i differs from the average of x. If the deviations $d_i = x_i - \bar{x}$ are all very small, our measurements are all close together and presumably very precise. If some of the deviations are large, our measurements are obviously not precise. Now that we have determined the "best value" for the measurement $\bar{x}$, and then we need to estimate the error in this value. The random deviation of multiple independent measurements data can be approximated as a normal distribution. We start by defining one way in which the spread of data about the mean value can be characterized.

The standard deviation σ is defined as

$$\sigma = \sqrt{\frac{1}{N-1}\sum_{i=1}^{N}(x_i - \bar{x})^2} \qquad (2-3-3)$$

If the standard deviation is small, then the spread in the measured values about the mean is small, hence the precision in the measurements is high. Note the standard deviation is always positive and it has the same units as the measured values. σ_m, which is defined to be

$$\sigma_m = \frac{S}{\sqrt{n}} \qquad (2-3-4)$$

Where σ is the standard deviation and n is the total number of measurements.

The result to be reported is then

$$\bar{x} \pm \sigma_m \qquad (2-3-5)$$

The interpretation of equation is that the measured value probably lies in the range from $\bar{x} + \sigma_m$ to $\bar{x} - \sigma_m$.

2.3.2 Normal distribution

If the measurements of x are subject to many small random deviations but negligible systematic deviations, for a very large number of measurement, the normal, or Gaussian distribution is the theoretical distribution of measured values x about the mean value $\bar{x}$. If the measurements are carried out with high precision, then σ will be small and the normal or Gaussian distribution will be sharply peak at the mean value $\bar{x}$ See Fig. 2-3-1.

$$f(x) = \frac{1}{\sigma\sqrt{2\pi}} e^{-(x-\bar{x})^2/(2\sigma^2)} \qquad (2-3-6)$$

Where $\bar{x}$ is the best estimate value of x, center of distribution, mean value after many measurements, and σ is width parameter of distribution, standard deviation after many measurements.

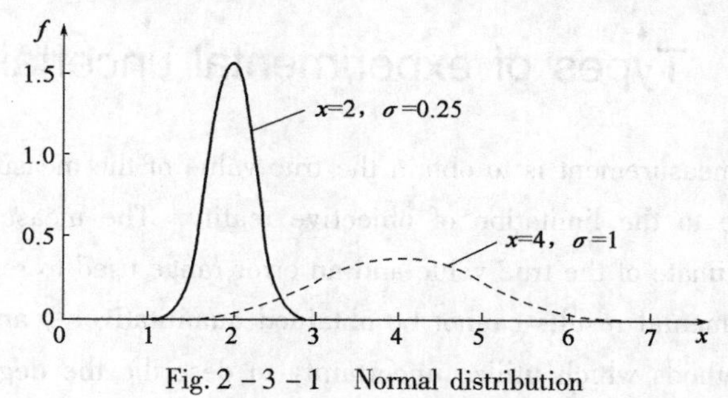

Fig. 2-3-1 Normal distribution

The confidence probability of obtaining the value x as a result of any single measurement is given by $f(x)$. Note that the most probable value resulting from any single measurement is the mean value $\bar{x}$. The normal distribution has the property that 68% of the measurements will fall within the range from $\bar{x}-\sigma$ to $\bar{x}+\sigma$, and 95% of the measurements will fall within the range from $\bar{x}-2\sigma$ to $\bar{x}+2\sigma$.

The probability of a single measurement falling within t standard deviations of x is

$$prob(\text{within} - t\sigma) = \frac{1}{\sqrt{2\pi}}\int_{-t}^{t} e^{-z^2/2}dz \quad (2-3-7)$$

The formula (2 – 3 – 7) can be obtained when the number of measurements is large, the data distribution can be considered as a condition of the normal distribution. In fact, the number of measurements is often less 10 times, and in this case, the measurement results are$t$ – distribution, for the t – distribution , the confidence factor is $t_P(n)$, Tab. 2 – 3 – 1 illustrates the relationship between the factors and the different times.

Tab. 2 – 3 – 1 t – distribution $t_P(n)$

P \ n	2	3	4	5	6	7	8	9	10	15	20	∞
0.997	235.8	19.21	9.22	6.62	5.51	4.90	4.53	4.28	4.09	3.64	3.45	3.00
0.950	12.70	4.30	3.18	2.78	2.57	2.45	2.36	2.31	2.26	2.14	2.09	1.96
0.900	6.31	2.92	2.35	2.13	2.02	1.94	1.90	1.86	1.83	1.76	1.72	1.65
0.683	1.84	1.32	1.20	1.14	1.11	1.09	1.08	1.07	1.06	1.04	1.03	1.00

2.3.3 Accuracy and Precision

Accuracy and precision are commonly used synonymously in daily life, but in experimental measurements there is an important distinction. The accuracy of a measurement signifies how close it comes to the true value, that is how nearly correct it is.

Precision refers to the agreement among repeated measurements, which is the spread of the measurements or how close they are together. The more precise a group of the measurements the closer together they are. However, a good precision does not necessarily imply good accuracy

Obtaining greater accuracy for an experimental value depends in general on minimizing systematic errors. Obtaining greater precision for an experimental value depends on minimizing random errors.

2.4 Types of experimental uncertainty

The purpose of the measurement is to obtain the true value of the measured value, but the true value is not accurate due to the limitation of objective reality. The measurement result can only obtain an approximate estimate of the true value and an error range used to represent the approximate degree, so that the experimental results cannot be obtained quantitatively, and possess uncertainty.

According to the methods which utilize uncertainty to describe the degree of reliability of the measurement [(INC - 1) 1980] [which was recommended and promoted by the Bureau International des Poids et Mesures (BIPM) in 1980], China's measurement departments have also developed a series of norms which are consistent with it. These norms cover almost all areas of the metering system, and they generally use an uncertainty representation system.

The measurement uncertainty is derived from a number of factors, so it consist of multiple components. Some of these components can be evaluated through a series of statistical analyses of the results of multiple measurements under the same conditions. This method is known as type A Uncertainties, symbolized u_A. Other components are evaluated by non - statistical methods, known as type B uncertainties, symbolized μ_B.

2.5 Type A evaluation of uncertainty

Uncertainties are classified into two groups: the random uncertainties (Type A uncertainty), which can be treated statistically, and the systematic uncertainties (Type B uncertainty). If $x_1, x_2, \cdots, x_n$ are the results of n measurements of the same quantity x, then, as we have seen, our best estimate for the quantity x is their mean $\bar{x}$. We have also seen that the standard deviation (σ_x characterizes the average error of the separate measurements $x_1, x_2, \cdots, x_n$). Our answer, $x_{best} = \bar{x}$, however, represents a judicious combination of all n measurements, and we express type A uncertainty by

$$u_A = t_P \cdot \sqrt{\frac{\sum_{i=1}^{n} (X_i - \overline{X})^2}{n(n-1)}} \qquad (2-5-1)$$

Where t_P is variate with measuring times and confidence probability.

This integral is often called the error function or the normal error integral.

2.6 Type B evaluation of uncertainty

In physical experiments, many of the experiments are conducted only once and the Type A uncertainty is zero. Here, we introduce the Type B standard uncertainty, or simply Type B uncertainty, to evaluate the uncertainty of the measurement. The uncertainties which are not in accordance with the statistic principles are called the Type B uncertainty, symbolized u_B. There are many components resulting in the Type B uncertainty. In this course, we only consider the instrument error and the reading error, which are denoted by Δ_{ins} and Δ_{read}, respectively.

The Type B uncertainty resulting from the instrument error is shown here

$$u = \frac{\Delta_{ins}}{C} \qquad (2-6-1)$$

where, C is the confidence probability. In this book, $\sqrt{3}$ is adopted for C. Therefore, we use the following formula to calculate Type B uncertainty resulting from the instrument error.

$$u_B = \frac{\Delta_{ins}}{\sqrt{3}} \qquad (2-6-1)$$

2.7 Combined standard uncertainty

Combining the Type A standard uncertainty u_A and the Type B standard uncertainty u_B, the uncertainty of a direct measurement u is found as

$$u = \sqrt{u_A + u_B} \tag{2-7-1}$$

2.8 Expression of the measurement result

The measurement result for one direct measurement quantity or one indirect measurement quantity can be reported in the form:

$$x = \bar{x} \pm 2u \quad (P = 95\%) \tag{2-8-1}$$

Where, $\bar{x}$ is the best estimate value, which also refers to the mean or average of the measurement, and $2u$ is the uncertainty of the measurement.

The physical meaning of the final expression of the measurement is that the probability of the true value lying in the range from $\bar{x} + 2u$ to $\bar{x} - 2u$ is 95%.

2.9 Propagation of uncertainties

Propagation of uncertainties is simply a method to determine the uncertainty in a value where the value is calculated using two or more measured values with known estimated uncertainties. This method will be discussed separately for addition and subtraction of measurements, and for multiplication and division of measurements.

2.9.1 Addition and subtraction of measurements

Suppose x, y and z are three measured values and estimated uncertainties are u_x, u_y and u_z. The results of the three measurements would be reported in the form ($x \pm u_x$, $y \pm u_y$, $z \pm u_z$), where each estimated uncertainty may be the smallest division of the measuring instrument.

If W, a value to be calculated from the measurements, is defined to be,

$$W = x \pm y \pm z \tag{2-9-1}$$

Statistical analysis shows that a better approximation is to write u_W as the square root of the sum of estimated uncertainties squared,

$$u_W = \sqrt{u_x^2 + u_y^2 + u_z^2} \tag{2-9-2}$$

Express the final result,

$$W = \overline{W} \pm 2u_W \quad (P = 95\%) \tag{2-9-3}$$

2.9.2 Multiplication and division of measurements

The area A of a rectangle of with W and height h is $W \times h$. Suppose we measure the width and height, and estimate their uncertainties. Then we know $W \pm u_W$ and $h \pm u_h$, and we want to calculate $A \pm u_A$.

To determine u_A, we first use differential calculus to obtain the differential area dA,

$$dA = h dW + W dh \tag{2-9-4}$$

Dividing both sides by A, $A = W \times h$,

$$\frac{u_A}{A} = \frac{u_W}{W} + \frac{u_h}{h} \tag{2-9-5}$$

As before, statistical analysis shows that a better estimation of the fractional in area, u_A/A, is

$$\frac{u_A}{A} = \sqrt{\left(\frac{u_W}{W}\right)^2 + \left(\frac{u_h}{h}\right)^2} \tag{2-9-6}$$

When measurements are multiplied or divided, the largest fractional uncertainty in the measurements becomes the fractional uncertainties in the result.

Express the final result,

$$A = \bar{A} \pm 2u_A \quad (P = 95\%) \tag{2-9-7}$$

2.10 Significant figures

The significant figures in a number are all figures that are obtained directly from the measuring process and exclude those zeros which are included solely for the purpose of locating the decimal point. This definition will be illustrated with a number of example(Tab. 2 – 10 – 1).

Tab. 2 – 10 – 1 Significant figures examples

Number	Number of significant figures	Remarks
2	1	Implies = 25% precision
2.0	2	Implies = 2.5% precision
2.00	3	Implies = 0.25% precision
0.1256	4	Leading zero is not necessary, but it does make the reader notice the decimal point
2.123	4	
2.123×10^3	4	
510	3	
5.10×10^2	3	
5.1×10^2	2	

A measurement and its experimental uncertainty should have their last significant digits in the same location(relative to the decimal point). Examples:

$$54.1 \pm 0.1, \quad 121 \pm 4, \quad 8.764 \pm 0.003, \quad (7.83 \pm 0.05) \times 10^2.$$

Uncertainties are usually only known roughly, so quote the uncertainties to no more than two significant figures. One significant figure is usually sufficient. Thus, if some calculation yields acceleration of gravity $g = 9.82 m/s^2$ and its uncertainty $u_g = 0.04385 m/s^2$, the uncertainty should be rounded to $u_g = 0.04 m/s^2$, and the conclusion should be rewritten as

$$g = (9.82 \pm 0.04) m/s^2$$

Properly, the correct number of significant figures to which a result should be quoted is obtained via uncertainty analysis. However, uncertainty analysis takes time, and frequently in actual laboratory practice it is postponed. In such a case, one should retain enough significant figures that round off error is no danger, but not so many as to constitute a burden, here is an example:

$0.91 \times 1.23 = 1.1$ Wrong

In this case, the numbers 0.91 and 1.23 are known to about 1%, where as the result, 1.1 is defined to about 10%. In this extreme case, the accuracy of the result is reduced by almost a factor of ten, due to round off error. Now, a factor of ten in accuracy is usually precious and expensive, and it must not be thrown away by careless data analysis.

$0.91 \times 1.23 = 1.1193$ Wrong

The extra digits, which are not really significant, are just a burden, and in addition they carry the incorrect implication of a result of absurd accuracy.

$0.91 \times 1.23 = 1.12$ Okey

$0.91 \times 1.23 = 1.192$ Less good, but still acceptable

In multiplication or division it is often acceptable to keep the same number of significant figures in the product or quotient as are the least precise factor.

Handle significant figure in addition and subtraction will be illustrated with examples:

$51.4 - 1.67 = 49.73 \to 49.7$, $7146 - 12.8 = 7133.2 \to 7133$,

$20.8 + 18.72 + 0.851 = 40.371 \to 40.4$, $1.4693 + 10.18 + 1.062 = 12.7113 \to 12.71$.

2.11 The rounding off rule

The mean of the seven measurements was calculated by dividing 51.567m by 7, yielding on calculator 7.366714286m. Recording that number as the mean would be improper because it contains more than one uncertain digit. In fact, it is discovered that the uncertainty in the mean is 0.001m. the last six digits (714286) are insignificant and meaningless. Consequently, if a computation yields a number with more digits than are justified, the number should be rounded off. To drop insignificant digits from a number, use the following rules:

(1) Drop all digits after the desired number of significant digits.

(2) If the leftmost digit to be dropped is greater than 5, or 5 followed by other digits, increase the last retained digit by one.

(3) If less than 5, leave the last retained digit unchanged.

(4) If exactly 5, increase the last retained digit by one if it is odd and leave it unchanged if it is even.

The reason for rule (4) is that consistently incrementing the least significant digit if the leftmost digit to be dropped is exactly 5 would lead to a systematic error. In this way, the least significant digit is increased approximately half the time and left unchanged the other half.

2.12 Graphical analysis

It is often desirable to find the relationship between measured variables. One good way to do this is to plot a graph of the data and then analyze The graph. The guideline should be followed in plotting your data. The example is illustrated in Fig. 2 – 12 – 1.

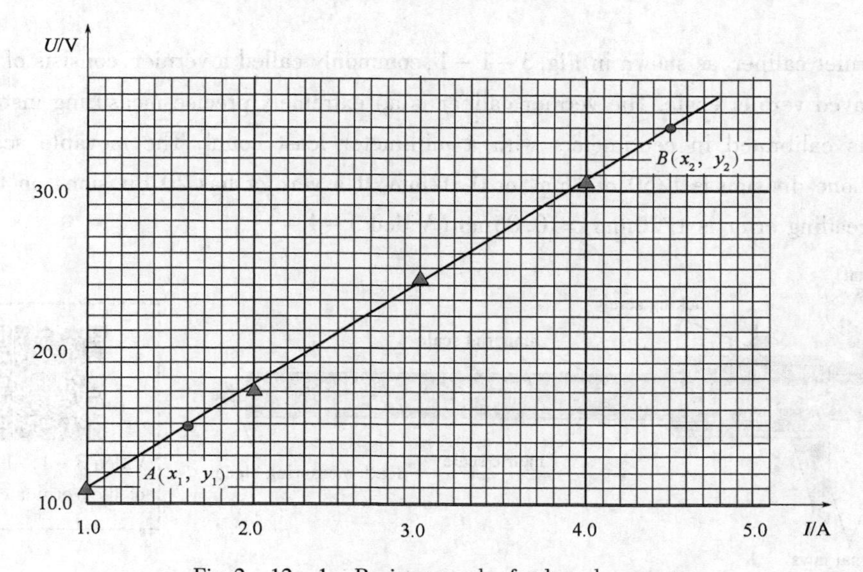

Fig. 2 – 12 – 1 Resistor graph of volt and current

(1) Use a sharp pencil or pen. A broad tipped pencil or pen will introduce unnecessary inaccuracies.

(2) Draw your graph on a full page of your notebook. A compressed graph will reduce the accuracy of graphical analysis.

(3) Give the graph a concise title.

(4) The dependent variable should be plotted along the vertical (y) axis and the independent variable (x) axis.

(5) Label axes and include units.

(6) Select a scale for each axil and start each axil at zero, if possible.

(7) Use error bars to indicate errors in measurements, e. g.

(8) Draw a smooth curve through the data points. If the errors are random, then about 1/3 of the point will not lie within their error range of the beat curve.

Chapter 3 Introduction of basic instruments

3.1 Vernier caliper

The vernier caliper, as shown in Fig. 3 – 1 – 1, commonly called a vernier, consists of a rule with a main engraved vernier scale. The vernier caliper is an extremely precise measuring instrument; the main scale is calibrated in centimeters with a millimeter least count. The movable scale has 50 division and one division is 1/50 of 1mm, or 0.02mm. If a vernier has 20 divisions on the Vernier scaler. The reading error is 1/20mm = 0.05mm (Video 3 – 1 – 1).

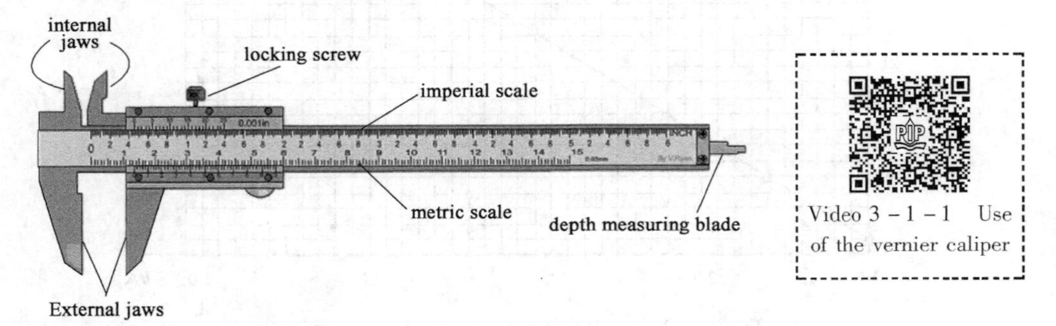

Video 3 – 1 – 1 Use of the vernier caliper

Fig. 3 – 1 – 1 Structure of vernier caliper

If you are measuring something with a round cross section, make sure that the axis of the object is perpendicular to the caliper. This is necessary to ensure that you are measuring the full diameter and not merely a chord. Notice that there is a fixed scale and a sliding scale. The boldface numbers on the fixed scale are centimeters. The tick marks on the fixed scale between the boldface numbers are millimeters. There are 50 tick marks on the sliding scale. The left-most tick mark on the sliding scale will let you read from the fixed scale the number of whole millimeters that the jaws are opened (Fig. 3 – 1 – 2).

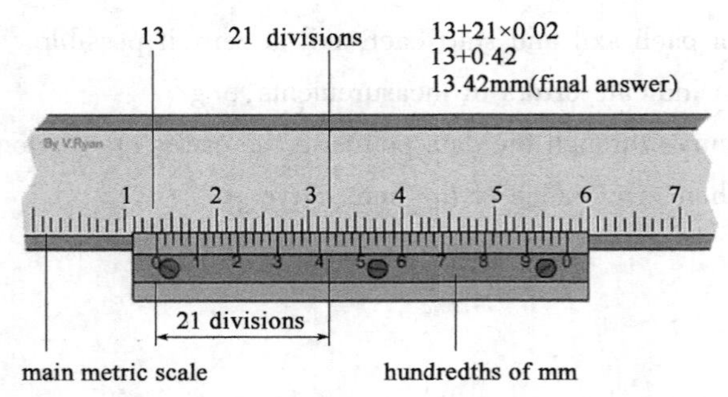

Fig. 3 – 1 – 2 Example of reading the vernier scale

In the example below, the zero scale line on the sliding scale is between 13 mm and 14 mm, so the number of whole millimeters is 13.

Next we find the 50ths of millimeters. Notice that the 50 tick marks on the sliding scale are the same width as 49 ticks marks on the fixed scale. This means that at most one of the tick marks on the sliding scale will align with a tick mark on the fixed scale; the others will miss.

The number of the aligned tick mark on the sliding scale tells you the number of 50ths of millimeters. In the example above, the 21th tick mark on the sliding scale is in coincidence with the one above it, so the caliper reading is (13.42 ± 0.02) mm.

3.2 Micrometer screw-gauge

Micrometer screw-gauge is an accurate instrument used for measuring the diameter of a thin wire or the thickness of a sheet of metal.

It consists of a U-shaped frame fitted with a screwed spindle which is attached to a thimble, as shown Fig. 3 – 2 – 1.

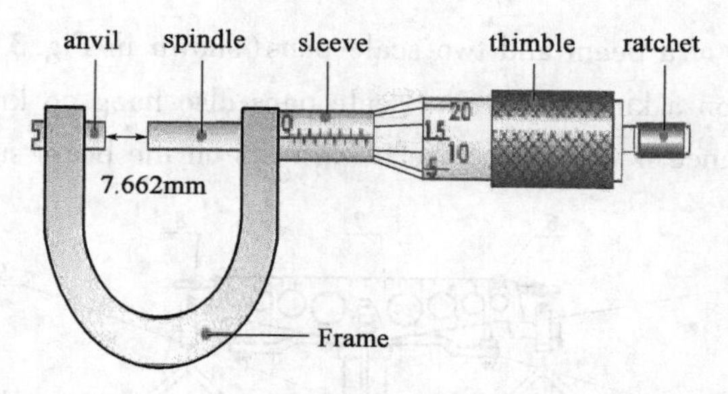

Fig. 3 – 2 – 1 Structure of Screw gauge

The screw has a known pitch such as 0.5mm. Pitch of the screw is the distance moved by the spindle per revolution. Hence in this case, for one revolution of the screw the spindle moves forward or backward 0.5mm. This movement of the spindle is shown on an engraved linear millimeter scale on the sleeve. On the thimble there is a circular scale which is divided into 50 or 100 equal parts.

When the anvil and spindle end are brought in contact, the edge of the circular scale should be at the zero of the sleeve (linear scale) and the zero of the circular scale should be opposite to the datum line of the sleeve. If the zero is not coinciding with the datum line, there will be a positive or negative zero error as shown in Fig. 3 – 1 – 2 (Video 3 – 2 – 1).

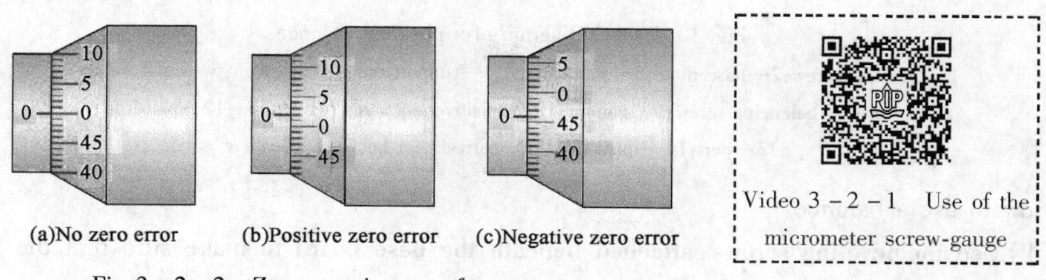

Fig. 3 – 2 – 2 Zero error in case of screw gauge

Video 3 – 2 – 1 Use of the micrometer screw-gauge

While taking a reading, the thimble is turned until the wire is held firmly between the anvil and the spindle.

The least count of the micrometer screw can be calculated using the formula given below:

Least count = Pitch/Number of divisions on the circular scale = 0.5 mm/50 = 0.01 mm

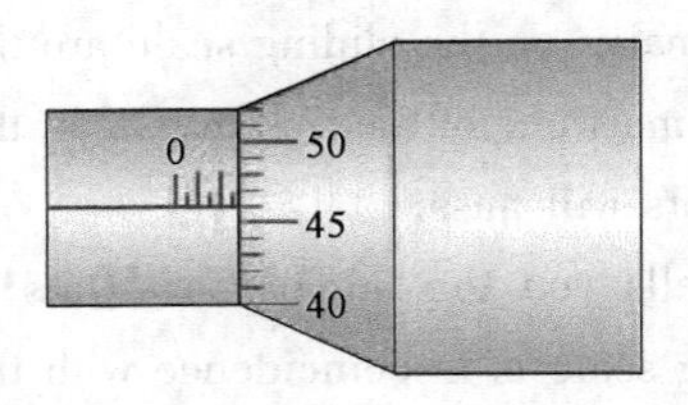

Fig. 3-2-3 Measured sample

Suppose measure diameter of a wire, the wire whose thickness is to be determined is placed between the anvil and spindle end, the thimble is rotated till the wire is firmly held between the anvil and the spindle. The ratchet is provided to avoid excessive pressure on the wire. It prevents the spindle from further movement. The thickness of the wire could be determined from the reading as shown in Fig. 3-2-3.

Reading = Linear scale reading + (Coinciding circular scale × Least count)

= 2.5 mm + (460 × 0.01) = (2.5 + 0.460) mm = 2.960 mm

3.3 Physical balance

This balance consists of a beam and two scale pans (shown in Fig. 3-3-1), the beam being balanced at its mid point on a knife-edge. The scale pans also hang on knife edges and rest on the base board. When the balance is not in use the beam rests on the beam support.

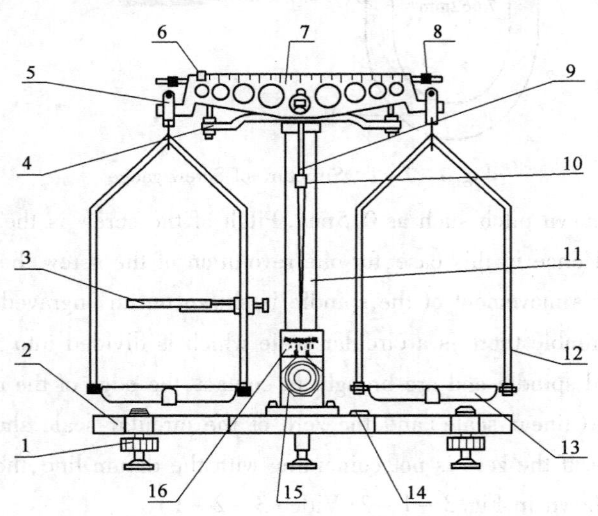

Fig. 3-3-1 Diagram of a physical balance

1—levelling screw; 2—base plate; 3—stand frame; 4—support frame; 5—stirrup; 6—small weight; 7—beam; 8—balancing screw; 9—pointer; 10—sensitiveness screw; 11—pillar; 12—pan frame; 13—pan; 14—spirit level; 15—arrestment knob; 16—scale

How to use a balance?

(1) Use the leveling screws, attached beneath the base board to make sure that the beam is horizontal. It can be verified with the help of the plumb-line provided shown in the diagram.

(2) Use the arrestment knob to raise the beam and the adjusting screw at the two ends of the beam, to bring the pointer to the middle or zero mark on the scale.

(3) Lower the beam using the arrestment knob again.

(4) Place the body to be weighed on the left scale pan and put weights on the right hand scale pan to balance the beam (when pointer is at zero).

3.4 Traveling microscope

A traveling microscope is an optical instrument which used to measure small distances or changes in small distances, such as hole diameter, and line width. A traveling microscope is composed of mechanical part and optical part. Its structure is shown in Fig. 3 – 4 – 1. The optical part is a long focal length microscope mounted on a sliding table driven by a screw which mounted on the base. The sliding table and microscope can be moved forward, up and down, left and right. The mechanical part is manufactured according to the micrometer principle. There are 100 sectors on a scale connected with a screw with 1mm, and the measuring precision of a traveling microscope is 0.01mm.

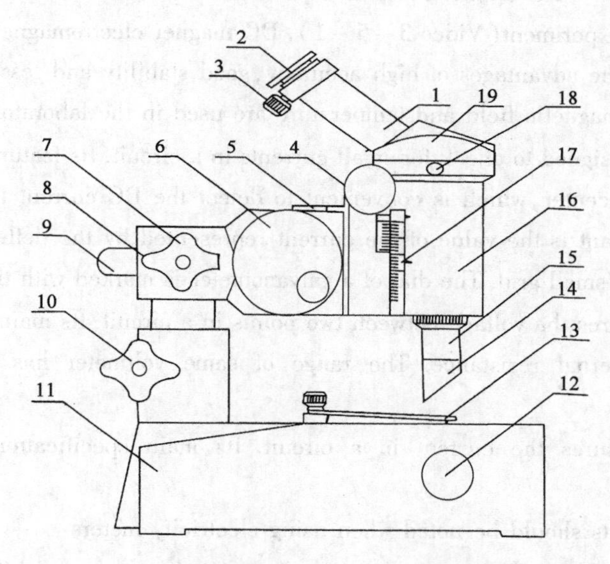

Fig. 3 – 4 – 1 Structure of a traveling microscope

1—eyepiece tube; 2—eyepiece; 3—lock screw; 4—focus wheel; 5—scale; 6—measurement wheel; 7—lock wheel I; 8—shaft; 9—square shaft; 10—lock wheel Ⅱ; 11—base; 12—mirror wheel; 13—pressing piece; 14—half reflection mirror; 15—objective lens; 16—len tube; 17—scale; 18—lock screw; 19—prism chamber

When measuring with a traveling microscope, the measured object is first placed on the carrier and fixed with a pressing piece firstly, then the traveling microscope is adjusted and aligned with the measured object. Adjust the eyepiece of the traveling microscope to make the fork filament clear. Secondly, adjust the focusing condition of the microscope or move the whole instrument to make the measured object image clear and eliminate parallax. Then align the fork hairs to a point (or a line) on the subject and take the reading; Turn the wheel, aim at another point, and record the reading. Finally, the difference between the two readings is the length of the measured object.

Precautions for using a traveling microscope:

(1) The focusing process should be adjusted so that the eyepiece can be focused slowly from bottom to top. It is prohibited to adjust the focusing of the eyepiece in reverse.

(2) When using a traveling microscope, the direction of the microscope and the measured points should be parallel.

(3) When measuring, the wheel should be rotated in the same direction. When the moving fork hairs exceeds the target, it is necessary to turn back a little more, and then rotate the wheel in the same direction again to align the target to avoid the backlash error caused by changing direction.

(4) The objective lens and eyepiece should not be touched by hand. Only use lens paper to clear.

3.5 Electric meters

Video 3 – 5 – 1 Use of the electric meters

The ammeter, voltmeter and galvanometer are commonly used in physical experiment (Video 3 – 5 – 1). DC magnet electromagnetic meters which have the advantages of high accuracy, good stability and less influence by external magnetic field and temperature are used in the laboratory.

A galvanometer designed to check for small currents in a circuit. Its feature is that the zero point of the pointer is in the center, which is convenient to detect the DC current in different directions. The galvanometer constant is the value of the current represented by the deflection of a small grid, generally about 10^{-5} A/small grid. The dial of a galvanometer is marked with the letter "G" usually.

A voltmeter measures the voltage between two points in a circuit. Its main specifications include the range and the internal resistance. The range of same voltmeter has the different internal resistance.

An ammeter measures the current in a circuit. Its main specifications include range and resistance.

The following points should be noted when using electricity meters:

(1) Ammeter connection, the ammeter must be connected in series, and the voltmeter should be connected in parallel with both ends of the measured voltage.

(2) Direction current, it is necessary to pay attention to the "+" mark of the terminal column on the electricity meter to indicate the current flowing in, and the "–" mark to indicate the current flowing out.

(3) Selection of measuring range, during measurement, the value to be measured should be estimated in advance. Choose a larger measuring range and try to measure it.

[Supplementary explanation]

Electrical instrument accuracy (accuracy level), according to the national standard, instrument accuracy is classed the following seven accuracy level respectively: 0.1, 0.2, 0.5, 1.0, 1.5, 2.5, 5.0.

The maximum absolute error of measurement under specified conditions with a one way scale is:

$$\Delta_{max} = x_n S\% \qquad (3-5-1)$$

Where, x_n is the measuring range of the instrument, and S is the accuracy level. The maximum relative error of a given value x in measurement is:

$$k = \frac{\Delta_{\max}}{x} = \frac{x_n S\%}{x} \tag{3-5-2}$$

It can be seen that the precision of this kind of instrument is close to the precision of the instrument only when it is near the full range. Therefore, when used, it is generally necessary to make the pointer work between 2/3 and full scale of the dial.

3.6 Sliding rheostat

The structure of the sliding rheostat is shown in Fig. 3 – 6 – 1. The resistance wire is a constantan wire that has been oxidized and insulated. The wire is tightly winded on the porcelain tube, and a metal protection bracket is installed outside. In Fig. 3 – 6 – 1, A and B are the two fixed ends, and C is the sliding end, which can be used for current limiting, voltage distributer and variable resistor to replace the undetermined resistance in the circuit. The main technical indicators of sliding rheostat are total resistance and rated current (power). It should be selected according to the size and adjustment requirements of the external load. It note especially that the current through any part of the rheostat is not allowed to exceed its rated current (Video 3 – 6 – 1).

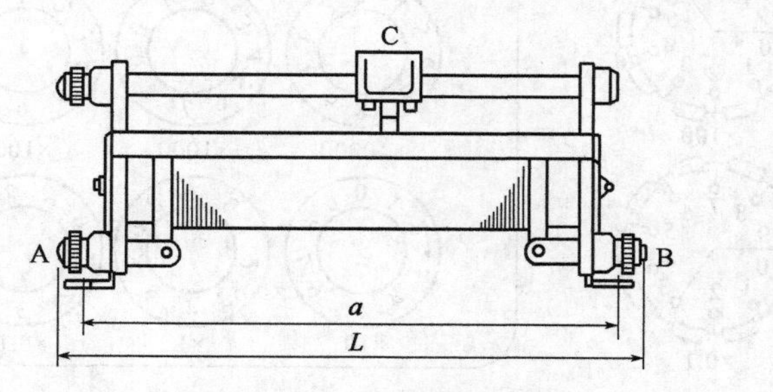

Fig. 3 – 6 – 1 Sliding rheostat structure graph

Video 3 – 6 – 1 Use of teh sliding rheostat

Sliding rheostat is commonly used in the laboratory to change the current or voltage in the circuit, and is connected to a current – limiting circuit and a voltage distributer circuit respectively. Fig. 3 – 6 – 2 shows the current limiting circuit. When sliding C, the entire circuit resistance is changed and the current is changed accordingly, so it is called the current limiting circuit. Before switching on the power supply, C should generally slide to the B end to make the current minimize. After switching on the power supply, the resistance can be gradually reduced to increase the current to the required value. As shown in Fig. 3 – 6 – 3, the voltage at both ends of the load changes when the position of sliding end C is changed, so it is called the voltage distributer circuit. Similar to the current limiting circuit, in order to ensure safety, before switching on the power supply, C should generally slide to the B end to make the output voltage minimize. After switching on the power supply, the resistance can be gradually increased to increase the voltage to the required value.

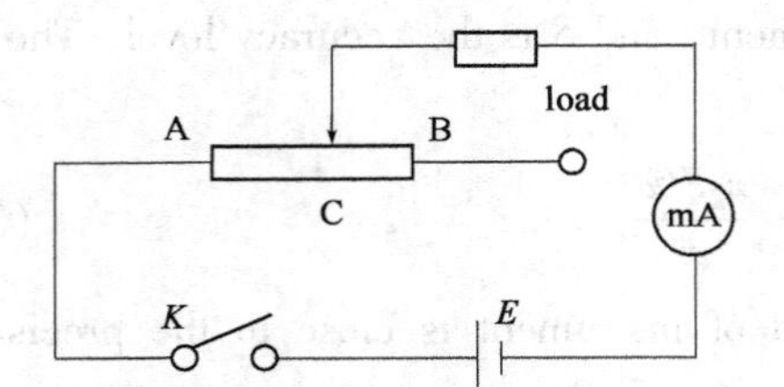

Fig. 3 – 6 – 2 Current-limiting circuit

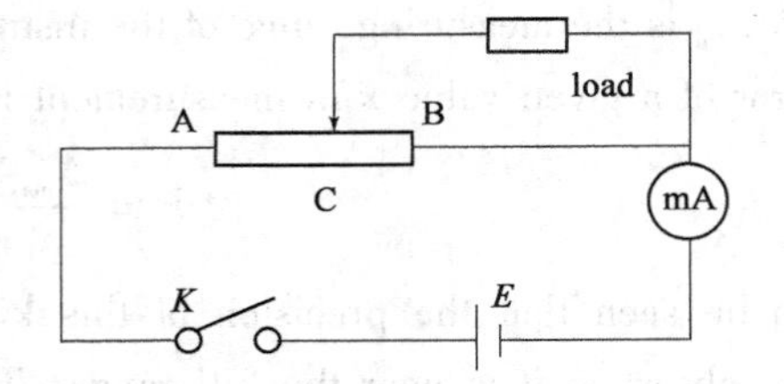

Fig. 3 – 6 – 3 Voltage distributor circuit

3.7 Resistance box

Video 3 – 7 – 1 Use of the resistance box

Fig. 3 – 7 – 1 and Fig. 3 – 7 – 2 are the inside circuit schematic diagram of the rotary resistance box and its Outline picture. The resistance is made of high stable manganese-copper alloy wire with low zero resistance. The resistance value between two terminals can be changed by turning the rotary knob, which can be adjusted in the DC circuit (Video 3 – 7 – 1).

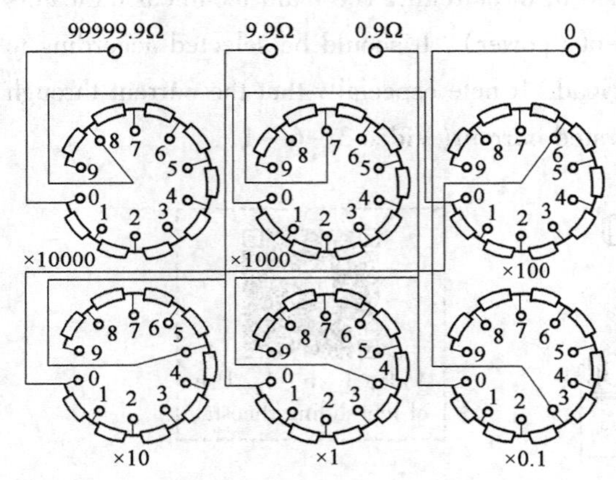

Fig. 3 – 7 – 1 Inside circuit schematic diagram of resistance box

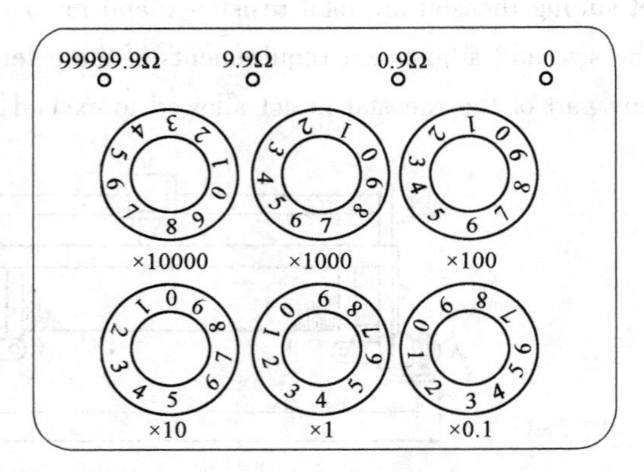

Fig. 3 – 7 – 2 Outline picture of resistance box

Chapter 4 Experiment

Exp. 1 Determination of the density of a solid

1.1 Objectives

(1) Learn how to use vernier calipers, micrometer and physical balance.

(2) Learn the method of determination of the density of a solid.

(3) Master how to deal with data correctly.

1.2 Apparatus

Physical balance, weights, beaker, vernier caliper, micrometer screw-gauge, regular shape cylinder, irregular shape solid. etc.

1.3 Principles

In order to classify and identify materials of a wide variety, scientists use numbers called physical constants (e. g. density, melting point, boiling point, index of refraction) which are characteristic of the material in question. These constants do not vary with the amount or shape of the material, and are therefore useful in positively identifying unknown materials. Standard reference works have been complied containing lists of data for a wide variety of substances. The chemist makes use of this in determining the identity of an unknown substance, by measuring the appropriate physical constants in the laboratory, consulting the scientific literature, and then comparing the measured physical constants with the values for known materials. This experiment illustrates several approaches to the measurement of the density of solids.

Density is a measure of the "compactness" of matter within a substance and is defined by the equation:

$$\text{Density} = \text{Mass/Volume} \tag{S1 - 1}$$

The standard metric units in use for mass and volume respectively are grams and milliners or cubic centimeters. Thus, density has the unit grams/milliner (g/mL) or grams/cubic centimeters (g/cm^3). The literature values are usually given in this unit. Density may be calculated from a separate mass and volume measurement, or, in the case of liquids, may be determined directly by the use of an instrument called hydrometer.

Volume measurements for liquids or gases are made using graduated containers, for example, a graduated cylinder. For solids, the volume can be obtained either from the measurement of the dimensions of the solid or by displacement. The first method can be applied to solids with regular geometric shapes for which the mathematical formulas can be used to calculate the volume of the solid from the dimensions of the solid. Alternatively, the volume of any solid object, irregular or regularly shaped, can be measured by displacement. The solid is submerged in a liquid in which it is not soluble, and the volume of liquid displaced measured.

The hydrometer measures density directly. An object that is less dense than a liquid will float in that liquid density to a depth such that the mass of the object submerged equals the mass of the liquid displaced (Archimedes' Principle). Since mass equals density multiplies volume, an object floated in liquids of different densities will displace different volumes of liquid. A hydrometer is a tube of constant mass that has been calibrated to measure density by floating the hydrometer in liquids of known densities and recording on a scale the fraction of the hydrometer submerged. Any hydrometer can be used over a limited range of densities because the hydrometer must float in the liquid being studied and the hydrometer level must be sufficiently submerged to obtain an on scale reading. Hydrometers may be calibrated in g/mL or some other unit of density.

$$V\rho_0 g = (m_1 - m_2)g \tag{S1-2}$$

In this experiment the determination of the volume with a balance is introduce. Determination of the solid density is by way of static weighting. The irregular solid is not dissolved in water and its mass is m_1, the weight of the solid submerged in water is m_2, as shown in Fig. S1 - 1. Let the density of the water is ρ_0 and the volume of the solid is V. According to Archimedes' law, the loss in weight of a body submerged in water is equal to the weight of the water in volume of the body, we have:

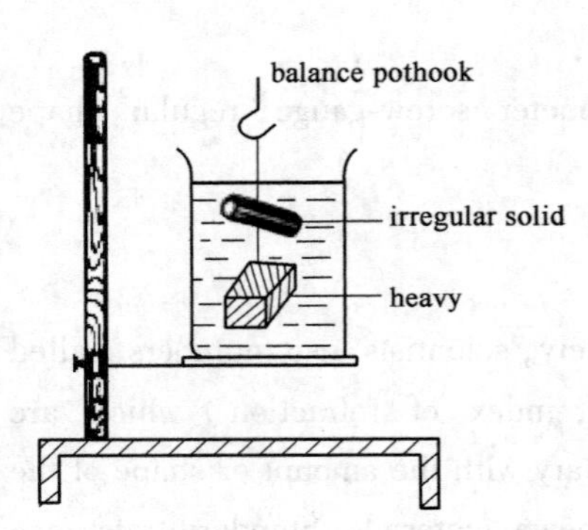

Fig. S1 - 1 Diagram of determining irregular solid

Where g is the acceleration of free fall, ρ_0 is the density of the water. Obviously the volume of the solid is

$$V = \frac{m_1 - m_2}{\rho_0} \tag{S1-3}$$

Thus, the density of the irregular shape solid is

$$\rho = \frac{m_1}{m_1 - m_2}\rho_0 \tag{S1-4}$$

In the following experiment, We will measure the densities of regular cylinder and irregular solid.

1.4 Procedures

1.4.1 Measuring the density of a regular cylinder

Weigh and record the mass of an unknown metal cylinder (Fig. S1 - 2). Also record the identity of the unknown metal cylinder. Calculate the volume of the metal cylinder by measuring (in cm) the height (h) and diameter (d) of the metal cylinder and then applying the formula: *volume* = h ×

$0.785d^2$. Also, measure the volume of the metal cylinder by displacement of water in a graduated cylinder. Calculate the density of the metal cylinder for each method of measuring volume and identify the metal by comparing the value obtained with the literature values for various metals.

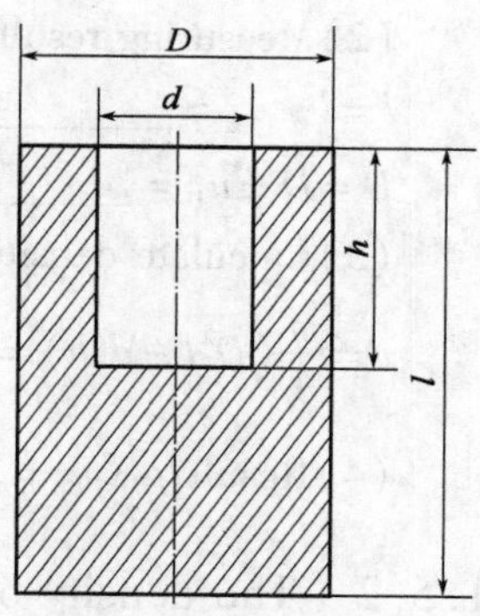

Fig. S1 - 2 Metal cylinder

1.4.2 Measuring the density of an irregular object by water displacement

(1) Obtain approximately 15 ~ 20cm³ of the unknown sample assigned by your instructor.

(2) Weigh the of Mass of dry irregular solid m_1.

(3) Fill the graduated cylinder about half full of water.

(4) Weigh the irregular solid submerged in water is m_2.

(5) Record the temperature of the water and find the value of ρ from the appendix.

(6) Calculate the density of the unknown solid by formula (S1 - 4).

Based on your assignment of the identity of the unknown sample, calculate the relative error for your experimental density. If your sample densities vary significantly (relative error > 5%), you may need to obtain additional samples of this unknown and make step (2) or more additional trials. Consult your instructor if you are unsure.

1.5 Data and calculations

1.5.1 The density of cylinder (Tab. S1 - 1、Tab. S1 - 2)

Tab. S1 - 1 The length, depth, inner diameter of cylinder

Vernier caliper (precision) ______ mm, Range ______ mm, Zero ______ mm

Position	l/mm			h/mm			d/mm		
Trial number	Direct values	Emendation zero l_i	$\Delta l_i = \bar{l} - l_i$	Direct values	Emendation zero h_i	$\Delta h_i = \bar{h} - h_i$	Direct values	Emendation zero d_i	$\Delta d_i = \bar{d} - d_i$
1									
2									
3									
4									
5									
Average		$\bar{l}$ =	σ_L =		$\bar{h}$ =	σ_h =		$\bar{d}$ =	σ_d =

Tab. S1 - 2 The outside diameter of the cylinder

Micrometer screw - gauge (precision) ______ mm, Range ______ mm, Zero ______ mm

Trial number	1	2	3	4	5	Average
Direct values						
Emendation zero D_i						
$\Delta D_i = \bar{D} - D_i$						

(1) Mass of the cylinder: m = ______.

(2) Measuring result:

$l = \bar{l} \pm u_L =$ ______; $h = \bar{h} \pm u_h =$ ______; $d = \bar{d} \pm u_d =$ ______;

$D = \bar{D} \pm u_D =$ ______; $m = \bar{m} \pm u_m =$ ______.

(3) Calculate density by the formula (S1 − 1).

$v = \frac{\pi}{4}(D^2 l - d^2 h) =$ ______; $\rho = 4m/(\pi D^2 l - d^2 h) =$ ______.

(4) Result: $\bar{\rho} \pm u_\rho =$ ______; $E = \frac{u_\rho}{\bar{\rho}} \times 100\% =$ ______.

1.5.2 The density of irregular shape solid (Tab. S1 − 3)

Tab. S1 − 3 the density of irregular shape solid

Temperature $t =$ ______ ℃, Balance precision: ______ g

A irregular solid mass in air m_1/kg	
A irregular solid mass in water m_2/kg	
The density of water on t℃ ρ_0/kg · m^{-3}	

Calculate the density: $\rho = \frac{m_1}{m_1 - m_2}\rho_0 =$ ______ kg · m^{-3} ($\rho > \rho_0$).

Result: $\rho \pm u_\rho =$ ______ kg · m^{-3}.

1.6 Precautions

(1) Take weights by tweezers and put them back into the weight box immediately after measurement.

(2) Adjust the floor level and carefully adjust the three horizontal screws on the base until the bubble in the level is in the middle.

(3) The load of the balance must not exceed the maximum weight of the balance.

Exp. 2 Measurement of the moment of inertia of a rigid body

2.1 Objectives

(1) Understand concept of a moment of inertia of a rigid body.

(2) Measure the moments of inertia of several objects by studying their accelerating rotation under the influence of unbalanced torque.

(3) Learn how to deal with data by plotting.

2.2 Apparatus

Measuring instrument of the moment of inertia, Stopwatch, Meter ruler, etc.

2.3 Principles

If we apply a single unbalanced force F to an object, the object will undergo a linear acceleration a, which is determined by the force and the mass of the object. The mass is a measure of an object's inertia, or its resistance to being accelerated. The mathematical relationship is $F = ma$. If we consider rotational motion, we find that a single unbalanced torque M = (Force) × (lever arm) produces an angular acceleration, which depends on not only the mass of the object but also the mass distribution. The equation which is analogous to $F = ma$ for an object that is rotationally accelerating is

$$M = I\alpha \tag{S2-1}$$

Where M is the torque in N · m; α is the angular acceleration in rad/s^2; I is the moment of inertia in kg · m^2. The moment of inertia is a measure of the way the mass is distributed on the object and determines its resistance to rotational acceleration.

Every rigid object has a definite moment of inertia about any particular axis of rotation. Here are a couple of examples of the expression for two special objects.

In our case the rigid body consists of two cylinders, which are placed on the metallic platform at varying radii from the axis of rotation (Fig. S2 - 1). We cannot ignore the mass of the platform and the supporting structure in our measurements, so their moment of inertia is not equal to zero. However, we can remove the masses from the apparatus and measure the moment of inertia for the supporting structure and the platform alone, $I_{support}$. We can then place the masses back on the support structure and measure the moment of inertia of the entire system, $I_{measured}$. If we then wanted just the moment of inertia of the masses, I_{mass}, we would subtract I_{mass} from $I_{measured}$.

To set up your rigid body, wrap the string around the axle several times, run it over the pulley to a known weight as shown in Fig. S2 - 1 and Fig. S2 - 2.

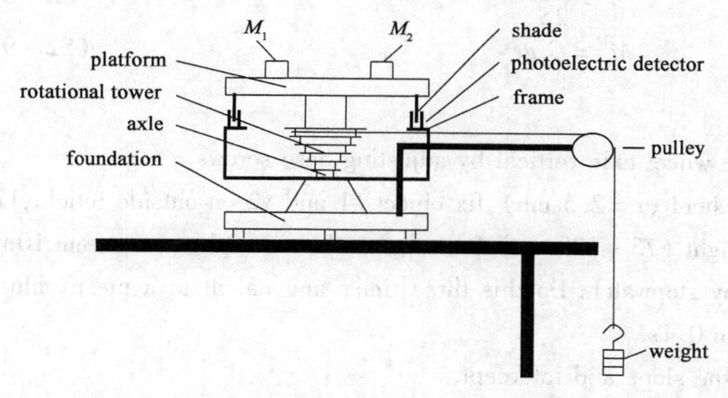

Fig. S2 - 1 Instrument of the moment of inertia of a rigid body

Fig. S2 - 2 Top view of platform

Consider the following steps:

If we release the weights from quiescence and measure how long it takes to fall a distance then from

$$S = \frac{at^2}{2} \tag{S2-2}$$

We can solve for a linear acceleration of the weight, the string and a point on the side of the axle. Using

$$\alpha = \text{angular acceleration} = \frac{\text{Linear acceleration}}{\text{Radius of axle}} = \frac{a}{r} \tag{S2-3}$$

We obtain the angular acceleration. The torque acting on the axle is given by

$$M = \text{Force} \times \text{lever arm}$$

$$M = T \times r = (mg - ma) \times r \tag{S2-4}$$

Since we now have a and M, when $a \ll g$, we can calculate I from equation (S2 - 4). Before class, be sure you know how to use equation (S2 - 4) and the above three steps to obtain the expression:

$$mgr - M_{\mu} = \frac{2SI}{rt^2} \tag{S2-5}$$

Which allows calculation of I from measurements of time with no intermediate steps. In this experiment compared to mgr, the friction torque M_{μ} is always quite small, therefore can be ignored when calculating I.

Then, we can get the following equation from equation (S2 - 5) with $M_{\mu} \ll mgr$:

$$mgr = \frac{2SI}{rt^2} \tag{S2-6}$$

Make a series of measurements of t, the time of the entire rigid body falls S distance, with the mass m_1 and m_2 placed an equal distance $r(r_1 = r_2)$ from the axis of rotation, and keep S constant. We can get from the formula (S2 - 5):

$$m = \frac{2SI}{gr^2} \cdot \frac{1}{t^2} + \frac{M_{\mu}}{gr} = k\frac{1}{t^2} + C_1 \tag{S2-7}$$

From equation (S2 - 7), there is linear relationship between m and $1/t^2$. Make a plot of your measurements of m vs. $1/t^2$. Calculate the slope and intercept of this data, you can get the moment of inertia I and friction torque M_{μ}.

$$I = \frac{kgr^2}{2S} \tag{S2-8}$$

$$M_{\mu} = C_1 gr \tag{S2-9}$$

2.4 Procedures

(1) Arrange the apparatus make wheel axis vertical by adjusting base screws.

(2) Circle the line around the wheel ($r = 2.5$ cm), fix object #1 and #2 on outside reticle, Let the weight m held still from some height (75 ~ 100 cm) fall to ground, the weight varies from 10mg to 40mg, record the time of falling by stopwatch. Do this three times and calculate a mean value. Make sure the every error is less than 0.1s.

(3) Plot the line and calculate the slope and intercept.

2.5 Data and calculations

2.5.1 Data records (Tab. S2 - 1)

Tab. S2 - 1 The data of falling time for different weight

m/g	t_1/s	t_2/s	t_3/s	$\bar{t}$/s	$\bar{t}^2$/s^2	$\frac{1}{\bar{t}^2}$/s^{-2}
10						
15						
20						
25						
30						
35						
40						

$S =$______, $R =$______.

2.5.2 Calculations

The moment of inertia of the rigid body is obtained by drawing $m : \frac{1}{t^2}$ plots. Measurement of the moment of inertia of a rigid body is shown in Video S2.

Video S2 Measurement of the moment of inertia of a rigid body

2.6 Precautions

(1) When adjusting different radii, the position of the pulley must be perpendicular to the axis of rotation, and the pulley must be in the same height with the string.

(2) The friction force should be kept constant as far as possible, and the adjusted device should not be changed arbitrarily in the experiment.

(3) The initial speed must be zero when the weight begins to fall.

Exp. 3 Determination of Young's modulus of a metallic wire

3.1 Objectives

(1) Understand the principle of the optical lever.

(2) Understand the basic physics concepts and measure method of Young's modulus.

3.2 Apparatus

Young's modulus apparatus, telescope, the optical lever mirror, micrometer screw gauge, vernier calipers, tapeline, masses, metal wire, etc.

3.3 Principles

Young's modulus is named after Thomas Young, the 19th century British scientist. In solid mechanics, Young's modulus is a measure of the stiffness of an isotropic elastic material. It is also known as the Young's modulus, modulus of elasticity, elastic modulus (though Young's modulus is actually one of several elastic modulus such as the bulk modulus and the shear modulus) or tensile modulus. It is defined as the ratio of the uniaxial stress over the uniaxial strain in the range of stress in which Hooke's Law holds.

The experimental apparatus is shown in Fig. S3 - 1 and Fig. S3 - 2. If a metal wire of cross-sectional area S is pulled by a force F at the end, the wire stretches from its original length L to a new length L_n. The stress is the quotient of the tensile force divided by the cross-sectional area, or F/S. The strain or relative deformation is the change in length, $\Delta L = L_n - L$, divided by the original length, or $\Delta L/L$. Thus Young's modulus may be expressed mathematically as:

$$\frac{F}{S} = E\frac{\Delta L}{L} \tag{S3-1}$$

If the diameter of metal wire is d, formula (S3 - 1) may be expressed mathematically as:

$$E = \frac{4FL}{\pi d^2 \Delta L} \qquad (S3-2)$$

where E—Young's modulus, usually expressed in Pascal, Pa;

F—the force of compression or extension;

S—the cross-sectional surface area or the cross-section perpendicular to the applied force;

ΔL—the change in length (negative under compression; positive when stretched).

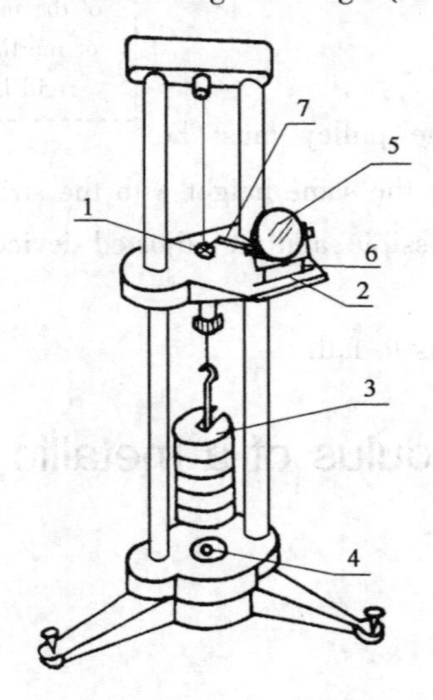

Fig. S3 - 1 Young's modulus measuring apparatus

1—the cylinder holding the wire collet; 2—groove of base plate; 3—weights; 4—spirit level; 5—optical lever mirror; 6—the two front metal toes of the optical lever mirror; 7—rear metal toe of the optical lever mirror

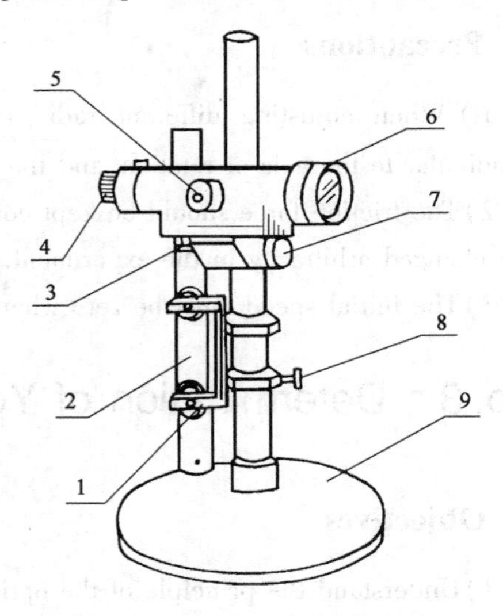

Fig. S3 - 2 Ruler & Telescope device

1—ruler fixing clip; 2—millimeter ruler; 3—height screw; 4—eye focus screw; 5—object focus screw; 6—telescope; 7, 8—fixing screw; 9—base board

When the tip of mirror changes the distance ΔL, the mirror will rotate an angle θ (Fig. S3 - 3). Where θ is a very small quantity and $\Delta L \ll b$, we can get

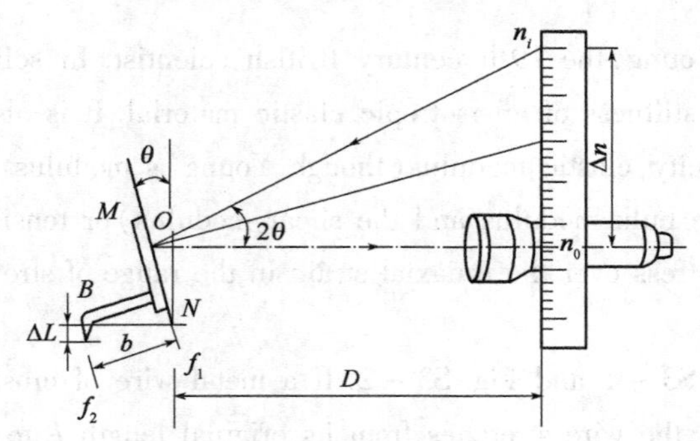

Fig. S3 - 3 Principle of optical lever

$$\Delta L/b = \tan\theta \approx \theta$$

and

$$\Delta n/D = \tan 2\theta \approx 2\theta$$

We can get from above formula

$$\Delta L = \frac{b}{2D}\Delta n \qquad (S3-3)$$

Where, $\frac{2D}{b}$ is the magnifying index of the optical Lever. Young's modulus can be expressed mathematically as:

$$E = \frac{8mgDL}{\pi d^2 b \Delta n} \qquad (S3-4)$$

3.4 Procedures

We will measure the change in length of a wire as it supports an ever increasing load. In the

Young's Modulus apparatus illustrated in Fig. S3 – 1, a wire is suspended from the top of the frame. Masses are added to the weights-hanger attached to the bottom of the wire in order to stretch the wire. To ensure a well-functioning apparatus, you will need to familiarize yourself with the spirit level and micrometer assembly. Figure out how it works! In particular, figure out how the various parts move relative to one another and why.

(1) Adjust the foot screw of the Young's modulus Apparatus so that the cylinder holding the wire collet moves freely within the hole through the base plate (below the pivoting plate) and that it is not catching on the pin that keeps the cylinder from turning.

(2) Adjust the "optical lever mirror" and "Ruler & Telescope device". The basic principle of adjustment is to coarse adjustment should be followed before fine adjustment. The standard of adjustment is a clear and parallax free scale reading from the telescope. As described below:

①The purpose of adjustment is to place the bracket of optical lever mirror horizontally, the mirror surface is perpendicular to the table, the telescope is horizontally aligned with the mirror, and the ruler is vertical to the telescope.

②The center of telescope and the plane mirrors on the same axis, and the height is the same. After that, align the telescope horizontal fork wire near the zero scale of the scale.

③Focus the telescope, so that the ruler can be seen through the eyepiece, the imaging is clear and there is no parallax.

(3) Starting with a suspended mass of 1kg (just the large mass hanger), measure and record the wire extension as a function of suspended mass in increments of 1kg, until the suspended mass reaches 8kg, ($n_1, n_2, n_3, \cdots, n_8$). Then you will remove mass in turn, from 8kg to 1kg and record the reading of the ruler for each weight removed, ($n_8, n_7, n_6, \cdots, n_1$).

(4) Measure the wire diameter with micrometer screw gauge. Make several measurements at various positions along the wire and find the average.

(5) Measure the distance (D) between the ruler and the optical lever mirror with the steel tape.

(6) Measure the total length (L) of the metal wire with the steel tape.

(7) Use Vernier caliper to measure the distance from the two front metal toes of optical lever mirror to the back metal toe.

3.5 Data and calculations

3.5.1 Data records (Tab. S3 – 1 ~ Tab. S3 – 3)

Tab. S3 – 1 The stretch of steel wire vs. weight unit: cm

n	Loading	Unloading	$\overline{n_i}$	n	Loading	Unloading	$\overline{n_{i+4}}$	Δn_i
1				5				
2				6				
3				7				
4				8				

$$\Delta \overline{n_i} = \frac{\sum_{i=1}^{4} \Delta n_i}{4} = \_\_\_\_\_\_ \text{cm}.$$

Tab. S3 – 2 The diameter of steel wire

$d_0 =$ ______ mm unit: mm

Position	Up			Middle			Bottom		
Trial number	1	2	3	4	5	6	7	8	9
$d = d_i - d_0$									

$$\overline{d} = \frac{\sum_{i=1}^{9} d_i}{6} = \_\_\_\_\_\_ \text{ mm.}$$

Tab. S3 – 3 The distance of the optical lever leg spacing

$\Delta_{\text{instru.}} = 0.02\text{mm}$ unit: mm

Trial number	1	2	3	4	5	6
$b_i = b - b_0$						

$$\overline{b} = \frac{\sum_{i=1}^{6} b_i}{6} = \_\_\_\_\_\_ \text{ mm.}$$

Other data:

$D =$ ______ cm, $L =$ ______ cm, $m \pm \sigma_m = (4.000 \pm 0.004)$ kg.

3.5.2 Calculations

$$E = \frac{8mgDL}{\pi \overline{d}^2 \overline{b} \Delta n_i} = \_\_\_\_\_\_.$$

3.6 Precautions

(1) Do not touch the surface of mirror and telescope lens with your hands.

(2) Take care of the mirror, avoid breaking it.

(3) The weight should be steadily suspended or removed to avoid impact.

Video S3 Determination of Young' s modulus of a metallic wire

Determination of Young' s modulus of a metallic wire is shown in Video S3.

Exp. 4 Measurement of the gravitational acceleration g using a simple pendulum

4.1 Objectives

(1) Learn the principle of measuring the acceleration of gravity with a simple pendulum.

(2) Measure the gravitational acceleration by means of a simple pendulum.

4.2 Apparatus

Simple pendulum, Calculagraph, Tapeline, Vernier caliper.

4.3 Principles

A simple pendulum (Fig. S4 - 1) displaced through a small angle $\theta(\theta<5°)$, will oscillate back and forth about its equilibrium position with period T. T is the time the pendulum takes to make one complete back - and - forth motion. The bob (the diameter is denoted as d) is hung from a rigid support on a string of length l. The mass of the bob is denoted as m. The tangent force forced on the bob is mgsin θ, according to Newton law, kinematical equation is as followed:

$$ma_{\tau} = -mg\sin\theta \quad (S4-1)$$

$$mL\frac{d^2\theta}{dt^2} = -mg\sin\theta \quad (S4-2)$$

Here $\sin\theta \approx \theta$ as the angle θ is less than 5°. So

$$\frac{d^2\theta}{dt^2} = -\frac{g}{L}\theta \quad (S4-3)$$

$$\theta(t) = A\cos(\omega t + \phi) \quad (S4-4)$$

$$\omega = \frac{2\pi}{T} = \sqrt{\frac{g}{L}} \quad (S4-5)$$

Fig. S4 - 1 Simple pendulum

Where, L includes the length l of the string and the radius $d/2$ of bob.

For oscillations where the angle θ is small, the period T is related to the length L of the string and the gravitation constant g by

$$T = 2\pi\sqrt{\frac{L}{g}} \quad (S4-6)$$

If one measures the gravitational accelerating g as a function of the length of the string and the period T, we can obtain the formula:

$$g = 4\pi^2\frac{L}{T^2} \quad (S4-7)$$

In experiment, the measurement error of a single period is very large, so we will measure the time t continuous oscillating of periods.

$$g = 4\pi^2\frac{n^2L}{t^2} \quad (n \text{ is the number of periods}) \quad (S4-8)$$

4.4 Procedures

(1) The measurement of the length l of the string and the radius $d/2$ of bob. The length l of the string is one - off measured using a tapeline. The radius $d/2$ of bob is determined by a Vernier caliper, and repeated five times along different diameter orientations to obtain average values $\overline{L} = l + \frac{\overline{d}}{2}$.

(2) The measurement of the period T. In experiment, the measurement error of a single period is very large, so we will measure the time t continuous oscillating 30 periods, and repeated five times with angle θ less than 5° to obtain average values t.

4.5 Data and calculations

4.5.1 Data records(Tab. S4 - 1)

The diameter of the ball: $d =$ ______ mm.

The length of the string: $l =$ ______ cm.

The time continuous oscillating 30 periods: $t =$ ______ ms.

Tab. S4 - 1 30 periods time data unit: s

	1	2	3	4	5	6
$t = 30T$						

4.5.2 Calculation

The average gravitation constant: $\bar{g} = 4\pi^2 \frac{n^2 \bar{L}}{t^2} (n = 30)$.

Relative error: $E = \frac{|\bar{g} - g_0|}{g_0} \times 100\% \ (g_0 = 9.8\text{m/s}^2)$.

The evaluation of uncertainty u_g.

$$g = \bar{g} \pm 2u_g$$

Video S4 Measurement of the gravitational acceleration g using a simple pendulum

4.6 Precautions

(1) The angle of simple pendulum is strictly controlled($\theta < 5°$).

(2) It is forbidden to vibrate the test table during measuring.

Measurement of the gravitational acceleration g using a simple pendulum is shown in Video S4.

Exp. 5 Measurement of the coefficient of viscosity of a liquid by falling ball method

5.1 Objectives

(1) Master the working principle of how to measure the viscosity of liquid using Stokes' Law.

(2) Understand the impact of temperature on the viscosity of liquid.

5.2 Apparatus

FB328B type variable temperature viscosity tester, several small steel spheres(about 1 ~ 2mm in diameter, more than 20), Micrometer, Stopwatch, Ruler, Liquid sample, Balance, Tweezers.

5.3 Principles

Not all liquids are the same. Some are thin and flow easily. Others are thick and gooey. Honey or corn syrup will pour more slowly than water. In order to cause motion of a fluid, a stress must be applied to it. Resistance to motion varies considerably from liquid to liquid and is very much greater than for gases. Some liquids, such as honey or syrup, offer considerable resistance. These are called viscous liquids. Others, such as water and alcohol, are much less viscous. Viscosity coefficient is one of the important properties of liquid.

An object moving in a fluid is acted on by a frictional force in the opposite direction to its direction of travel. The magnitude of this force depends on the geometry of the object, its velocity, and the internal friction of the fluid. The falling ball method is one of the earliest and simplest methods to determine the viscosity of fluid. In this method, a ball is allowed to fall freely a measured distance through an infinitely extended liquid medium of dynamic viscosity η, and its velocity is determined. The viscous drag of the falling ball results in the creation of a frictional force f, described by Stokes' Law:

$$f=6\pi\eta rv \tag{S5-1}$$

Where, r is the radius of the ball, v is the terminal velocity of the falling ball.

If the ball falls vertically in the fluid, after a while, it will move at a constant velocity v, and all the forces acting on the ball will be in equilibrium (Fig. S5 - 1): the frictional force f which acts upwards, the buoyancy force F which also acts upwards and the downward acting gravitational force mg.

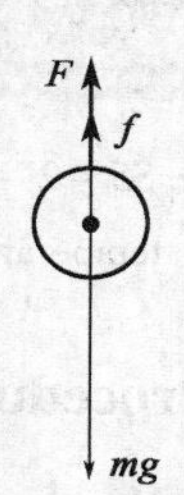

Fig. S5 - 1 Force analysis diagram

The buoyancy force F is given by

$$F=\frac{4}{3}\pi r^3\rho_0 g \tag{S5-2}$$

Where ρ_0 is the density of the fluid, g is the gravitational acceleration.

The equilibrium between these three forces can be described by:

$$mg-\frac{4}{3}\pi r^3\rho_0 g-6\pi\eta rv=0 \tag{S5-3}$$

Then the viscosity can be determined by the velocity v of falling ball:

$$\eta=\frac{\left(m-\frac{4}{3}\pi r^3\rho_0\right)g}{6\pi rv} \tag{S5-4}$$

Where v can be determined by measuring the fall time t over a given distance $L/v = L/t$. The viscosity then becomes:

$$\eta=\frac{\left(m-\frac{4}{3}\pi r^3\rho_0\right)gt}{6\pi rL} \tag{S5-5}$$

Since we cannot let a ball falling in an infinitely extended liquid, and the liquid sample is contained in a burette in practical experiment, the Stokes' Law must be corrected considering the effect of the container. The viscosity is, therefore, calculated by:

$$\eta=\frac{\left(m-\frac{4}{3}\pi r^3\rho_0\right)gt}{6\pi rL\left(1+2.4\frac{r}{R}\right)} \tag{S5-6}$$

Where, R is the radius of the burette. Since the mass of the ball, $m=\frac{4}{3}\pi r^3\rho$ and the diameters of the ball and the inner diameter of the burette can be directly measured in experiment, the liquid viscosity can be calculated by the following formula:

$$\eta = \frac{(\rho - \rho_0) g d^2 t}{18L\left(1 + 2.4\dfrac{d}{D}\right)} \tag{S5-7}$$

in which, ρ and ρ_0 are the densities of the steel ball and castor oil, respectively. d and D are the diameters of the ball and burette, respectively.

Furthermore, the viscosity is dependent on the temperature. The liquid viscosity tends to decrease (or alternatively, its fluidity tends to increase) as its temperature increases. The experimental devices were arranged as shown in Fig. S5-2, which includes the glass tube containing the liquid to be measured, a tubular jacket for thermal control, a water reservoir with a heater, a temperature control system, and a pump for circulation.

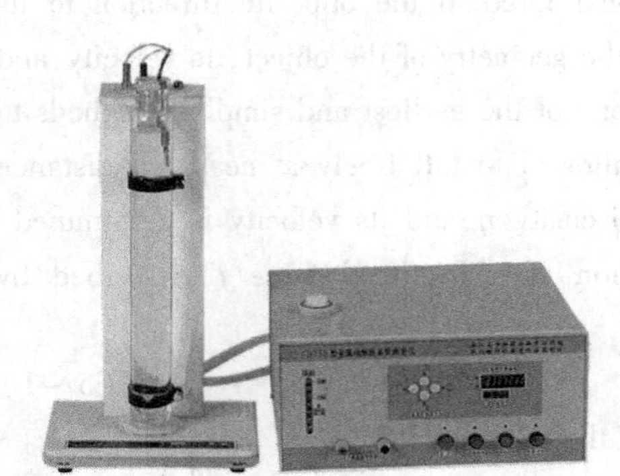

Fig. S5-2 FB328B type variable temperature viscosity tester

5.4 Procedures

(1) Measure the diameter d of the ball with a micrometer, measure 6 balls in total, and find the average value of the diameter of the steel ball.

(2) According to the instructions, you can adjust the variable temperature viscosity tester in use.

(3) In order to make the surface of the ball is completely covered by the test liquid. Use tweezers to pick up the ball and soak it in the liquid in advance. Then make the ball drop from the center of glass tube along axis and determine the position where the ball begins to fall in uniform speed. Within the range of the ball falling with a uniform speed, mark a location A where we start the timer. Usually the mark line A is selected at a slightly lower position when the ball starts to move at a constant speed. Select the mark line B below the mark line A as the timing stop position. The distance L between A and B can be measured with a tape measure.

(4) Start the thermostat at room temperature. Use a stopwatch to measure the time t taken for the ball to fall between the mark lines A and B.

(5) Repeat step 4 for two times, record the data in the Table.

(6) Set the temperature at 30℃, 35℃, 40℃, respectively. Repeat the steps 2-4, record the corresponding fall time of the balls to the Table.

(7) After the experiment, use a magnet to take out the small steel balls from the oil cylinder and place it properly.

5.5 Data and calculations

5.5.1 Data records (Tab. S5-1)

The diameter of the ball: d = ______ m;

The inner diameter of the burette: $D =$ ______ m;

The density of castor oil: $\rho_o = 957\text{kg/m}^3$;

The density of steel ball: $\rho = 7800\text{kg/m}^3$;

The timed falling distance: $L =$ ______ m.

Tab. S5 - 1 The data of falling time for the ball at different temperatures.

Temperature/℃	Room temperature: ______	30	35	40
Falling time of ball 1/s				
Falling time of ball 2/s				
Falling time of ball 3/s				
Average falling time t/s				
Terminal velocity of ball $v_0/\text{m}\cdot\text{s}^{-1}$				
Viscosity η				

5.5.2 Calculations

$$\eta = \frac{(\rho_o - \rho) g d^2 \bar{t}}{18L\left(1 + 2.4\frac{d}{D}\right)}$$

Using the above formula, the viscosity of castor oil can be obtained at different temperature.

5.6 Precautions

(1) The little balls should fall along the central axis of the glass cylinder, and do not release the ball close to the tube wall.

(2) The liquid in the tube should be free of bubbles, and the surface of the ball should be smooth and free of oil.

(3) The dynamic viscosity of the liquid is related to the temperature. Avoid touching the glass tube with your hands during the experiment to ensure that the liquid temperature does not change.

(4) When the ball is falling, the liquid should remain still, and there should be a certain time interval between placing two balls.

Exp. 6 Determination of the melting heat of ice by mixture method

6.1 Objectives

(1) Learn calorimetry and method of noting temperature.

(2) Master measurement of the melting heat of ice.

(3) Learn how to correct the effect of heat loss.

6.2 Apparatus

Copper Calorimetric container and stirrer, Thermometer, Physical balance, graduated cylinder, beaker, etc.

6.3 Principles

The word calorimetry is derived from latin word calor, that means heat. It refers to the measurement of the heat of chemical reactions and/or thermodynamic phase changes. This heat does not contribute to the temperature change but it is rather used for changes in the molecular structure of matter. It is often called a latent heat, because it causes no apparent change. Temperature of an object at its melting temperature t_m does not respond to further heating until the object melts throughout its entire volume. Heat that is consumed by the melting object follows the equation:

$$Q = m\lambda \tag{S6 - 1}$$

where λ is a specific heat of melting(melting heat). Numerically, λ corresponds to a heat that turns 1kg solid object at its melting temperature t_m into a liquid without the change in temperature. It is expressed in units of J/kg.

On the other hand, temperature of a matter that is away from phase transition region changes from temperature t_A to temperature t_B upon matter's heating. This process follows the equation:

$$Q = mC(t_B - t_A) \tag{S6 - 2}$$

Where C is a specific heat capacity(specific heat) of a particular matter, heat capacity, ratio of heat absorbed by a material to the temperature change. Numerically, it represents a heat that is consumed by 1kg of matter for temperature increase of 1℃. Its units are J/(kg · K).

In a case of object that consists of several substances, we define a heat capacity C as a heat consumed by the entire object for the temperature increase of 1℃. Heat that is consumed by the melting object follows another equation:

$$Q = C(t_B - t_A) \tag{S6 - 3}$$

Heat capacity is expressed in units of J/K.

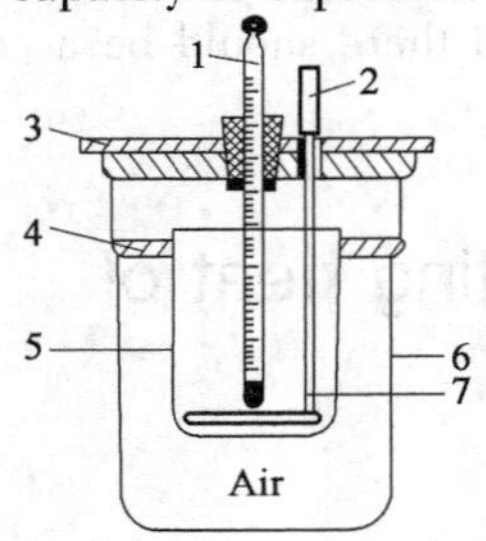

Fig. S6 - 1 Schematic diagram of calorimeter
1—thermometer; 2—insulated handle; 3—insulated cover; 4—insulated frame; 5—inner container; 6—outer container; 7—stirrer

A simple calorimeter consists of a thermometer attached to an insulated container (Fig. S6 - 1). For example, a calorimetric container with stirrer and thermometer is going to be used in this assignment. We fill this calorimeter, which is characterized by the heat capacity C_1, with water of the weight m_1 and temperature T_1. If we add some ice of the weight m and temperature 0℃, temperature inside calorimeter continuously decreases down to the value $T_2 > 0℃$. Steady temperature means the completion of ice melting and homogeneous distribution of temperature through the entire system. Now, let's have a closer look at the whole process in detail. The heat Q_1 was necessary to melt the ice [see the equation (S6 - 1)]:

$$Q_1 = m\lambda \tag{S6-4}$$

Where λ is a specific melting heat of ice, that we are going to determine. After absorbing this heat, added ice changed itself into liquid water of the same weight m and temperature 0℃. However, the final temperature $T_2 > 0℃$ indicates that this water received additional heat Q_2 [see the equation (S6 - 2)] can be described by:

$$Q_2 = mC_1(T_2 - T_0) = mC_1 T_2 \tag{S6-5}$$

Where $C_1 = 4.18\text{kJ} \cdot \text{K}^{-1} \cdot \text{kg}^{-1}$ is a specific heat capacity of water. At the same time, add some mass ice (m) to the calorimeter at the beginning, released the heat Q_3 while its temperature decreased from T_1 to T_2 [see the equation (S6 - 2)]:

$$Q_3 = m_1 C_1(T_1 - T_2) \tag{S6-6}$$

In addition to temperature change of water, Calorimetric container and stirrer of the weight m_2 underwent the same temperature change and thus released the heat Q_4 [see the equation (S6 - 3)]:

$$Q_4 = m_2 C_2(T_1 - T_2) \tag{S6-7}$$

where $C_2 = 0.38\text{kJ}/(\text{kg} \cdot \text{K})$ is a heat capacity of calorimeter. Since the whole system is insulated from the surroundings, released heat is equal to the received one, which can be described by:

$$Q_1 + Q_2 = Q_3 + Q_4 \tag{S6-8}$$

Combining equations (S6 - 7) yields:

$$m\lambda + mC_1 T_2 = m_1 C_1(T_1 - T_2) + m_2 C_2(T_1 - T_2) \tag{S6-9}$$

And equation for λ is then expressed as:

$$\lambda = \frac{1}{m}(m_1 C_1 + m_2 C_2)(T_1 - T_2) - C_1(T_2 - T_0) \tag{S6-10}$$

6.4 Procedures

(1) Use graduated cylinder, fill the calorimeter with 200mL of hot water which the temperature is 10℃ above room temperature. Calculate the water weight m_1, supposing its density is 1g/cm^3. Close the container and turn on the mixer. Record the temperature T_1 of the hot water before adding the ice.

(2) Weight out about 30 ~ 50g of ice (m) and add it quickly to the calorimeter. Cover the container and start stirring. Wait until the temperature becomes steady for at least 3 minutes and write down the final temperature T_2.

(3) Weigh water m_1 and ice m, record initial T_1 and final T_2 temperature of water, and calculate specific melting heat of ice λ.

(4) Pour water out of the calorimeter, dry it and repeat the measurement two more times.

(5) Use the equation (S6 – 10) to calculate specific melting heat of ice for each measurement. Determine average value and compare it to the value listed in the handbook of physics ($\lambda_0 = 332.9$ kJ/kg).

6.5 Data and calculations

6.5.1 Data records (Tab. S6 – 1)

Tab. S6 – 1. The data of temperature, mass and specific melting heat of ice

Trial number	m	m_1	m_2	T_1	T_2	λ
1						
2						
3						

6.5.2 Calculations

$$\lambda = \frac{1}{m}(m_1 C_1 + m_2 C_2)(T_1 - T_2) - C_1(T_2 - T_0)$$

Using the above formula, the melting heat of ice can be obtained, then calculate the percentage error with the standard value ($\lambda_0 = 332.9$ kJ/kg).

6.6 Precautions

(1) Avoid splashing water when casting ice and stirring, and avoid dripping water in the process of taking out the thermometer.

(2) Keep stirring when measuring, and don't touch the thermometer and calorimeter.

(3) The thermometer should not be in contact with the calorimeter and ice cubes, and should be suspended in water.

Video S6 Determination of the melting heat of ice by mixture method

(4) The fine thermometer is large and easy to break, so be careful when using it.

(5) During the experiment, do not directly touch the calorimeter with your hands, do not conduct experiments under sunlight or where the air flows too fast, and the calorimeter should be far away from the heat source.

Determination of the melting heat of ice by mixture method is shown in Video S6.

Exp. 7 Determination of the thermal conductivity of a poor conductor

7.1 Objectives

(1) Understand the physical process of heat conduction phenomena.

(2) Master the principle and method of measuring thermal conductivity of poor conductors by steady state method.

(3) Learn how to determine the thermal conductivity of rubber disc.

7.2 Apparatus

FD – TC – B thermos conductivity measurer, vernier caliper, LD100 – 1 electronic scales.

7.3 Principles

Thermal conductivity, is an important physical quantity reflecting the material's ability to transfer heat. The thermal conductivity of the material is not only closely related to the species of the material, but also related to its microstructure, temperature and other factors. The steady state method is used to measure the thermal conductivity. The upper position of the instrument is a heating plate C, the middle is a poor conductor rubber disc B, and the lower position of the instrument is a radiating plate A (Fig. S7 – 1). The sample rubber's side area is much smaller than its cross-sectional area. It is assumed that the heat can only be transmitted vertically from the top towards the bottom, and the heat that radiated from the side area is negligible.

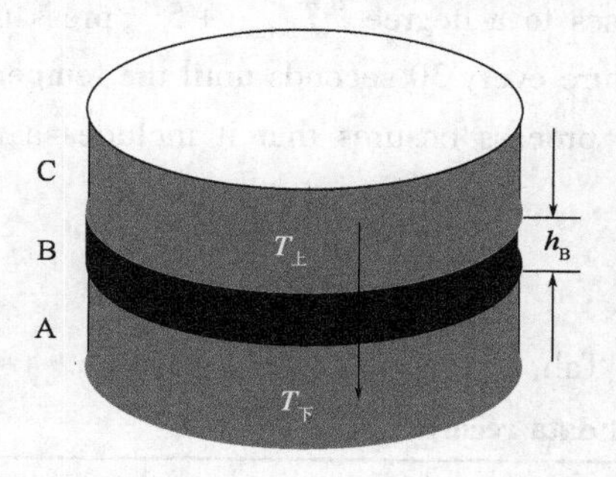

Fig. S7 – 1 Schematic diagram of thermal conductivity

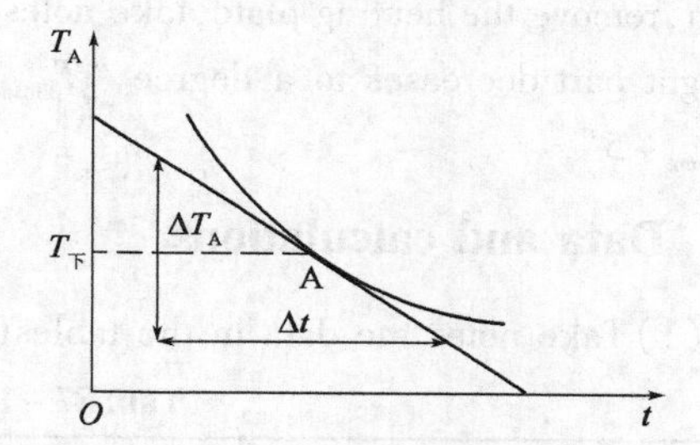

Fig. S7 – 2 Cooling curve of radiating plate A

According to the principle of heat conduction, it can be shown that the dQ of the dS area in the dt time is proportional to the temperature gradient inside the object as below (Fig. S7 – 2):

$$\frac{dQ}{dt} = -\lambda \frac{dT}{dx} dS \tag{S7 – 1}$$

Where, $\frac{dQ}{dt}$ is the heat transfer rate, the symbol " – " means that the heat transfers from the high temperature region to the low temperature region, $\frac{dT}{dx}$ is the temperature gradient along the vertical direction x, and λ is the thermal conductivity of the object, its unit is $W \cdot m^{-1} \cdot K^{-1}$.

According to the theory derivation, the sample rubber's thermal conductivity can be shown as

$$\lambda = -\frac{2m_{copper}c_{copper}h_B(D_A + 4h_A)}{\pi D_B^2(T_{top} - T_{bottom})(D_A + 2h_A)} \times \left.\frac{dT}{dt}\right|_{T=T_{bottom}} \tag{S7 – 2}$$

Where, m_{copper} is the mass of the radiating plate A; $c_{copper} = 0.380 kJ \cdot kg^{-1} \cdot K^{-1}$, it is the specific heat of the radiating plate A; h_B is the sample rubber B's thickness; D_A is the radiating plate A's diameter; h_A is the radiating plate A's thickness; D_B is the sample rubber B's diameter; When the system reaches a stable state, T_{top} is the stable temperature of plate C; T_{bottom} is the stable temperature of plate A; $T_{top} - T_{bottom}$ is their temperature difference between the two surfaces of the sample rubber; $\left.\frac{dT}{dt}\right|_{T=T_{bottom}}$ is the slope of the temperature curve at the point $T = T_{bottom}$.

7.4 Procedures

(1) Turn on the power. Open the fan. Press the button "increase temperature". At the left part, it displays "Bxx. x", and set it to 50 degrees. Then press the button "OK". Now, at the left part, it should display "Axx. x", and it means the heating plate A's temperature at this time.

(2) As time goes by, the temperature at the left part reaches 50 degrees and hold on for a while. The temperature at the right part increases to a degree (about 34℃ or others) and hold on for a while. Let the temperature of the right part maintains for nearly 10 minutes, and it is the balance temperature, write down T_{top} and T_{bottom}.

(3) Press the reset button, remove the rubber, put the heating plate to the radiating plate directly, press the button "increase temperature", set it to 50 degrees, and press the button "OK".

(4) When the temperature at the right part increases to a degree "T_{bottom} + 5", press the reset button, remove the heating plate, take notes the temperature every 30 seconds until the temperature at the right part decreases to a degree "T_{bottom} − 5". This process ensures that it includes a range of "T_{bottom} + 5".

7.5 Data and calculations

(1) Take notes the data in the tables (Tab. S7 − 1、Tab. S7 − 2).

Tab. S7 − 1 Experiment data record

m_{copper}		D_A		$\left.\frac{dT}{dt}\right\|_{T=T_{bottom}}$	
c_{copper}	$0.380 kJ \cdot kg^{-1} \cdot K^{-1}$	h_B		T_{top}	
h_A		D_B		T_{bottom}	

Tab. S7 − 2 Temperature data

30″	60″	90″	120″	150″	180″	210″	240″	270″	300″
330″	360″	390″	420″	450″	480″	510″	540″	570″	600″
630″	660″	690″	720″	750″	780″	810″	840″	870″	900″

(2) With time as horizontal axis and temperature as vertical axis, plot the curve, connect the points, and draw its slope $\left.\frac{dT}{dt}\right|_{T=T_{bottom}}$.

(3) Calculate λ.

7.6 Precautions

(1) Pay attention to connect the instruments correctly. The two sensors of the heating plate and the radiating plate should be corresponded and not be interchangeable.

(2) Some silicon oil is needed before inserting the temperature sensor into the hole, and make sure the sensor and copper plate have good thermal contact.

(3) The fan under the copper plate of the thermal conductivity meter is used for convection heat transfer, which can reduce the heat release ratio between the side and the bottom of the sample and increase the temperature gradient inside the sample, thus reducing the experimental error. Therefore, make sure the fan is turned on during the experiment.

(4) After the measurement, cut off the power supply and pay attention to prevent scalding when taking out the sample.

Determination of the thermal conductivity of a poor conductor is shown in Video S7.

Video S7 Determination of the thermal conductivity of a poor conductor

Exp. 8 Study on the characteristics of the power control circuit

8.1 Objectives

(1) Grasp the basic principles of current-limiting control circuit.

(2) Grasp the basic principles of voltage dividing control circuit.

8.2 Apparatus

DC power supply, galvanometer, two resistance boxes, milliammeter, switch, slide-wire rheostat etc.

8.3 Principles

The circuit of electrical experiment is usually composed of power supply, control circuit and measurement circuit. The measurement circuit often needs the voltage or current provided by the power supply to be adjustable in a certain range, and the output of the power supply is generally a fixed value. In order to solve this contradiction and meet the requirements of measurement circuit, the change of voltage and current in different control circuits is controlled by slide-wire rheostat of different specifications.

There are two basic connection methods of control circuit: current-limiting connection and voltage dividing connection. The performance index of these two connection methods can be characterized by adjustment range, fine adjustment degree and linearity degree. The synthesis of these three factors is the output characteristics of power control circuit, which can be expressed by the output characteristic curve of control circuit.

8.3.1 Current-limiting control circuit

The current-limiting control circuit is shown in Fig. S8 – 1, ε is the power supply, and the internal resistance of the power supply is generally very small and can be ignored. S is the power switch, R_L is the load resistance, R_0 is the full resistance of the slide-wire rheostat, R_2 is the resistance in the connected circuit, and R_1 is the resistance not connected. When the load R_L is connected with the point C of the slide-wire rheostat, the resistance value R_2 between A and C can be continuously changed by moving the position of sliding end C of slide-wire rheostat, so as to change the current of the whole circuit and control the load current to meet the requirements.

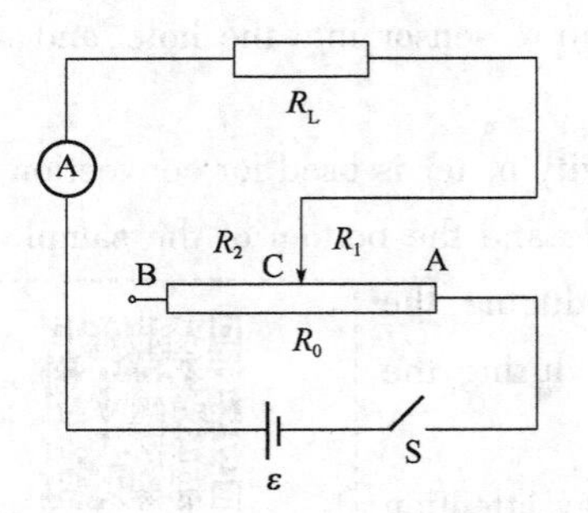

Fig. S8 – 1 Current-limiting control circuit

The output characteristic curve of the current-limiting circuit is the curve that the ratio of the output current I of the control circuit and its maximum value I/I_{max} varies with the position of the sliding end of the rheostat. It can directly observe the change rule of the output current of the control circuit in the regulation range. Take the ratio of load resistance and total resistance of slide-wire rheostat as the circuit characteristic coefficient K, and $K = R_L/R_0$. x is the relative position of sliding end C on the slide-wire rheostat, and $x = R_1/R_0$. Then

$$\frac{I}{I_{max}} = \frac{\varepsilon}{R_L + R_2} \cdot \frac{R_L}{\varepsilon} = \frac{R_L}{R_L + R_2} = \frac{R_L/R_0}{R_L/R_0 + R_2/R_0} = \frac{K}{1 + K - x} \tag{S8 – 1}$$

The current-limiting characteristic curves under different K values can be drawn by experiments.

8.3.2 Voltage dividing control circuit

The voltage dividing control circuit is shown in Fig. S8 – 2, which ε is the power supply, and the internal resistance of the power supply is generally very small and can be ignored. R_L is the load resistance and R_0 is the full resistance of slide-wire rheostat. The two fixed terminals A and B of the slide-wire rheostat are respectively connected with the positive and negative poles of the power supply through the power switch S, one end of the load R_L is connected with the sliding end C, and the other end is connected with the fixed terminal A. The resistance of slide-wire rheostat is separated into two parts. The resistance R_1 is parallel to R_L. The other part of resistance is R_2. By moving the position of sliding end C of slide-wire rheostat, the voltage value between A and C can be changed continuously, so that the load voltage can be controlled to meet the requirements.

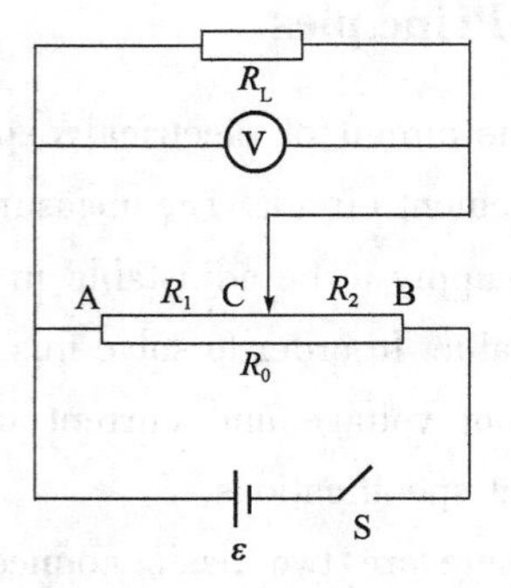

Fig. S8 – 2 Voltage dividing control circuit

U/U_m is the ratio of the output voltage of control circuit U and its maximum value U_m. The output characteristic curve of voltage dividing control circuit is a curve that U/U_m varies with the position of the rheostat sliding end.

Take the ratio of load resistance and total resistance of slide-wire rheostat as the circuit characteristic coefficient K, and $K = R_L/R_0$. x is the relative position of sliding end C on the slide-wire rheostat, and $x = R_1/R_0$. Then

$$\frac{U}{U_{max}} = \frac{\varepsilon}{\frac{R_L R_1}{R_L + R_1} + R_2} \cdot \frac{R_L R_1}{R_L + R_1} \cdot \frac{1}{\varepsilon} = \frac{R_L R_1}{R_L R_1 + (R_0 - R_1)(R_L + R_1)} \cdot \frac{R_0^2}{R_0^2}$$

$$= \frac{Kx}{Kx + K + x - Kx - x^2} = \frac{Kx}{K + x - x^2} \qquad (S8-2)$$

The voltage dividing characteristic curves under different K values can be drawn by experiments.

8.4 Procedures

(1) Connect the circuits according to the schematic diagram respectively.

(2) $x = R_1/R_0$. Let $K = 0.1$ and $K = 1$ respectively. According to the total resistance of R_0, calculate the value of resistance of load R_L. Regulate the rheostat sliding end position. Increase R_1 from the left to the right with 1/10 of R_0 as an interval. Record the corresponding output current I. Compute the ratio I/I_m. Record the corresponding data in Tab. S8-1 and Tab. S8-2. Draw the relationship between I/I_m and x in the Fig. S8-3 coordinate system. Analyze the phenomena.

(3) $x = R_1/R_0$. Let $K = 0.1$ and $K = 1$ respectively. According to the total resistance of R_0, calculate the value of resistance of load R_L. Regulate the rheostat sliding end position. Increase R_1 from the left to the right with 1/10 of R_0 as an interval. Record the corresponding output voltage U. Compute the ratio U/U_m. Record the corresponding data in Tab. S8-3 and Tab. S8-4. Draw the relationship between U/U_m and x in the Fig. S8-4 coordinate system. Analyze the phenomena.

8.5 Data and calculations

8.5.1 Current-limiting control circuit (Tab. S8-1、Tab. S8-2、Fig. S8-3)

Tab. S8-1 The data of normalized current values at different positions (K = 0.1)

R_L = ______ Ω, R_0 = ______ Ω, $x = R_1/R_0$, I_m = ______ A

x	0	0.1	0.2	0.3	0.4	0.5	0.6	0.7	0.8	0.9	1
I/A											
I/I_m											

Table S8-2 The data of normalized current values at different positions (K = 1)

R_L = ______ Ω, R_0 = ______ Ω, $x = R_1/R_0$, I_m = ______ A

x	0	0.1	0.2	0.3	0.4	0.5	0.6	0.7	0.8	0.9	1
I/A											
I/I_m											

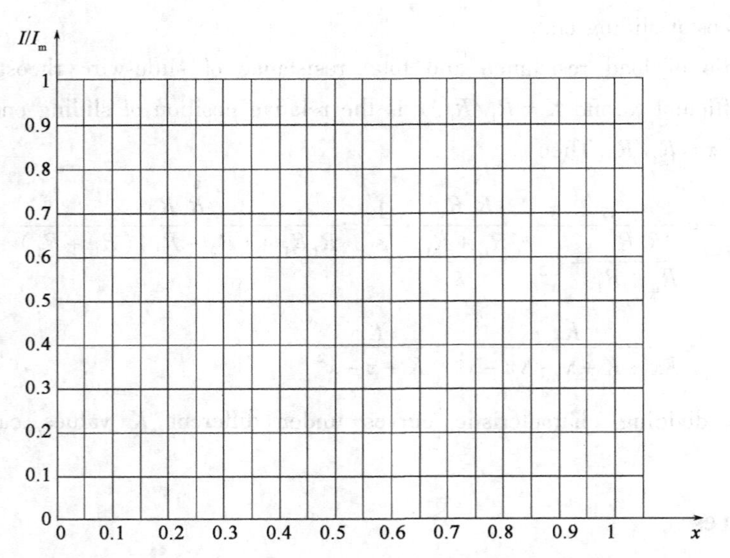

Fig. S8 – 3 Coordinate system for current-limiting control circuit

8.5.2 Voltage dividing control circuit (Tab. S8 – 3、Tab. S8 – 4、Fig. S8 – 4)

Tab. S8 – 3 The data of normalized voltage values at different positions (K = 0.1)

R_L = ______ Ω, R_0 = ______ Ω, $x = R_1/R_0$, U_m = ______ V

x	0	0.1	0.2	0.3	0.4	0.5	0.6	0.7	0.8	0.9	1
U/V											
U/U_m											

Tab. S8 – 4 The data of normalized voltage values at different positions (K = 1)

R_L = ______ Ω, R_0 = ______ Ω, $x = R_1/R_0$, U_m = ______ V

x	0	0.1	0.2	0.3	0.4	0.5	0.6	0.7	0.8	0.9	1
U/V											
U/U_m											

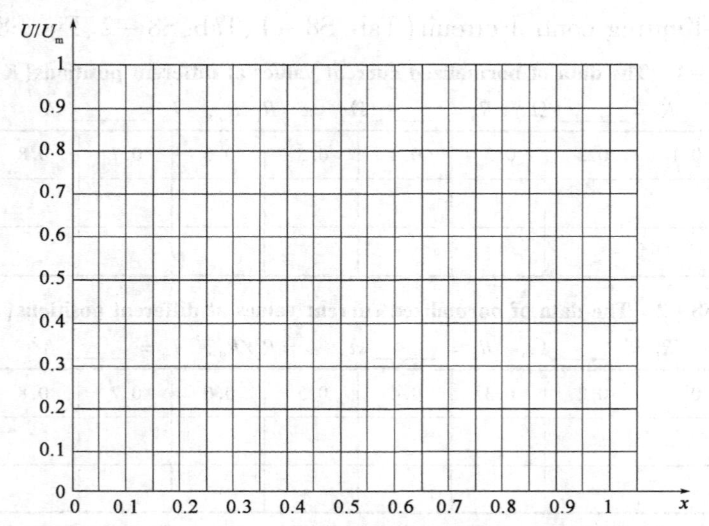

Fig. S8 – 4 Coordinate system for voltage dividing control circuit

8.6 Precautions

(1) Be careful not to mistake the positive and negative poles of ammeter and voltmeter.

(2) Be careful not to confuse the sliding end and the fixed end of the slide-wire rheostat.

(3) There should be a certain order of connection circuit to prevent wrong connection.

Study on the characteristics of the power control circuit is shown in Video S8.

Video S8 Study on the characteristics of the power control circuit

Exp. 9 Newton's rings

9.1 Objectives

(1) Understand the concept of interference.

(2) Learn to measure the radius of curvature of convex lens using the method of Newton's rings.

(3) Master how to use the travelling microscope.

9.2 Apparatus

Travelling microscope, Newton's rings device (plane convex lens, plane glass plate), sodium lamp.

9.3 Principles

In this experiment the physical property of interference of light will be used to determine the wavelength λ, of a light source. The interference fringe system here is a pattern of concentric circles, the diameter of which you will measure with a travelling microscope (which has a Vernier scale). If a clean plano-convex lens is placed on a clean glass plate (optically flat) and viewed in monochromatic light, a series of rings may be seen around the point of contact between the lens and the plate. These rings are known as Newton's rings and they arise from the interference of light reflected from the glass surfaces at the air film between the lens and the plate. The experimental set-up is shown in Fig. S9-1.

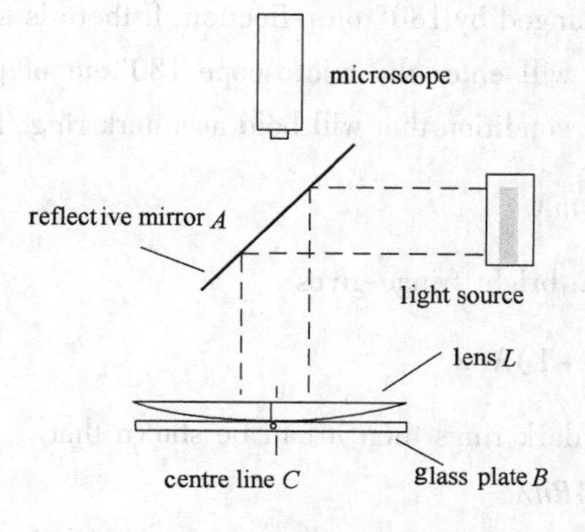

Fig. S9-1 Diagram of principle

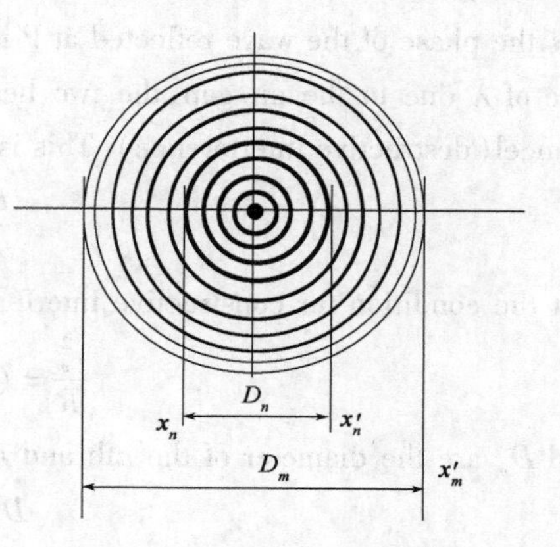

Fig. S9-2 Interference pattern

When a parallel beam of monochromatic light is incident normally on a combination of a plane convex lens L and a glass plate B, as shown in Fig. S9 – 1, a part of each incident light is reflected from the lower surface of the lens, and a part, after refraction through the air film between the lens and the plate, is reflected back from the plate surface. These two beams of reflected light are coherent, hence they will interfere and produce a system of alternate dark and bright rings with the point of contact between the lens and the plate as the center. These rings are known as Newton's ring (Fig. S9 – 2). The fact that the light wave is reflected from air to glass surface introduces a phase shift of P.

Let R be the radius of curvature of the lower surface of the lens. Let r_m be the radius of the mth dark ring, to be measured with the traveling microscope.

Let the corresponding thickness of the air gap at the point P be t. (See Fig. S9 – 3)

The path difference between the beams reflected at Q and P is approximately $2t$ (for vertical viewing, at small radius r).

From geometry (refer to Fig. S9 – 4),

$$r_m^2 = t(2R - t) \tag{S9 – 1}$$

If t is small, we can neglect terms the size of t^2. This gives,

$$t = \frac{r_m^2}{2R} \tag{S9 – 2}$$

Hence the path difference is

$$2t = \frac{r_m^2}{R} \tag{S9 – 3}$$

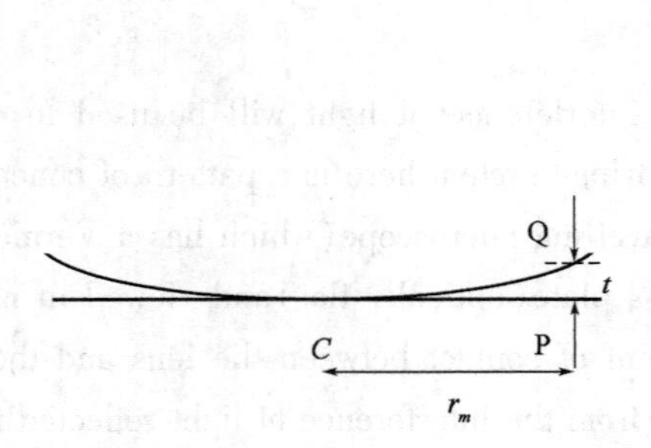

Fig. S9 – 3 Measuring thickness

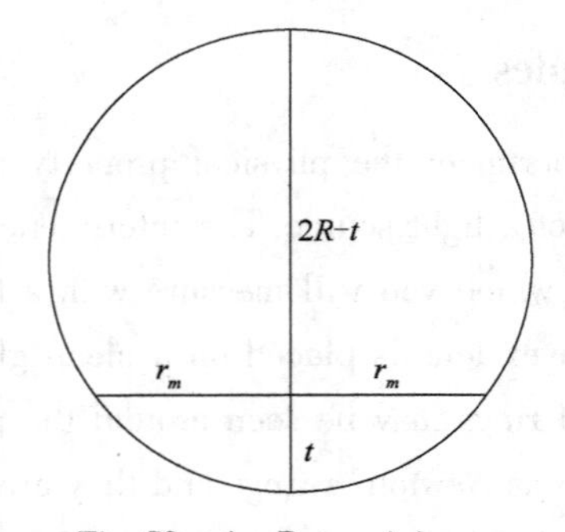

Fig. S9 – 4 Determining r_m

Now the phase of the wave reflected at P is changed by 180° on reflection. If there is also a path difference of λ due to the air gap, the two beams will enter the microscope 180° out of phase and hence cancel (destructive interference). This is the condition that will hold at a dark ring. Therefore,

$$\frac{r_m^2}{R} = m\lambda \tag{S9 – 4}$$

Note that the condition for constructive interference-bright fringe-gives

$$\frac{r_m^2}{R} = (2m + 1)\lambda/2 \tag{S9 – 5}$$

If D_n and D_m are the diameter of the nth and mth dark rings then it can be shown that

$$D_n^2 = 4Rn\lambda \tag{S9 – 6}$$

$$D_m^2 = 4Rm\lambda \tag{S9 – 7}$$

Where λ is the wave length of the light used and R is the radius of curvature of the lower surface of the lens.

From formula (S9 – 6) and formula (S9 – 7), we get

$$R = \frac{D_m^2 - D_n^2}{4(m-n)\lambda} \quad \text{(S9 – 8)}$$

This is the method used in industry for the determination of the radius of curvature of optical surfaces.

9.4 Procedures

In this experiment you will measure the diameter of 30 rings using the Vernier scale on the travelling microscope, and then use this data to measure the radius of curvature of convex lens

(1) Setup Newton's rings device as shown in Fig. S9 – 1. Turn on the sodium lamp. Adjust the angle of reflective mirror until you view the maximum brightness of the travelling microscope field.

(2) Turn the focused hand-wheel so that the eye lens moves slowly from bottom to top until clear interference rings can be seen in the eye lens.

(3) Focus the microscope on the fringes and align the cross-hair tangential to the central dark spot.

(4) Rotate the measurement wheel in one direction so that the microscope moves in the direction of the horizontal cross-hair, the vertical cross-hair will sweep the stripes one by one and count the number of moving dark rings m. When m is 35, turn the measurement wheel in the opposite direction to the 30th dark ring and align the cross-hair to the middle of the dark ring to record the reading (that is, the position reading of the ring). Rotate the measurement wheel and measure the readings of $m = 30, 29, 28, 27, 26$ and $n = 20, 19, 18, 17, 16$ in turn. Then continue to rotate the measurement wheel so that the cross-hair passes through the center of the Newton's rings to the other side, and then measure the readings $n = 16, 17, 18, 19, 20$ and $m = 26, 27, 28, 29, 30$ in turn. The readings filled in Table S9 – 1.

9.5 Data and calculations

The data of measuring the radius of curvature of convex lens see Tab. S9 – 1.

Tab. S9 – 1 The data of measuring the radius of curvature of convex lens

The order of rings	m	30	29	28	27	26
Position/mm	left					
	right					
Diameter D_m/mm						
The order of rings	n	20	19	18	17	16
Position /mm	left					
	right					
Diameter D_n/mm						
D_m^2/mm²						
D_n^2/mm²						
$(D_m^2 - D_n^2)$/mm²						
$\Delta(D_m^2 - D_n^2)$/mm²						

Note: $\Delta(D_m^2 - D_n^2) = \overline{D_m^2 - D_n^2} - (D_m^2 - D_n^2)$.

$\overline{D_m^2 - D_n^2} =$ ________. $\bar{R} = \dfrac{\overline{D_m^2 - D_n^2}}{4(m-n)\lambda} =$ ________. ($\lambda = 5896 \times 10^{-10}$m)

Result: $R = \bar{R} \pm \Delta\bar{R} =$ ________.

Video S9 Newton's rings

9.6 Precautions

(1) Don't touch the surface of optical instrument by hand.

(2) Don't turn on sodium light source frequently.

(3) For eliminate errors due to backlash error, during measurement travelling microscope must be moved one way.

Newton's rings is shown in Video S9.

Exp. 10 Adjustment and use of the spectrometer

10.1 Objectives

(1) Understand the structure of the spectrometer, learn the correct method of adjustment and use;

(2) Grasp the principle of angular cursor and the method of measuring angle by spectrometer;

(3) Determinate apical prism angle by reflection method.

10.2 Apparatus

Spectrometer, sodium discharge lamp, mirror, glass prism, etc.

10.3 Principles

The spectrometer is an instrument for measuring the light deflection angle. It is a kind of sophisticated goniometer. The spectrometer is a basic instrument in optical experiments since many parameters such as refractive index and wavelength can often be identified by measuring the deflection angle of light, Placing a dispersive prism or diffraction grating on the table of the spectrometer makes it a simple spectrometer; the addition of a photodetector to the spectrometer allows a quantitative study of the polarization of light. In order to ensure the accuracy of measurement, a spectrometer must be adjusted before use. There are some things in common among the adjustment of spectrometer and other optical instruments. So learning spectrometer adjustment method is also a basic training for using optical instruments.

10.3.1 The structure of the spectrometer

A schematic diagram of a prism spectrometer is shown in Fig. S10 – 1. It consists of a collimator, a telescope, a circular spectrometer table and Vernier dial. The collimator holds an aperture at one end that limits the light coming from the source to a narrow rectangular slit. A lens at the other end focuses the image of the slit onto the face of the prism. The telescope magnifies the light exiting the prism and focuses it onto the eyepiece. The vernier scale allows the angles at which the collimator and telescope are located to be read off.

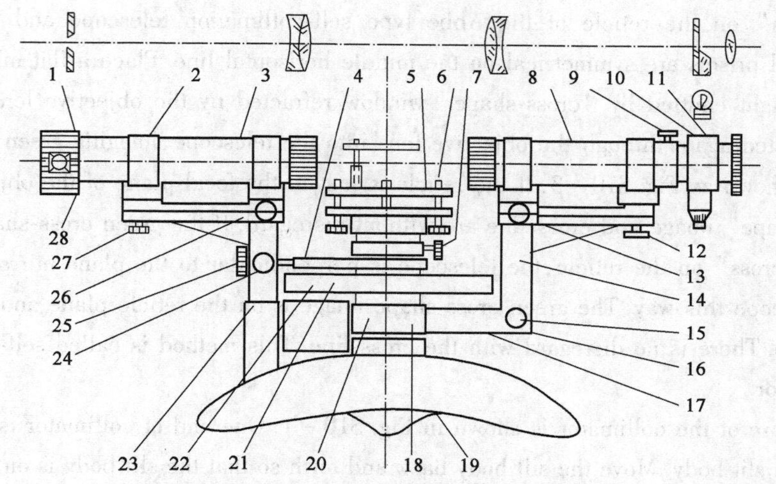

Fig. S10 - 1 Structure of a spectrometer

1—slit plate; 2—lock screw; 3—collimator; 4—braking bracing; 5—table; 6—table leveling screw; 7—table locking screw; 8—telescope; 9—telescope pipe locking screw; 10—Abbe self-collimating eyepiece; 11—adjustment screw of telescope eyeiece focus; 12—adjustment screw of telescope tilt; 13—horizontal axis adjustment screw of telescope; 14—support arm; 15—fine tuning screw; 16—stop screw of the main dial; 17—stop screw of the telescope; 18—braking bracing; 19—pedestal; 20—rotating shaft base; 21—vernier disk; 22—the main dial; 23—stand column; 24—fine tuning screw of vernier disk; 25—stop screw of vernier disk; 26—horizontal adjusting screw for optical axis of collimator; 27—adjustment screw of collimator tilt; 28—adjustment screw of slit

1. Self-collimation telescope

The telescope equipped with the Abbe eyepiece is called Abbe type self-collimation telescope, shown in Fig. S10 - 2. We use it to observe parallel light. Like ordinary telescopes, it consists of objective and an eyepiece. Changing the distance from the objective lens to the eyepiece, you can obtain clear imaging of distant objects at different distances. When the telescope focuses on infinity, the parallel light from infinity can be clearly imaged.

For measurement, a graduated board is attached between the objective lens and the eyepiece, and cross hairs are provided on the graduated board shown as Fig. S10 - 3. Distance between eyepiece and objective lens can be adjusted, cross hairs should be located in the eyepiece focal plane. The eyepiece is composed of a field lens and an eyepiece. The eyepiece of spectrometer is an Abbe eyepiece. There is a transparent total reflection small prism between eyepiece and cross-line. The shape of the prism is a "cross". The light emitted by small lights reflected by the prism makes the cross-line bright. A part of the reticle is covered by a small prism looking from the eyepiece, so only the other parts of the cross-line can be seen.

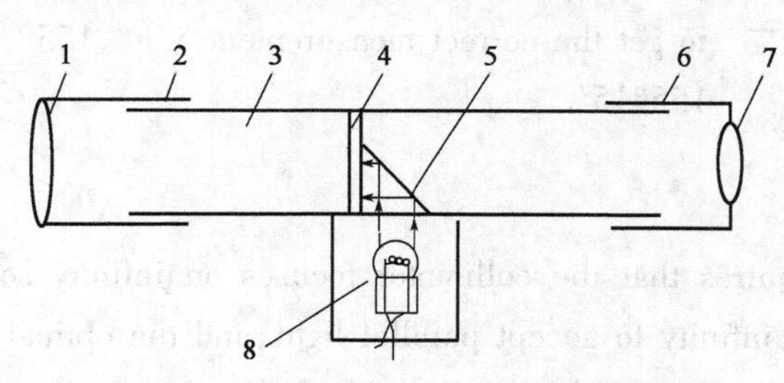

Fig. S10 - 2 Structure of a telescope

1—objective lens; 2—objective lens pipe; 3—fork wire plate tube; 4—graduated board; 5—45° right angle reflection prism; 6—eyepiece sleeve; 7—eyepiece; 8—lighting

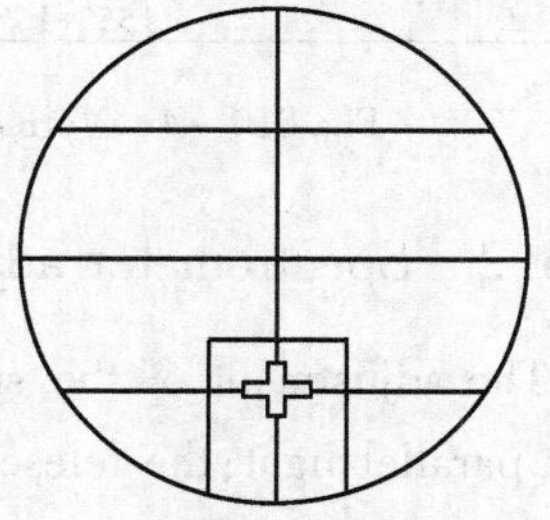

Fig. S10 - 3 Field of view

The " cross " on the reticle of the Abbe type self-collimation telescope and " cross-shape " window on small prisms are symmetrical on the middle horizontal line. Place a flat mirror in front of the telescope. Light emitted by " cross-shape " window refracted by the objective lens, plane mirror reflection, refracted again through the objective lens into the telescope, imaging green " cross-shape " on the reticle, shown as Fig. S10 – 3. If the reticle is just in the focal plane of the objective lens, the green " cross-shape " image and cross-line are without disregard. If the green cross-shape image right on the upper " cross " on the reticle, the telescope is perpendicular to the plane mirror. We can only adjust the telescope this way. The green cross-shape image is on the reticle plane and coincides with cross-shape line. There is no disregard with the cross-line. This method is called self-collimation.

2. Collimator

The structure of the collimator is shown in Fig. S10 – 1. One end of collimator is equipped with a variable width slit body. Move the slit body back and forth so that the slit body is on the focal plane of the parallel convex lens. At this moment, the light on the slit can become a parallel light after passing through the convex lens. The angle of the collimator can also be adjusted by screws on the mount.

3. Reading

A central shaft is fixed at the center of the pedestal, and the dial and the cursor sleeve are sleeved on the central shaft so as to rotate around the central shaft. A thrust bearing support is arranged at the lower end of the dial so that the rotation is light and flexible. A dial engraved with 720 points engraved lines, each grid division has a value of 30′. In order to eliminate we error of the dial, use two cursors A and B with difference of 180 degrees of reading.

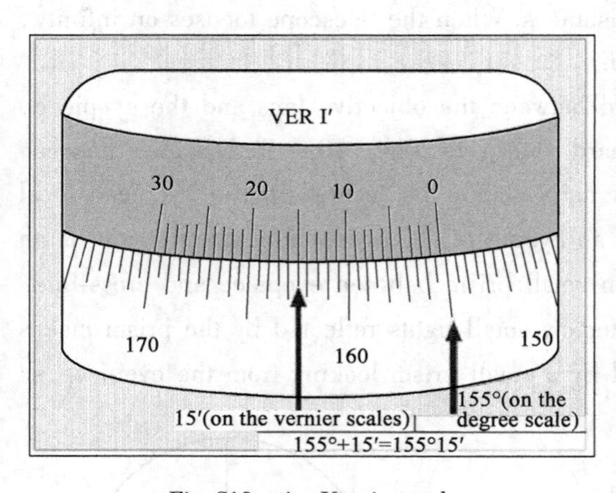

Fig. S10 – 4 Vernier scales

Note: To read the angle, first find where the zero point of the vernier scale aligns with the degree plate and record the value. In Fig. S10 – 4, the zero point on the vernier scale is between the 155° and 155°30′ marks on the degree plate, so the recorded value is 155°. Now find the line on the vernier scale that aligns most closely with any line on the degree scale. In the figure, this is the line corresponding to a measurement of 15 minutes of arc. Add this value to the reading recorded above to get the correct measurement, i. e., 155° + 15′ = 155°15′.

10.3.2 Spectrometer adjustment

The adjustment of the spectrometer requires that the collimator focuses on infinity so that it emits parallel light; the telescope focuses on infinity to accept parallel light; and the optical axes of the collimator and the telescope are perpendicular to the major axis of the spectrometer. Before adjusting, tighten the dial lock screw. In the adjustment, we first adjust roughly with a visual method, and then adopt the sub-four steps below for fine-tuning.

(1) Adjust the telescope with self-collimation so that it focuses on infinity.

Turn on the power, observe the reticle black cross-shape line from the eyepiece, and adjust the distance between the eyepiece and the reticle until the black cross-shape line is visible.

Adjust the table leveling screw so that the table is horizontal. Place a flat mirror on the table so that its mirror surface is perpendicular to the line connecting the two screws B_1 and B_2 on the table and through another screw B_1 as shown in Fig. S10 – 5. We put it this way, when adjusting the inclination of the plane mirror, just adjust B_1 or B_2 on the line. This has nothing to do with B_3.

Find the green cross-shape reflection image. Adjust the objective lens so that the green cross shape image is clear, and move the green cross-shape image to the upper cross line.

Orienting the reticle's cross-shape line rotates the table together with the mirror relative to the telescope. At this moment, the moving direction of the green cross-shape image is parallel to the horizontal line of the cross. After adjusting the above steps, lock the eyepiece tube.

(2) Adjust the optical axis of the telescope to be perpendicular to the spectrometer spindle and ensure that the other mirror plane of the plane mirror is facing the telescope. Look at the eyepiece to see if the green cross-shape image is visible. If not, adjust the screws "12" (do not adjust B_1 or B_2 at this time) so that the green cross-shape image appears in the field of view. Adjust so that the flat mirror on both sides of the reflection of the green cross-shape image appear in the field of view.

Use the "half-adjustment method" to make the telescope optical axis perpendicular to the spindle. Observe whether the reflected green cross-shape image is on the horizontal line. If not, we adjust the platform screws B_1 or B_2 so that the green cross-shape closer to the top line as close as half (Fig. S10 – 6). Adjust the telescope again so that the green cross-shape is on the horizontal line (Fig. S10 – 7). At this point, the optical axis of the telescope is perpendicular to the plane mirror (but not to say that the telescope optical axis is perpendicular to the spectrometer spindle). Then, turn the flat mirror 180 to observe whether the other side of the green cross-shape image is on the horizontal line (Fig. S10 – 7). If not, we use the "half-adjustment method" to repeatedly adjust several times (usually 3 times) until the two sides of the reflected green cross-shape image in the horizontal line. At this point, the optical axis of the telescope is perpendicular to the spectrometer spindle; the horizontal position of the telescope is fixed.

Note: The telescope adjustment screws cannot be moved anymore during the entire experiment.

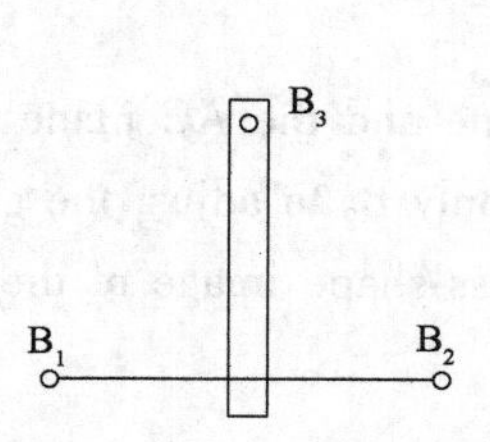

Fig. S10 – 5 Position of mirror

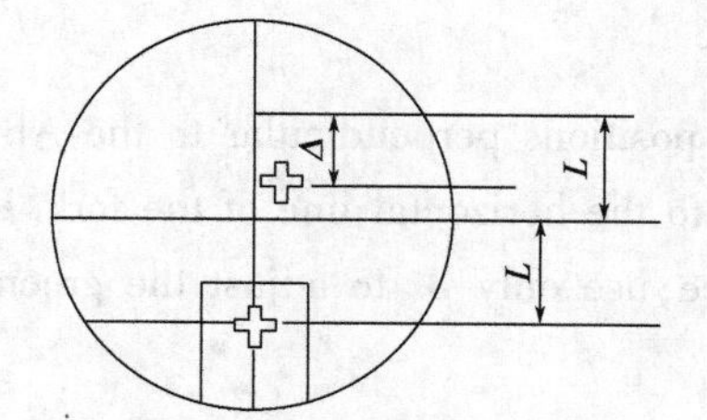

Fig. S10 – 6 Position of green cross

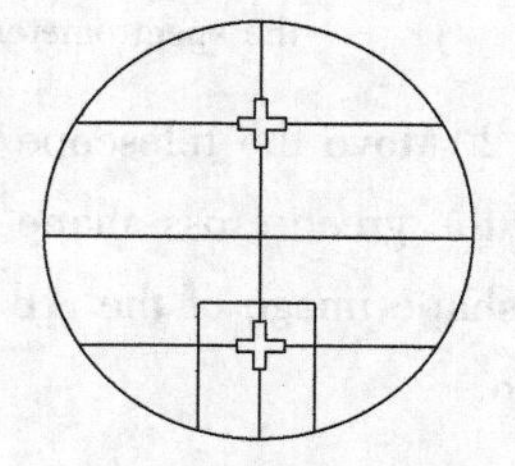

Fig. S10 – 7 Position of green cross

(3) Adjust the collimator so that it focuses on infinity.

Use the light parallel to the slit of the collimator and turn the adjusted telescope to the position of the collimator. Observe the slit image from the eyepiece and adjust the distance between the slit and the lens until the slit image is clearly visible without parallax. At this moment, the slit just happens to be in the focal plane of the collimator objective lens. The light emitted by the objective lens is parallel light. Slipped the slit object so that the slit is parallel to the vertical cross-shape line and then locked. Adjust the slit image width so that it is roughly the same width as the black cross line.

(4) Adjust the collimator perpendicular to the spectrometer spindle.

Adjust the level of parallel light pipe so that the length of the slit image is divided by the middle of the reticle horizontal line (when adjusting, we can offset the slit image from the vertical line of the cross-shape line so that all the slit images can be seen in the field of view). This is shown in Fig. S10 - 8. After adjusting, the horizontal position of the parallel light pipe is fixed.

10.3.3 Triangular prism main section adjustment

After the spectrometer adjustment is complete, place the prism on the table to be measured. Before the measurement, the two reflected faces of the prism must be perpendicular to the spindle of the spectrometer so that the telescope is perpendicular to the side of the prism. At this point, the telescope is perpendicular to the spectrometer spindle; we cannot adjust the telescope. The specific method of adjustment is as follows:

(1) Place the prism on the stage as shown in Fig. S10 - 9, where AB is perpendicular to B_1B_2, BC is perpendicular to B_2B_3, and AC is perpendicular to B_1B_3. The advantage of this placement is to adjust the AB plane perpendicular to the telescope. Tune only B_2. When adjusting the AC plane perpendicular to the telescope, adjust only B_3. Adjustment will affect both AB and AC surface, so do not touch B_1.

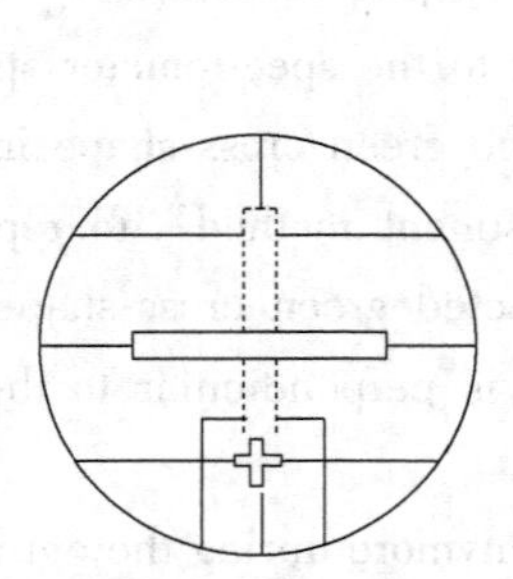

Fig. S10 - 8 Collimator perpendicular to the spectrometer spindle

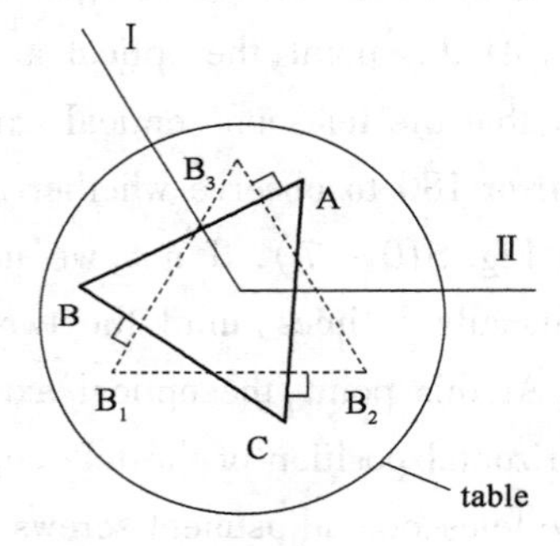

Fig. S10 - 9 Position of tri-angle prism

(2) Move the telescope to the positions perpendicular to the AB plane and the AC plane, and adjust the green cross-shape image to the horizontal line of the fork. Use only B_2 to adjust the green cross-shape image of the AB Surface; use only B_3 to adjust the green cross-shape image of the AC surface.

10.4 Procedures

Measure apical prism angle *A*.

(1) Rotate the table so that the apical of the prism faces the parallelized light tube as shown in Fig. S10 - 9. Rotate the telescope to one side to receive the cross light which is reflected from the

face AB of the prism. Carefully position the telescope until the cross image is centered on the cross-wires. Record the angular position of the telescope as θ_1 in Tab. S10 – 1.

(2) Rotate the telescope around to receive cross light reflected off the other face AC of the prism and record the angular position of the telescope as θ_2 in Tab. S10 – 1.

(3) Rotate the spectrometer table slightly by a few degrees. Repeat the above steps one more times to have a total of two trials in estimating the prism angle A.

(4) Record all your results in Tab. S10 – 1.

(5) Calculate the prism angle A.

$$\varphi = \frac{1}{2}(|\theta_1 - \theta_1'| + |\theta_2 - \theta_2'|)$$

$$A = 180° - \varphi$$

10.5 Data and calculations

Record the data in Tab. S10 – 1.

Tab. S10 – 1 Measurement of the prism angle A

Trial number	θ_1 (left)	θ_2 (left)	θ_1' (right)	θ_2' (right)	φ	$\overline{A}$
1						
2						

10.6 Precautions

(1) When reading, the slit image should be aligned with the vertical cross-line.

(2) Do not touch the optical surface of the optical element by hand.

(3) When adjusting the telescope, coarse adjustment should be followed before fine adjustment.

(4) Do not turn the telescope, parallel light tube, load table, reading plate and focusing system without loosening some fixing screws.

Adjustment and use of the spectrometer is shown in Video S10.

Video S10 Adjustment and use of the spectrometer

Exp. 11 Measurement of the refractive index of a prism

11.1 Objectives

(1) Understand configuration of spectrometer and the function of all parts, and the principle of measure the index of refraction of the prism.

(2) Determine the refraction index of a glass prism by measuring the angle of minimum deflection.

11.2 Apparatus

Prism, Spectrometer, Sodium discharge lamp, Mirror, etc.

11.3 Principles

Suppose that the size of the vertex A of prism is A and the refraction index of prism is n, shown in Fig. S11 - 1. which is incident on the AB surface of the prism an the incident angle α. After being refracted twice by the prism, the light ray emerges from the other surface AC at the angle β to go out. The total variation of the light propagation direction can be represented by the angle δ between the incident light and the extension line of the outgoing light after light is refracted twice through the prism. δ is called the deflection angle. From geometric relationships, we can get

$$A = \angle 1 + \angle 2 \tag{S11 - 1}$$

$$\delta = (\alpha - \angle 1) + (\beta - \angle 2) = \alpha + \beta - A \tag{S11 - 2}$$

From the law of refraction (Snell's law) gives an equation for the index of refraction of the prism, we get

$$\sin\alpha = n\sin\angle 1 \tag{S11 - 3}$$

$$n\sin\angle 2 = \sin\beta \tag{S11 - 4}$$

If the angle of incidence α is varied from some arbitrary starting value, either by moving the light source or rotating the prism, the deflection angle δ will become smaller until it reaches a minimum value, after which continued adjustment of α will cause the deflection angle to increase again. This special, minimum, value of δ is called the angle of minimum deflection which is denoted by δ_m. The angle of minimum deflection is important, because for this particular angle, a relatively simple relationship between δ_m and the index of refraction can be developed (Fig. S11 - 2).

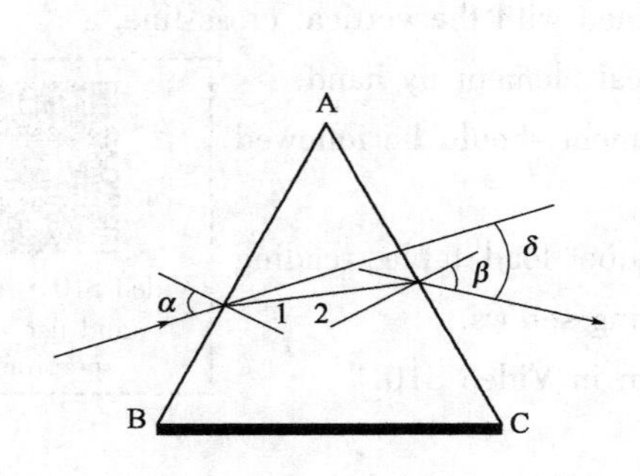

Fig. S11 - 1 Principle of the deflection angle

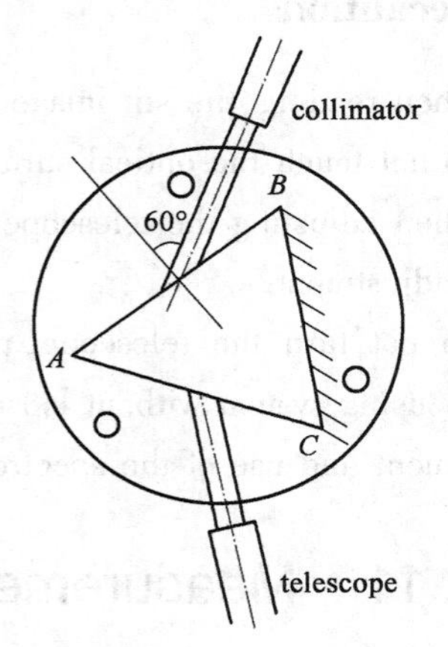

Fig. S11 - 2 Measuring the minimum deflection angle of prism

One can use differential calculus to show that the condition for δ to be minimum is

$$\angle 1 = \angle 2 \quad \text{and} \quad \alpha = \beta \tag{S11 - 5}$$

A light ray incidents on the prism at an angle α such that the above condition is satisfied will pass through the prism symmetrically. Further geometric manipulations reveal that when the minimum deflection conditions are satisfied, A and $\angle 1$ obey rather simple conditions

$$\angle 1 = \angle 2 = \frac{A}{2} \tag{S11 - 6}$$

$$\alpha = \angle 1 + A = \frac{A+\delta_m}{2} \quad (S11-7)$$

From equation (S11 − 3), equation (S11 − 4), equation (S11 − 6) and equation (S11 − 7), We can obtain that the refractive index of the prism is

$$n = \frac{\sin\left(\frac{A+\delta_m}{2}\right)}{\sin\frac{A}{2}} \quad (S11-8)$$

11.4 Procedures

(1) Adjust prism spectrometer (Refer to the adjustment methods of spectrometer in Experiment 10).

(2) The prism angle $A = 60°$ (Fig. S11 − 2).

(3) Measure the angle of minimum deflection:

①Set the slit to a suitable width, using the micrometer screw for slit width. The resolution increases with decreasing slit widths; however, the light intensity of the spectrum is correspondingly reduced.

②Release the telescope stop screw and move the table and telescope to the corresponding position as shown in Fig. S11 − 2.

③Slowly turn the prism table and observe the shift of the silt image with the telescope until silt image just passes through a reversing point (minimum setting).

④Fix the prism table and move telescope until the silt image align with the vertical line. Read the angular coordinate θ_A, θ_B.

⑤Remove the prism and rotate the telescope to incident light on the vertical line. Record the angular coordinate θ'_A, θ'_B.

Repeat the above steps to do three trials.

11.5 Data and calculations

11.5.1 Data records (Tab. S11 − 1)

Tab. S11 − 1 Minimum deflection angle data

Trial number	θ_A (left)	θ_B (left)	θ'_A (right)	θ'_B (right)	δ_{min}	$\overline{\delta_{min}}$
1						
2						
3						

11.5.2 Calculations

(1) Calculate the minimum deflection angle:

$$\delta_{min} = \frac{1}{2}[\,|\theta_B - \theta_A| + |\theta'_B - \theta'_A|\,]\delta_{min} = \frac{1}{2}[\,|\theta_B - \theta_A| + |\theta'_B - \theta'_A|\,]$$

(2) Reflection index: $n = \dfrac{\sin\left(\frac{A+\delta_m}{2}\right)}{\sin\frac{A}{2}}$.

(3) Standard deflection:

$$\frac{u_n}{n}=\left[\left(\frac{1}{2}\cot\frac{A}{2}-\frac{1}{2}\cot\frac{A+\delta_{\min}}{2}\right)^2u_A^2+\left(\frac{1}{2}\cot\frac{A+\delta_{\min}}{2}\right)^2u_A^2\right]^{\frac{1}{2}}$$

$$\sigma_A=\frac{\Delta_{\text{ins}}}{\sqrt{3}}$$

(4) Result: $n \pm u_n =$ ______.

11.6 Precautions

(1) When reading, the slit image should be aligned with the vertical cross-line.

(2) Do not touch the optical surface of the optical element by hand.

(3) When adjusting the telescope, coarse adjustment should be followed before fine adjustment.

(4) Do not turn the telescope, parallel light tube, load table, reading plate and focusing system without loosening some fixing screws.

Exp. 12 Measurement of wavelengths with grating spectrometer

12.1 Objectives

(1) Familiarize and use spectrometer.

(2) Understand the configuration and properties of diffraction grating.

(3) Measure the wavelengths of spectral lines of a Mercury (Hg) source using diffraction grating and a spectrometer.

12.2 Apparatus

Diffraction grating, Spectrometer, Mercury (Hg) source, Spirit level.

12.3 Principles

The diffraction phenomenon of light is an important characterization of the wave nature of light. In modern optical technology, light diffraction has become an important research method in many fields, such as spectral analysis, crystal analysis, optical information processing, etc. Therefore, studying diffraction phenomenon and its law are of great significance both in theory and in practice.

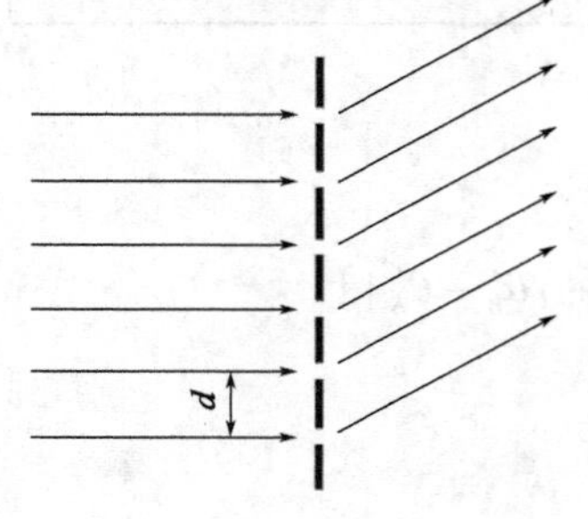

Fig. S12 - 1 Diffraction grating

Diffraction grating is an arrangement of large number of equidistant and parallel slits (Fig. S12 - 1). One of the techniques to manufacture diffraction grating is to rule the equidistant lines on glass plate. Typical diffraction gratings consist of $(1.5 \sim 2) \times 10^4$ lines per inch (this number can reach up to 10^5 lines per inch). The resolving power of grating depends upon number of slits and slit density. Using theory of diffraction to multiple slits, following grating equation can be derived

$$d\sin\theta = m\lambda \qquad (S12-1)$$

Where, d is grating constant; θ is angle of diffraction; m is diffraction order of spectrum; λ is wavelength of light. In equation (S12 – 1), d and m are constant. This implies that θ is proportional to λ. Thus, if a grating is exposed to light from polychromatic source, the colors are separated on account of their different wavelengths. Thus, diffraction grating can form the spectrum of the light. With respect to dispersive power and resolving power, grating is far better than prism. Further, if d and m are known and if θ is measured then λ, the wavelength of spectral lines can be calculated. Due to its ability to form well resolved spectrum and calculation of wavelengths, diffraction grating finds applications in spectrometers. Such spectrometers (Fig. S12 – 2) find applications in an important discipline called spectroscopy, a technique extremely useful in science and technology. Each source has its own characteristic spectrum. In spectroscopy the spectra of various atomic or molecular species are analyzed to evaluate the properties of the sources. A few applications of spectroscopy for understanding the structure and properties of atoms and molecules, detection of various elements in planets and stars, study of various effects such as Zeeman effect, Raman effect, Stark effect, etc.

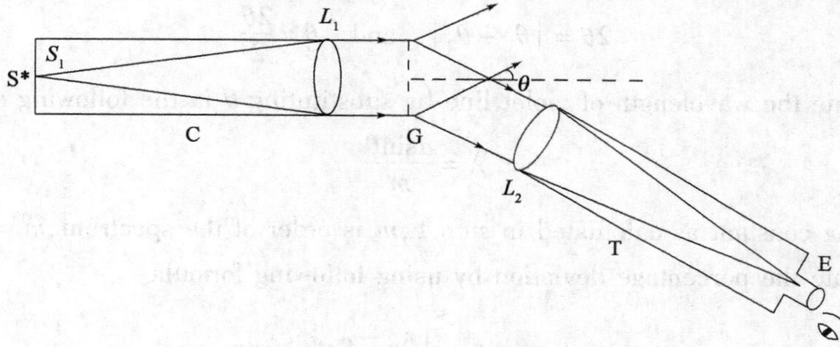

Fig. S12 – 2 Spectrometer

12.4 Procedures

(1) At first calculate the grating constant d of the grating by using following formula

$$d = \frac{1}{N}\text{inch} = \frac{2.54}{N}\text{cm} = \frac{2.54\times10^{8}}{N}\text{Å} \qquad (S12-2)$$

Where, N is number of slits per inch, which equal to 15000 slits per inch.

(2) Switch on the Mercury source, level the all parts of spectrometer such as telescope, collimator, grating table etc. using spirit level.

(3) Bring the slit of collimator in front of spectrometer. Adjust the slit width optimum value.

(4) Adjust the telescope and collimator for sharp images using prism and Schuster's method (Refer to the adjustment methods of spectrometer in Exp. 10).

(5) Mount the diffraction grating on the table such that its plane is exactly perpendicular to collimator axis as well as the table.

(6) Observe the central image of slit through telescope. This image is white, as colors cannot be separated in zero order. This is called as zero order spectrum. Make the image sharp by focusing the telescope and collimator.

(7) Unlike prism, grating produces multiple spectra. Move the telescope on both sides of the central image to observe the first as well as second order spectra on both the sides of the central

image. The second order spectrum is faint as compared to first order. So, consider first order spectrum for observations. Thus, the order of spectrum m in equation (S12 - 1) is 1. The first order spectrum consists of four prominent lines namely violet, green, yellow (doublet) and red. The other relatively faint lines are purple and orange.

(8) Move the telescope on left hand side and adjust the cross wire on violet line. Clamp the telescope. Measure the angular position θ_1 of the violet line, by using following procedure

$$\theta_1 = \text{MSR} + \text{VSR} \times \text{LC} \tag{S12-3}$$

Where MSR: main scale reading, a reading on the scale which coincides with the zero of the vernier scale. If no reading coincides then MSR is the reading on the main scale previous to zero of the vernier scale. VSR: Vernier scale reading is the sequence number of the division on the vernier scale which exactly coincides with the division on main scale. LC: Least count of the angular scale. LC = 1 minute.

(9) Now unclamp the telescope and move it on right side of the central image and focus the cross wire on the violet line. Measure its position θ_2 by using the procedure in step 8.

(10) Calculate θ by using following procedure

$$2\theta = |\theta_1 - \theta_2| \quad \text{and} \quad \theta = \frac{2\theta}{2} \tag{S12-4}$$

(11) Calculate the wavelength of violet line by substituting θ in the following equation.

$$\lambda = \frac{d\sin\theta}{m} \tag{S12-5}$$

Where d is grating constant as calculated in step 1, m is order of the spectrum, $m = 1$.

(12) Calculate the percentage deviation by using following formula

$$\text{deviation}(\%) = \left|\frac{\lambda_e - \lambda_s}{\lambda_s}\right| \times 100\% \tag{S12-6}$$

Where λ_e Experimental wavelength as calculated in step 11, λ_s Standard wavelength, given in the Tab. S12 - 1.

(13) Repeat the same procedure in step 9 to 13 for remaining spectral lines i. e. green, yellow and red.

(14) Tabulate your observations, calculations and results in Tab. S12 - 1.

12.5 Data and calculations

Record the data in Tab. S12 - 1.

Tab. S12 - 1 Observation, calculations and results

d = grating element = ______ Å m = order of the spectrum = 1

No.	Spectral line	Angular position		2θ (deg. min.)	Angle of diffraction θ (deg. min.)	Experimental wavelength λ_e/Å	Standard wavelength λ_s/Å	Deviation %
		θ_1 (left) (deg. min.)	θ_2 (right) (deg. min.)					
1	Violet						4387	
2	Green						5460	
3	Yellow						5790	
4	Red						6330	

12.6 Precautions

(1) Pay attention to adjust and used the Spectrometer.

(2) The grating is a precision optical device. It is forbidden to touch the optical surface by hand.

(3) Don't look directly at the ultraviolet light of the mercury lamp so as not to hurt your eyes.

Exp. 13 Use of the oscilloscope

13.1 Objectives

(1) Understand the construction of oscilloscope and the function of all components.

(2) Learn how to adjust and use the oscilloscope.

13.2 Apparatus

Oscilloscope, signal generator, BNC to BNC connecting wire, alligator clip to BNC connecting wire.

13.3 Principles

13.3.1 The construction of oscilloscope

The oscilloscope is a particularly useful instrument that can be used for testing and fault-finding a variety of electronic circuits, from logic circuits through analogue circuits to radio circuits.

The basic structure of the oscilloscope includes the oscilloscope tube, X and Y-axis amplifier, scanning circuit, synchronous trigger circuit, power supply and other parts, which are shown in Fig. S13 - 1.

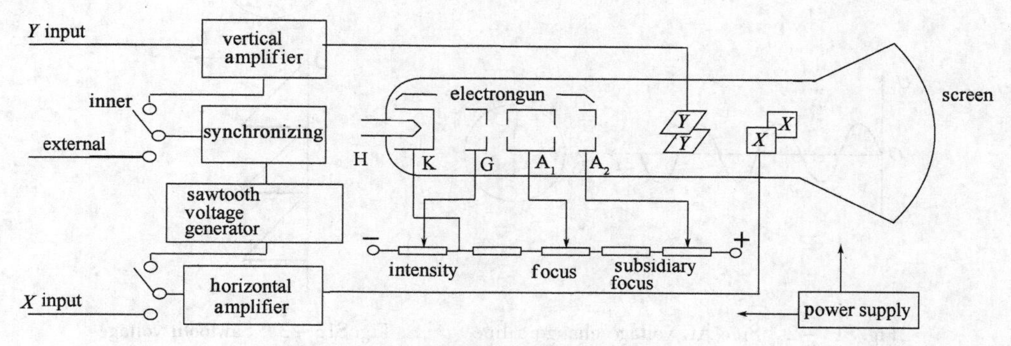

Fig. S13 - 1 Block diagram of an oscilloscope

The oscilloscope tube is the heart of the oscilloscope, which consists of four main parts: an electron gun, a time-base generator (that serves as a clock), two sets of deflection plates used to steer the electron beam, and a phosphorescent screen which lights up when struck by electrons. The electron gun, deflection plates, and the phosphorescent screen are all enclosed by a glass envelope which has been sealed and evacuated.

The key to the versatility of the cathode ray tube(CRT) is two sets of electron deflection plates which deflect the electron beam in the vertical and horizontal directions as a function of the voltage applied to each set of plates. By selecting the appropriate applied voltages to these plates it is possible to deflect the beam so that it strikes any point of the screen. In practice the applied voltage on the plates is usually varied as a function of time causing the beam to sweep in some pattern across the screen. The most common technique is to cyclically apply a linearly increasing voltage, i. e. , the sawtooth voltage, to the horizontal plates causing the beam to sweep from left to right across the screen. Without any vertical signal this appears as a continuous line across the screen, because of the slow decay of the fluorescent screen and the vertical plates beam will be deflected vertically as it sweeps left to right causing a wave pattern to form. Fig. S13 – 1 shows a block diagram of an oscilloscope. The functions of the sub-blocks are given as follows:

(1) The power supply provides the voltages necessary to produce the electron beam.

(2) The screen, the CRT(Cathode-Ray Tube) is often used. It displays in the principle which is the same as the TV. That is, it displays the signal wave form with the deflection of the electron beam (it moves to upper and lower either side).

(3) The vertical amplifier amplifies the input signal to the input signal to the vertical deflection plates so that small amplitude input signals produce an observable deflection of the electron beam (Fig. S13 – 2).

(4) Synchronizing is to make the wave-form of the input signal on the scope stationary, it is necessary to make the scan period of the horizontal axis synchronize with the input signal. Therefore, the way of making the scan period synchronize with the input signal becomes necessary.

(5) The horizontal amplifier amplifies the input signal or the sawtooth voltage which is then applied to horizontal deflection plates(Fig. S13 – 3).

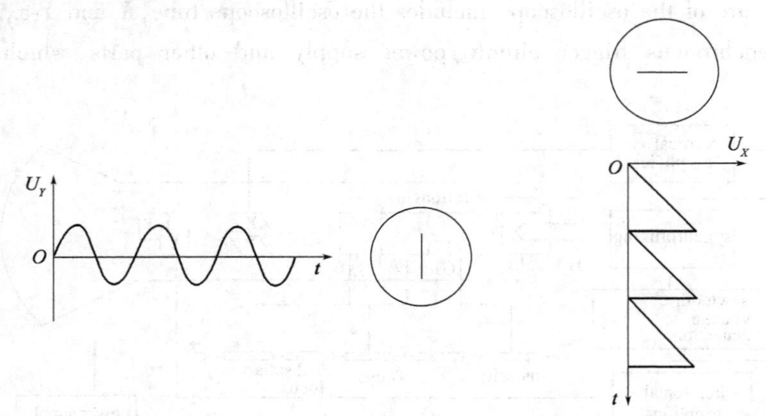

Fig. S13 – 2 Sine AC voltage change a line Fig. S13 – 3 Sawtooth voltage

(6) The sawtooth voltage generator provides the sawtooth voltage that causes the electron beam spot to move from left to right at a constant speed. The sawtooth voltage may be synchronized with a respective vertical input voltage.

13.3.2 The principle of oscilloscope display waveform

When the deflection plate is a certain voltage, the electron beam will deflect with effect of the electric field, and the distance of the spot moving on the screen is proportional to the voltage of the deflection plate. If the Y deflection plate inputs the sinusoidal voltage $U_Y = U_m \sin\omega t$, the x deflector without voltage, the spot light on the screen goes up and down the sine vibration. If the vibration frequency is fast, it looks like a vertical line (Fig. S13 – 2). If the X deflection plate inputs voltage $U_X = kt$ only, the spotlight on the screen goes left and right. If the charging frequency is fast, it looks like a horizontal line (Fig. S13 – 3). When the y deflection plate and the X deflection plate input signal at same time, the bright spot can be expended along the vibration of the Y direction, so as to display the relation curve of the signal voltage U and the time t on the screen. The principle of the wave is shown in Fig. S13 – 4. If the light spot moves evenly along the positive x axis for a uniform U_Y cycles, quickly rebounds to the original starting position, and then repeats the even motion along the positive x axis, the sinusoidal motion trace of the spot will coincide with the previous trace. Repeating the same movement in each cycle, the spot trace will be able to maintain a fixed position. When the repetition rate is larger, you can see a continuous curve the periodic waveform on the screen. Duo to the generation of oscilloscope scan circuit, the sawtooth wave's period T_X (or frequency $f_X = 1/T_X$) can be continuously adjusted by the circuit. When the period of the scan voltage T_X is n times of the period T_Y of the signal voltage, that is, $T_X = nT_Y$ or $f_Y = nf_X$, the screen will display n cycles the waveform. Even if the sawtooth signal made move without synchronizing with the signal, the wave-form which was stationary cannot be shown on the screen. It is necessary to make move at the period which is the same as the period of the input signal or its integer multiple period. To make the wave-form of the input signal on the scope stationary, it is necessary to make the scan period of the horizontal axis synchronize with the input signal. Therefore, the way of making the scan period synchronize with the input signal becomes necessary.

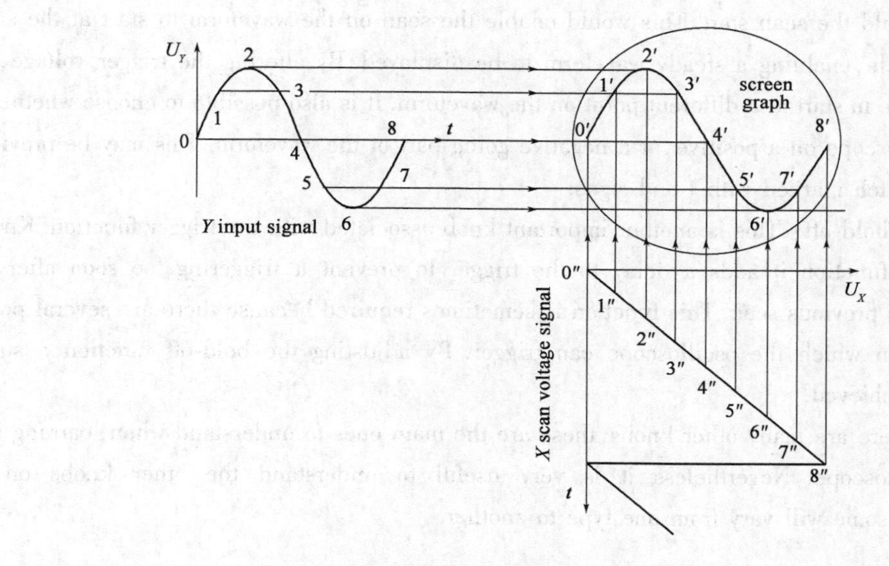

Fig. S13 – 4 Principle of the wave

13.3.3 The basic knob of an oscilloscope

There are a number of controls on any oscilloscope. A short overview of some of the controls is given below:

(1) Vertical sensitivity knob (V/cm): This knob on the oscilloscope alters the gain of the amplifier that controls the size of the signal in the vertical axis. It is generally calibrated in terms of a certain number of volts per centimeter. The vertical sensitivity is increased and the amplitude of the visible waveform on the screen is increased.

When using the oscilloscope, the vertical gain is normally set so that the waveform fills the vertical plane as best as possible, i. e. as large as possible without going outside the visible or calibrated area.

(2) Vertical position: This knob on the oscilloscope governs the position of trace when no signal is present. It is normally set to a convenient line on the graticule so that measurements above and below the "zero" position can be measured easily. It also has an equivalent horizontal position control that sets the horizontal position. Again this one should be set to a convenient position for making any timing measurements.

(3) Timebase (TIME/DIV) knob: The timebase knob sets the speed at which the screen is scanned. It is calibrated in terms of a certain time for each centimeter calibration on the screen. From this the period of a waveform can be calculated. This if a full cycle of a waveform too 10 microseconds to complete, this means that its period is 10 microseconds, and the frequency is the reciprocal of the time period, i. e. 1/10 microseconds = 100kHz.

Normally the timebase knob is adjusted so that the waveform or a particular point on the waveform under investigation can be seen at its best.

(4) Trigger knob: The trigger knob on the oscilloscope sets the point at which the scan on the waveform starts. On analogue oscilloscopes, only when a certain voltage level had been reached by the waveform would the scan start. This would enable the scan on the waveform to start at the same time on each cycle, enabling a steady waveform to be displayed. By altering the trigger voltage, the scan can be made to start at a different point on the waveform. It is also possible to choose whether to trigger the oscilloscope on a positive, or a negative going part of the waveform. This may be provided by a separate switch marked with + and-signs.

(5) Trigger hold-off: This is another important knob associated with the trigger function. Known as the "hold-off" function it adds a delay to the trigger to prevent it triggering too soon after the completion of the previous scan. This function is sometimes required because there are several points on a waveform on which the oscilloscope can trigger. By adjusting the hold-off function a stable display can be achieved.

Although there are many other knobs, these are the main ones to understand when learning how to use an oscilloscope. Nevertheless it is very useful to understand the other knobs on an oscilloscope, but some will vary from one type to another.

13.4 Procedures

13.4.1 Steps in using an oscilloscope

The first stage comes when turning on an oscilloscope and this is where knowing a few steps about how to use an oscilloscope can be very useful.

(1) Turn power on: This may appear obvious but is the first step. Usually the switch will be labelled "Power" or "Line". Once the power is on, it is normal for a power indicator or line indicator light to come on. This shows that power has been applied.

(2) Wait for oscilloscope display to appear: Although many oscilloscopes these days have semiconductor based displays, many of the older ones still use cathode ray tubes, and these take a short while to warm up before the display appears. Even modern semiconductor ones often need time for their electronics to "boot - up". It is therefore often necessary to wait a minute or so before the oscilloscope can be used.

(3) Find the trace: Once the oscilloscope is ready it is necessary to find the trace. Often it will be visible, but before any other waveforms can be seen, this is the first stage. Typically the trigger can be set to the center and the hold-off turned fully counter-clockwise. Also set the horizontal and vertical position controls to the center, if they are not already there. Usually the trace will become visible.

(4) Set vertical sensitivity knob: The next stage is to set the vertical sensitivity knob. This should be set so that the expected trace will nearly fill the vertical screen. If the waveform is expected to be 8 volts peak to peak, and the calibrated section of the screen is 10 centimeters high, then set the gain so that it is 1 volt/centimeter. This way the waveform will occupy 8 centimeters, almost filling the screen.

(5) Set the time base speed: It is also necessary to set the timebase speed on the oscilloscope. The actual setting will depend on what needs to be seen. Typically if a wave form has a period of 10ms and the screen has a width of 12 centimeters, then a timebase speed of 1ms per centimeter or division would be chosen.

(6) Input the signal: With the controls set approximately correctly the signal can be applied and an image should be seen.

(7) Adjust the trigger: At this stage it is necessary to adjust the trigger level and whether it triggers on the positive or negative going edge. The trigger level control will be able to control where on the waveform the timebase is triggered and hence the trace starts on the waveform. The choice of whether it triggers on the positive or negative going edge may also be important. These should be adjusted to give the required image.

(8) Adjust the knobs for the best image: With a stable waveform in place, the vertical sensitivity knob and timebase knobs can be re-adjusted to give the required image.

For a digital oscilloscope, all you need to know is the location of the "Reset button" - this will get you out of all trouble as you can set up digital oscilloscopes in many different ways and they often have options buried in the depths of the menu system.

Note: Never open the oscilloscope as these voltages are extremely dangerous they are high current and high voltage. Even 10mA at 250V(AC) can kill and voltages in the oscilloscope are far higher. The cathode ray oscilloscope (above) is the original oscilloscope and uses a high voltage cathode ray tube. Electrons are forced off a plate at one end using very high voltages and guided by electric fields to the phosphor screen that fluoresce when hit by an electron.

13.4.2 Voltage and frequency measurement

(1) Calibrate the oscilloscope: Connect the calibration signal to channel one or channel two by alligator clip to BNC connecting wire. Turn the knob of voltage calibration to locked position and the knob of time base calibration to locked position. Adjust oscilloscope make the signal stable. Check the voltage and frequency of calibration signal.

(2) Measure AC signals: An AC signal is simply Alternating Current and is more commonly used to describe an alternating voltage as well and the text book AC waveform is the sine wave.

Before you measure a steady AC signal, set the switch to GND and move the trace to the lowest horizontal graticule (black lines on screen). This sets the zero voltage position, now set the switch to AC and put the probe on the AC signal-adjust the channel amplifier to keep the signal on screen. Now set the trigger level and adjust the channel amplifier so that the signal fills the screen and is stable.

Count the number of divisions and multiply by the channel amplifier setting to read the voltage. Of course it's easy to select an easy voltage and amplifier setting to start with e. g. 5V with a 1V/division setting will make the trace move up 5 graticule divisions. For a signal period is same.

13.5 Data and calculations

13.5.1 Measuring AC signal voltage (Tab. S13 - 1)

Tab. S13 - 1 The data of AC voltage

Input signal voltage V	Vertical sensitivity V/cm	Number of divisions (peak to peak)	Measured voltage V

13.5.2 Measuring AC signal frequency (Tab. S13 - 2)

Tab. S13 - 2 The data of AC frequency

Input signal frequency Hz	Time base	Number of divisions (one period)	Period	Frequency
200				
1000				
5000				
25000				

13.6 Precautions

(1) The ground should not be confused with the signal, otherwise it will cause short circuit of the signal and damage to the instrument.

(2) The oscilloscope should be moderate brightness, light point should not stay for a long time, and otherwise it will damage the screen.

(3) Rotate the knob slowly, avoiding damaging the knob.

Use of the oscilloscope is shown in Video S13.

Video S13 Use of the oscilloscope

Exp. 14 Measurement of resistance with Wheatstone bridge

14.1 Objectives

(1) Understand the working principle of the Wheatstone bridge.

(2) Be able to use the Wheatstone bridge to measure resistance.

(3) Understand the concept of Wheatstone bridge's sensitivity.

14.2 Apparatus

QJ24 portable DC Wheatstone bridge, unknown resistance, resistance boxes, galvanometer, power-supply, Slide-wire rheostat, switch, and several wires.

14.3 Principles

14.3.1 The principle of Wheatstone bridge

We have learned how to measure the resistance using a voltmeter and an ammeter. The accuracy of such measurements is limited by the internal resistances of these meters. Ideally, we need the internal resistance of a voltmeter to be infinity, and that of an ammeter to be zero. But that is impossible in reality. All the measurements with these meters cannot avoid systematic errors.

A Wheatstone bridge is an electrical circuit used to measure an unknown electrical resistance by balancing two branches of a bridge circuit, one branch of which includes the unknown component. The primary benefit of a Wheatstone bridge is its ability to provide extremely accurate measurements. It was invented by Samuel Hunter Christie in 1833 and improved and popularized by Sir Charles Wheatstone in 1843. One of the Wheatstone bridge's initial uses was for the purpose of soils analysis and comparison.

The working circuit of Wheatstone bridge is shown in Fig. S14 - 1, in which, R_x is the unknown resistance to be measured, R_1, R_2 and R_s are there resistors, each of which called an arm of the bridge. The two nodes A and C are connected to the power supply and the other two nodes B and D are connected to a galvanometer G, The so-called "bridge" is for the BD branch, its function is to compare the potential U_B and U_D of the B and D two points. Generally, the potential of points B and D are not equal, the pointer of of G will deflect. When the G shows zero, the potentials of B, D are equal, the Wheatstone bridge

Fig. S14 - 1 The principle diagram of Wheatstone bridge.

reaches a balance. Let the currents flowing through the AB, AD, BC and DC branches be I_1, I_2, I_x and I_s. Then

$$I_1 = I_x, \quad I_2 = I_s \tag{S14-1}$$

Since $U_B = U_D$, we have

$$U_{AB} = U_{AD}, \quad U_{BC} = U_{DC}$$

So R_1, R_2, R_s and R_x should meet the following relationship (balance conditions)

$$I_1R_1 = I_2R_2, \quad I_xR_x = I_sR_s \tag{S14-2}$$

We can obtain

$$\frac{R_x}{R_1} = \frac{R_s}{R_2} \quad \text{or} \quad R_x = \frac{R_1}{R_2}R_s \tag{S14-3}$$

This is the balance equation for the Wheatstone bridge. In using this equation you must have the resistors labeled as in Fig. S14 – 1; if you change the labeling you must change the subscripts in equation (S14 – 3) accordingly.

Now take a closer look at the balance equation and you will notice that it involves neither the current through R_x nor the voltage applied to the bridge. This is quite desirable when measuring ohmic resistances since variations in the power supply voltage do not affect the balance conditions. The other side of the coin is that the bridge is not useful in measuring non-ohmic resistances since in that case R depends on I which is undetermined.

The Wheatstone bridge illustrates the concept of a difference measurement, which can be extremely accurate. Variations on the Wheatstone bridge can be used to measure capacitance, inductance, impedance and other quantities, such as the amount of combustible gases in a sample, with an explosimeter. In many cases, the significance of measuring the unknown resistance is related to measuring the impact of some physical phenomenon (such as force, temperature, pressure, etc.) which thereby allows the use of Wheatstone bridge in measuring those elements indirectly.

14.3.2 The sensitivity of Wheatstone bridge

When the bridge is balanced, there will be no current through the galvanometer. Since the sensitivity of the galvanometer is limited, it cannot distinguish the tiny unbalanced current resulting from a little change of R_s.

In order to describe the sensitivity of Wheatstone bridge, we define the relative sensitivity as:

$$S = \frac{\Delta d}{\Delta R_s / R_s} \tag{S14-4}$$

Where, ΔR_s is the change of resistance R_s, Δd is the deflection grids of the galvanometer causing by ΔR_s. Larger S means higher sensitivity of Wheatstone bridge and less error will be occurred in the measurement.

14.3.3 Exchange measurement

The accuracy of the measurement results from the Wheatstone bridge is mainly determined by two factors: one is R_1, R_2 and R_s, the other is the sensitivity of the whole bridge. By exchanging R_x and R_s, we can eliminate the error introduced by R_1 and R_2. Keep the R_1/R_2 ratio and other part of the circuit, exchange the position of R_s and R_x, and relabel R_s to R'_s. Once we find the bridge balance again, we can have

$$R_x = \frac{R_2}{R_1} R'_s \quad \text{(S14-5)}$$

From equation(S14 - 3) and equation(S14 - 5), we get

$$R_x = \sqrt{R_s R'_s} \quad \text{(S14-6)}$$

This method we use here is called the exchange measurement.

14.3.4 The error of portable Wheatstone bridge

$$\Delta = \pm k\% \left(CR_s + \frac{CR_N}{10} \right) \quad \text{(S14-7)}$$

Where, k is the accuracy class of the portable Wheatstone bridge; R is the measured value of the unknown resistance; $R_N = 5000\Omega$ is the benchmark value.

14.4 Procedures

14.4.1 Assemble a Wheatstone bridge and measure the unknown resistance R_x

(1) In accordance with the circuit diagram, connect the circuit. R_1, R_2 and R_s are decade-resistance-boxes, let $R_1 = R_2 = 100\Omega$. In order to protect the circuit, the slide-wire rheostat R_E should be placed at maximum at the beginning.

(2) The power should be turned on after the circuit being checked carefully. Observe the galvanometer in coarse state and let it shows zero by adjusting R_s.

(3) Adjust the slide rheostat R_E for decreasing the output resistance, the sensitivity of the bridge is increased. If the galvanometer does not show zero, adjust R_s so that the galvanometer shows zero again. Repeat the procedure of R_E's adjustment until the output resistance of the R_E is minimum. At this time, the bridge is at the highest sensitivity and the error is the minimum. Then write down the value of R_s.

(4) Measure the sensitivity of the bridge: in the case of balance, changing the value of R_s to make galvanometer's pointer deflect (less than 5 grids), then record the change amount of R_s (ΔR) and the deflecting grids of the galvanometer's pointer (Δd).

14.4.2 Measure the unknown resistance R_x with the box-type bridge.

(1) Roughly measure the unknown resistance R_x with a multimeter.

According to the resistance value range, select the appropriate ratio $C = R_1/R_2$. In order to ensure there are four significant figures for the measurement data, the ratio selection is defined as: if the measured resistance is thousand magnitude resistances select "1", hundred resistances select "0.1", and so on.

(2) Connect the measured resistance between terminals X_1 and X_2. Set the adjustable resistor R for the rough measurement value of multimeter. Press the power-supply switch B_0 and turn on G_2, descending adjust each gear until the galvanometer pointer close to zero when G_1 is turned to coarse position. Then turn G_1 to fine position, carefully adjust the R, so that galvanometer shows zero.

(3) Record the value of the adjustable resistor R, and the measured resistance value can be obtained by multiplying the showing value of R by the ratio C, that is $R_x = CR$.

(4) Measure the sensitivity of the box-type bridge: in the case of the bridge is balanced, changing the value of turntable resistor R makes galvanometer ' s pointer deflecting (less than 5 grids), then record the change amount of R_s (ΔR) and the deflecting grids of the galvanometer ' s pointer(Δd).

14.5 Data and calculations

Record the data in Tab. S14 – 1、Tab. S14 – 2.

Tab. S14 – 1 The data of Assemble Wheatstone bridge and measuring R_x

R_1/Ω	R_2/Ω	R_s/Ω	R_s'/Ω	R_x/Ω	$\Delta R_s/\Omega$	Δd	S
100.0	100.0						

$$R_x = \sqrt{R_s R_s'} = \_\_\_\_\_\_; \quad S = \frac{\Delta d}{\Delta R_s / R_s} = \_\_\_\_\_\_$$

Tab. S14 – 2 Data of measuring R_x with box-type Wheatstone bridge.

C	R	R_x	ΔR	Δd	S	Ω

$$R_x = CR = \_\_\_\_\_\_; \quad S = \frac{\Delta d}{\Delta R / R_x} = \_\_\_\_\_\_; \quad \Delta = \pm k\% \left(CR + \frac{CR_N}{10} \right) = \_\_\_\_\_\_.$$

14.6 Precautions

Video S14 Measurement of resistance with Wheatstone bridge

(1) In the self-assembly bridge experiment, before closing the circuit switch, it is necessary to check whether R_E is in a safe position.

(2) When using the box-type bridge, the B_0 and G buttons cannot be pressed for a long time. When measuring resistance, first press B_0 and then G; when disconnecting, release G first then B_0.

Measurement of resistance With wheatstone bridge is shown in Video 14.

Exp. 15 Measurement of induced magnetic field

15.1 Objectives

(1) Master to measure magnetic field by induced wire.

(2) Investigate the magnetic field of a current carrying coil.

(3) Learn the characteristic of Helmholtz coils.

(4) Check theory of magnetic field piling up.

15.2 Apparatus

Magnetic field measurement box, Helmholtz coils, detecting coil , etc.

15.3 Principles

15.3.1 Magnetic field of a circular loop

A wire coil that is carrying a current I produces a magnetic field $\boldsymbol{B}(\boldsymbol{r})$, where $\boldsymbol{r}$ is the distance from the center of the coil to the field point. The magnetic field $\boldsymbol{B}$ is proportional to the current I in the coil. The strength and direction of the magnetic field depend on $\boldsymbol{r}$. For large distances from the coil ($r \gg R$, where R is the radius of the coil), the shape of the magnetic field of a coil is identical to the electric field produced by a point electric dipole. For large distances both fields fall off as $1/r^3$. In this experiment you will measure the magnetic field of a circular coil at distances that are fairly close to the coil. The large distance approximation is not valid. A constant magnetic field can be measured in many ways. You can use a compass, a Hall Probe, a rotating coil of wire, or nuclear magnetic resonance. In this experiment the magnetic field will not be constant but will vary sinusoidal with time. Such a varying magnetic field will induce a varying voltage in a small coil which will be called the "Detecting coil." The Detecting coil will be used to measure the magnetic field produced by a larger coil called the "field" coil. The current in the field coil will be varied sinusoidally with time and produce a sinusoidally varying magnetic field.

In Fig. S15 – 1, Along the coil axis, if the origin of the coordinates is taken at the center of the coil and if the x axis is taken along the coil axis, the magnitude of the magnetic field B, which points in the x direction, is given by

$$B_x = \frac{\mu_0 N I R^2}{2\left(R^2 + x^2\right)^{\frac{3}{2}}} \qquad \text{(S15 – 1)}$$

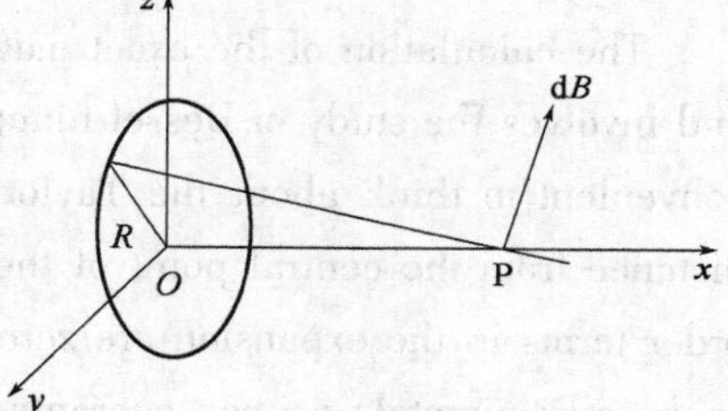

Fig. S15 – 1 Magnetic field on the axis of circular loop

where, B is in tesla; μ_0 is the vacuum permeability; $\mu_0 = 4\pi \times 10^{-7}$ H/m; N is the number of turns of the field coil, $N = 600$; I is the current in the wire, in amperes; R is the radius of the coil in meters, $R = 0.1$ m; x is the axial distance in meters from the center of the coil.

As $x = 0$, we can get the magnetic field B in the y direction,

$$B_y = \frac{\mu N I}{2R}\left[\frac{1}{\left(1 - \frac{y^2}{R^2}\right)} + \frac{\frac{y^2}{R^2}}{4\left(1 - \frac{y^2}{R^2}\right)^{3/2}} + \cdots\right] \qquad \text{(S15 – 2)}$$

15.3.2 Magnetic field of Helmholtz coils

The term Helmholtz coils refers to a device for producing a region of nearly uniform magnetic field. It is named in honor of the German physicist Hermann von Helmholtz.

A Helmholtz pair consists of two identical circular magnetic coils that are placed symmetrically one on each side of the experimental area along a common axis, and separated by a distance h equal to the radius R of the coil (Fig. S15 – 2). Each coil carries an equal electrical current I flowing in the same direction.

Setting $h = R$, which is what defines a Helmholtz pair, minimizes the nonuniformity of the field at the center of the coils, in the sense of setting $\frac{d^2B}{dx^2}=0$, but leaves about 6% variation in field strength between the center and the planes of the coils. A slightly larger value of h reduces the difference in the field between the center and planes of the coils, at the expense of worsening the field's uniformity in the region near the center, as measured by $\frac{d^2B}{dx^2}=0$.

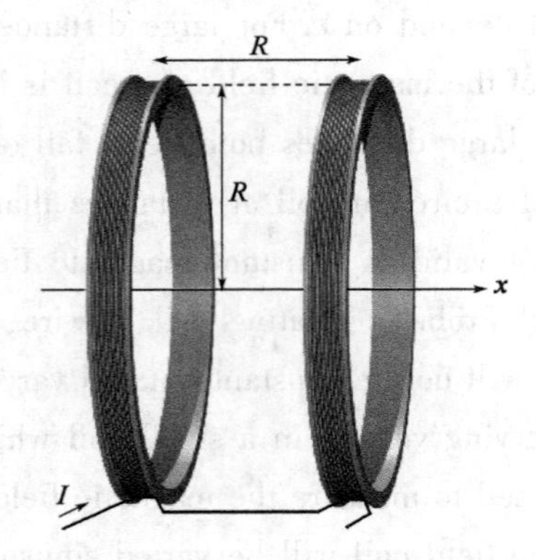

Fig. S15 - 2 Composition of Helmholtz coils

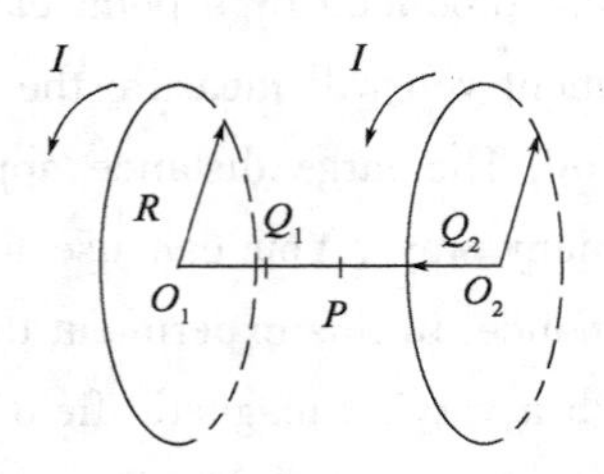

Fig. S15 - 3 Magnetic field of Helmholtz coils

The calculation of the exact magnetic field at any point in space has mathematical complexities and involves the study of Bessel functions. Things are simpler along the axis of the coil-pair, and it is convenient to think about the Taylor series expansion of the field strength as a function of x, the distance from the central point of the coil-pair along the axis (Fig. S15 - 3). By symmetry the odd order terms in the expansion are zero. By separating the coils so that $x=0$ is an inflection point for each coil separately we can guarantee that the order x^2 term is also zero, and hence the leading non-uniform term is of order x^4. One can easily show that the inflection point for a simple coil is $R/2$ from the coil center along the axis; A simple calculation gives the correct value of the field at the center point. If the radius is R, the number of turns in each coil is N and the current flowing through the coils is I, then the magnetic flux density, B at the O_1 and O_2 points will be given by

$$B_{O_1}=B_{O_2}=\frac{\mu NI}{2R}+\frac{\mu NIR^2}{2\,(R^2+R^2)^{3/2}}=0.677\frac{\mu NI}{R} \tag{S15 - 3}$$

On the x-axis P point magnetic field B is

$$B_P=2\frac{\mu NIR^2}{2\left[R^2+\left(\frac{R}{2}\right)^2\right]^{3/2}}=0.724\frac{\mu NI}{R} \tag{S15 - 4}$$

On the x-axis Q_1 and Q_2 point(distance from P point equal to $R/4$) magnetic field B is

$$B_{Q_1}=B_{Q_2}=\frac{\mu NIR^2}{2\left[R^2+\left(\frac{R}{4}\right)^2\right]^{3/2}}+\frac{\mu NIR^2}{2\left[R^2+\left(\frac{3}{4}R\right)^2\right]^{3/2}}=0.712\frac{\mu NI}{R} \tag{S15 - 5}$$

Summarizing, Helmholtz coils produces a uniform magnetic field .

15.3.3 Measurement of magnetic field by electromagnetic induction

The voltage generated in the coil is due to electromagnetic induction. Assume that the Detecting coil is small and that the magnetic field at a given instant of time is approximately uniform over the area of the coil. For this situation the flux Φ of the vector $\boldsymbol{B}$ through the Detecting coil is defined as the product of the area A of the coil times the component of $\boldsymbol{B}$ normal to the plane of the detecting coil.

A coil carrying an alternating current produces an alternating magnetic field. Supposed a Detecting coil is put in the field, Faraday's law of induction then gives for the voltage induced in one turn of the Detecting coil as $-\mathrm{d}\Phi/\mathrm{d}t$. If n is the number of turns in the Detecting coil, the voltage ε induced in the Detecting coil is

$$|\varepsilon| = \frac{\mathrm{d}\Phi}{\mathrm{d}t} = n\omega SB_m\cos\theta\cos\omega t \tag{S15-6}$$

As $\cos\theta = 1$, the maximal magnetic field is

$$B_m = \frac{\sqrt{2}\,\varepsilon_m}{n\omega S} \tag{S15-7}$$

The effective value is

$$B = \frac{\varepsilon_{Em}}{n\omega S} \tag{S15-8}$$

Summarizing, in the alternating magnetic field, by an alternating voltmeter the maximal effective induced magnetic field of the detecting coil is

$$B = \frac{108\varepsilon}{13\pi n d^2 \omega} \tag{S15-9}$$

Where, n is the turns of the detecting coil, $n = 1200$ turns; d is the outer diameter of the detecting coil, $d = 12.8\mathrm{mm}$; ω is the angular-velocity of the alternating current, $\omega = 2\pi f$, f is 400 Hz.

To determine the direction of the magnetic field, it is more accurate to rotate the coil so that a maximum is seen rather than to look for a zero or minimum signal. Add or subtract 90 degree to the orientation of the detecting coil to find the direction of the field.

15.4 Procedures

Fig. S15-4 is main panel for magnetic field measurement box.

(1) Connect excitation voltage to the coil one by wire.

(2) Set measuring conversion knob to position-current.

(3) Adjust the excitation adjusting knob to obtain suitable working current I.

(4) Connect the detecting coil to the detecting coil terminals of the magnetic field measurement box by wires.

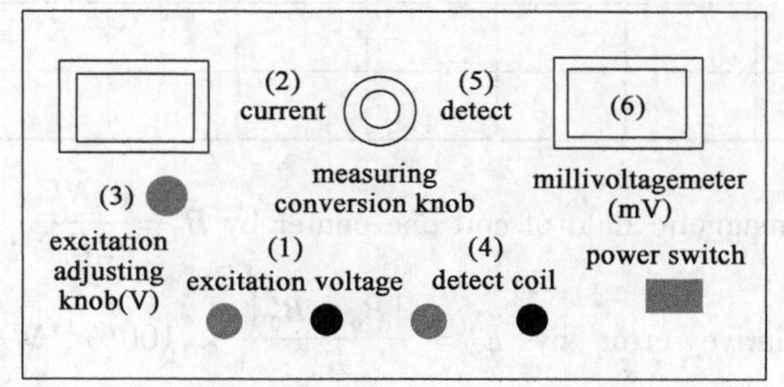

Fig. S15-4 Main panel for magnetic field measurement box

(5) Set measuring conversion knob to detect.

(6) Rotate the direction of the detecting coil and measure the maximum inductive electromotive force of the detecting coil in the axial and radial direction of the coil one.

(7) Connect excitation voltage to the coil two by wire and keep current I constant. Measure the maximum inductive electromotive force of the detecting coil in the axial direction of the coil two.

(8) Connect excitation voltage to the Helmholtz coils by wires and adjust current I to the same value as before. Measure the maximum inductive electromotive force of the detection coil in the axial direction of the Helmholtz coils.

15.5 Data and calculations

(1) Data records (Tab. S15 - 1 ~ Tab. S15 - 4).

Tab. S15 - 1 The magnetic field data of coil one on the radial

Current I = ______ mA

Y/cm	0	1	2	3	4
ε/mV					
$B/10^{-5}$T					

Tab. S15 - 2 The magnetic field data of coil one on the axis

Current I = ______ mA

X/cm	0	1	2	3	4	5	6	7	8	9	10
ε/mV											
$B/10^{-5}$T											

Tab. S15 - 3 The magnetic field data of coil two on the axis

Current I = ______ mA

X/cm	0	1	2	3	4	5	6	7	8	9	10
ε/mV											
$B/10^{-5}$T											

Tab. S15 - 4 The magnetic field data of Helmholtz coils

Current I = ______ mA

X/cm	0	1	2	3	4	5	6	7	8	9	10
ε/mV											
$B/10^{-5}$T											

(2) Calculating the magnetic field of coil one center by $B_0 = \frac{\mu_0 NI}{2R}$.

(3) Calculating relative error by $E = \frac{|B_0 - B_0'|}{B_0} \times 100\%$, Where B_0' is experimental measurement value.

15.6 Precautions

Video S15 Measurement of induced magnetic field

(1) Excitation voltage output prohibit short.

(2) During magnetic field description, strong electromagnetic interference should be eliminated, otherwise the inductive electromotive force of the detecting coil would be unstable or wrong.

Measurement of induced magnetic field is shown in Video S15.

Exp.16 Use of the multi-meter

16.1 Objectives

(1) Understand the working principle of multi-meter.

(2) Master how to use the multi-meter.

16.2 Apparatus

Multi-meter, measured circuit board.

16.3 Principles

A multi-meter is an instrument for measuring the properties of an electrical circuit (such as resistance, voltage, or current). This instrument allows you to check for the presence of voltage on a circuit, etc. This experiment will discuss analog meters.

Become familiar with the parts of a Multi-meter. Inspect the meter. Starting from the top and working to the bottom (Fig. S16 - 1).

The dial (Fig. S16 - 2). The pointer indicates values read from the scale. The pointer or needle, this is the thin black line at the left-most position in the dial face window in the image. The needle moves to the value measured. Arc shaped lines or scales on the meter dial face. May be different colors for each scale but will have different values. These determine the ranges of magnitude.

Fig. S16 - 1 Multi-meter

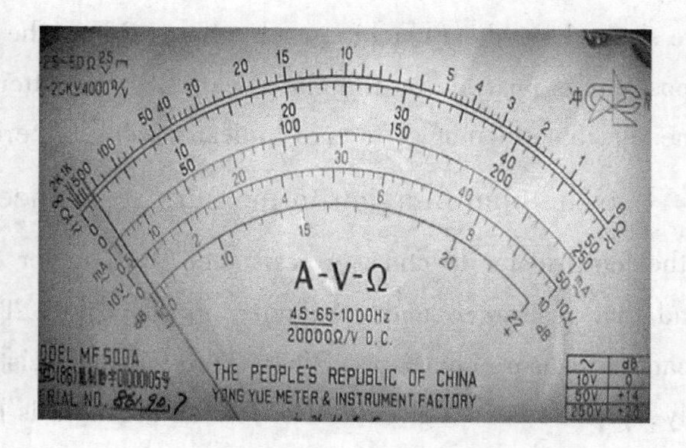

Fig. S16 - 2 The dial of the multi - meter

A wider mirror. The mirror is used to help reduce parallax viewing error by lining up the pointer with its reflection before reading the value the pointer is indicating. In the image, it appears as a wide gray strip between the red and black scales.

A selector switch or knob. This allows changing the function (volts, ohms, amps) and scale (×1, ×10, etc.) of the meter. Many functions have multiple ranges. It is important to have both set correctly, otherwise serious damage to the meter or harm to the operator. Most meters employ the knob type like the one shown in the image, but there are others. Regardless of the type, they work similarly. Some meters (like the one in the image above) have an "Off" position on this selector switch while others have a separate switch to turn the meter off. The meter should be set to off when stored.

Jacks or openings in the case to insert test leads. Most multi-meters have several jacks. The normal type has just two. One is usually labeled "COM" or (–) for common and negative. This is where the black test lead is connected. It will be used for nearly every measurement taken. The other jack(s) is labeled "V(+) and the Omega symbol" (an upside down horseshoe) for Volts and Ohms respectively and positive. The + and – symbols represent the polarity of probes when set for and testing DC volts. If the test leads were installed as suggested, the red lead would be positive as compared to the black test lead. This is nice to know when the circuit under test isn't labeled + or –, as is usually the case. Many meters have additional jacks that are required for current or high voltage tests. It is equally important to have the test leads connected to the proper jacks as it is to have the selector switch range and test type (volts, amps, ohms) set. All must be correct. Consult the meter manual if unsure which jacks should be used.

Test leads. There should be two test leads or probes. Generally, one is black and the other red.

Battery and fuse compartment. Usually found on the back, but sometimes on the side. This holds the fuse (and possibly a spare) and the battery that supplies power to the meter for resistance tests. The meter may have more than one battery and they may be of different sizes. Sometimes there is more than one fuse. Fully charged batteries will be required for resistance / continuity tests.

Zero position adjustment. This is a small knob usually located near the dial that is labeled "Ohms Adjust", "0 Adj", or similar. This is used only in the ohms or resistance range while the probes are shorted together (touching each other). Rotate the knob slowly to move the needle as close to the 0 position as possible on the ohms scale. If new batteries are installed, this should be easy to do – a needle that will not go to zero indicates weak batteries that should be replaced.

16.3.1 Use the ohm function to measure resistance

Set the multi-meter to ohms or resistance (turn meter on if it has a separate power switch). Understand that resistance and continuity are opposites. The multi-meter measures resistance in ohms, it cannot measure continuity. When there is little resistance there is a great deal of continuity. Conversely, when there is a great deal of resistance, there is little continuity. With this in mind, when we measure resistance we can make assumptions about continuity based on the resistance values measured. Observe the meter indication. If the test leads are not in contact with anything, the needle or pointer of an analog meter will be resting at the left most position. This is represents an infinite amount of resistance, or an "open circuit"; it is also safe to say there is the no continuity. Careful inspection of the dial should reveal the OHM scale. It is usually the top-most scale and has values that are highest on the left of the dial (a sideways "∞" for infinity) and gradually reduce to 0 on the right. This is opposite of the other scales; they have the lowest values on the left and increase going right.

Connect the black test lead to the jack marked "Common" or " – "

Connect the red test lead to the jack marked with the Omega (Ohm symbol) or letter "R" near it.

Set the range (if provided) to $R \times 100$.

Hold the probes at the end of the test leads together. The meter pointer should move fully to the right. Locate the "Zero Adjust" knob and rotate so that the meter indicates "0" (or as close to "0" as possible). Note that this position is the "short circuit" or "zero ohms" indication for this $R \times 1$ range of this meter. Always remember to "zero" the meter immediately after changing resistance ranges.

Replace batteries if needed. If unable to obtain a zero ohm indication, this may mean the batteries are weak and should be replaced. Retry the zeroing step above again with fresh batteries.

Measure resistance of something like a known-good light bulb. Locate the two electrical contact points of the bulb. They will be the threaded base and the center of the bottom of the base. Have a helper hold the bulb by the glass only. Press the black probe against the threaded base and the red probe against the center tab on the bottom of the base. Watch the needle move from resting at the left and move quickly to 0 on the right.

Change the range of the meter to $R \times 1$. Zero the meter again for this range. Repeat the step above. Observe how the meter did not go as far to the right as before. The scale of resistance has been changed so that each number on the R scale can be read directly. In the previous step, each number represented a value that was 100 times greater. Thus, 150 really was 15000 before. Now, 150 is just 150. Had the $R \times 10$ scale been selected, 150 would have been 1500. The scale selected is very important for accurate measurements. With this understanding, study the R scale. It is not linear like the other scales. Values at the left side are harder to accurately read than those on the right. Trying to read 5 ohms on the meter while in the $R \times 100$ range would look like 0. It would be much easier at the $R \times 1$ scale instead. This is why when testing resistance, adjust the range so that the readings may be taken from the middle rather than the extreme left or right sides.

Test resistance between hands. Set the meter to the highest $R \times$ value possible. Zero the meter. Loosely hold a probe in each hand and read the meter. Squeeze both probes tightly. Notice the resistance is reduced. Let go of the probes and wet your hands. Hold the probes again. Notice that the resistance is lower still. For these reasons, it is very important that the probes not touch anything other than the device under test. A device that has burned out will not show "open" on the meter when testing if your fingers provide an alternate path around the device, like when they are touching the probes.

16.3.2 Use the volts function to measure voltage

Set the meter for the highest range provided for AC or DC volts. Usually, the voltage to be measured has a value that is unknown. For this reason, the highest range possible is selected so that the meter circuitry and movement will not be damaged by voltage greater than expected. If the meter were set to the 50 volt range and a common electrical outlet were to be tested, the 120 volts present could irreparably damage the meter. Start high, and work downward to the lowest range that can be safely displayed.

Insert the black probe in the "COM" or " – " jack.

Insert the red probe in the "V" or " + " jack.

Locate the Voltage scales. There may be several Volt scales with different maximum values. The range chosen the selector knob determines which voltage scale to read. The maximum value scale should coincide with selector knob ranges. The voltage scales, unlike the Ohm scales, are linear. The scale is accurate anywhere along its length. It will of course be much easier accurately reading 24V on a 50V scale than on a 250V scale, where it might look like it is anywhere between 20V and 30V.

Test a common electrical outlet. In China you might expect 220V. In other places, 110V, 240V or 380V might be expected. Press the black probe into one of the straight slots. It should be possible to let go of the black probe, as the contacts behind the face of the outlet should grip the probe, much like it does when a plug is inserted. Insert the red probe into the other straight slot. The meter should indicate a voltage very close to 220V (depending on type outlet tested). Remove the probes, and rotate the selector knob to the lowest range offered, that is greater than the voltage indicated (220V). Reinsert the probes again as described earlier. The meter may indicate between 210V and as much as 230V this time. The range of the meter is important to obtain accurate measurements. If the pointer did not move, it is likely that DC was chosen instead of AC. The AC and DC modes are not compatible. The correct mode MUST be set. If not set correctly, the user would mistakenly believe there was no voltage present. This could be deadly. Be sure to try both modes if the pointer does not move. Set meter to AC volts mode, and try again. Whenever possible, try to connect at least one probe in such a way that it will not be required to hold both while making tests. Some meters have accessories that include alligator clips or other types of clamps that will assist doing this. Minimizing your contact with electrical circuits drastically reduces that chances of sustaining burns or injury.

16.3.3 Use the amps function to measure amperes

Determine if AC or DC by measuring the voltage of the circuit as outlined above.

Set the meter to the highest AC or DC amp range supported. If the circuit to be tested is AC but the meter will only measure DC amps (or vice-versa), stop. The meter must be able to measure the same mode (AC or DC) Amps as the voltage in the circuit, otherwise it will indicate 0.

Be aware that most Multi-meters will only measure extremely small amounts of current, in the μA and mA ranges. 1μA is 0.000001A and 1mA is 0.001A. These are values of current that flow only in the most delicate electronic circuits, and are literally thousands (and even millions) of times smaller than values seen in the home and automotive circuits that most homeowners would be interested testing. Just for reference, a typical 100W/120V light bulb will draw 0.833A. This amount of current would likely damage the meter beyond repair. A "clamp-on" type ammeter would be ideal for the typical homeowner requirements, and does not require opening the circuit to take measurements (see below). If this meter were to be used to measure current through a 4700Ω resistor across 9V DC, it would be done as outlined below:

Insert the black probe into the "COM" or " – " jack.

Insert the red probe into the "A" jack.

Shut off power to the circuit.

Open the portion of the circuit that is to be tested (one lead or the other of the resistor). Insert the meter in series with the circuit such that it completes the circuit. An ammeter is placed IN SERIES with the circuit to measure current. It cannot be placed "across" the circuit the way a voltmeter is used (otherwise the meter will probably be damaged). Polarity must be observed. Current flows from the positive side to the negative side. Set the range of current to the highest value.

Apply power and adjust range of meter downward to allow accurate reading of pointer on the dial. Do not exceed the range of the meter, otherwise it may be damaged.

16.3.4 Error

$$\text{Error} = \text{precision} \times \text{range}$$

Multimeter is considering a digital meter instead of the older analog types. Digital meters usually offer automatic ranging and easy to read displays. Since they are electronic, the built-in software helps them withstand incorrect connection and ranges better than the mechanical meter movement in analog types.

16.4 Procedures

(1) Measure AC voltage. The Experiment board diagram is Fig. S16 – 3.

(2) Measure DC voltage. Choose appropriate range, measure volt (U_{ad}, U_{ab}, U_{bc}, U_{cd}), refer to experiment board, calculate $U_{ab} + U_{bc} + U_{cd}$

(3) Measure DC current, Open the switch K, Serial the Multi-meter to aa′ port.

(4) Switch off power supply, measure the resistor on the board.

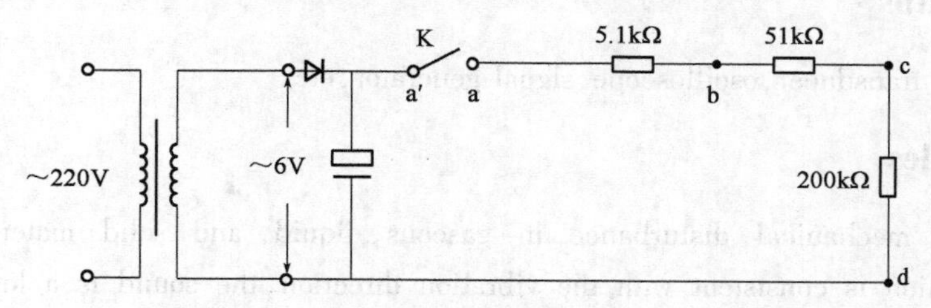

Fig. S16 – 3 The diagram of measurement circuit

16.5 Data and Calculations

(1) AC voltage.

(2) DC voltage.

U_{ad} = ________, $U'_{ad} = U_{ad} + u_U$ = ________.

U_{ab} = ________, $U'_{ab} = U_{ab} + u_U$ = ________.

U_{bc} = ________, $U'_{bc} = U_{bc} + u_U$ = ________.

U_{cd} = ________, $U'_{cd} = U_{cd} + u_U$ = ________.

(3) DC current.

I'_{aa} = ________, $I' = I + u_I$ = ________.

(4) Resistor.

R_{ab} = ________, R_{bc} = ________, R_{cd} = ________.

16.6 Precautions

(1) When you are going to check any part for continuity, you must remove the power. Meters supply their own power from an internal battery. Leaving power on while testing resistance will damage the meter.

(2) Always check meters on known good voltage sources to verify operational status before using. A broken meter testing for volts will indicate 0 volts, regardless of the amount present.

(3) Never connect the meter across a battery or voltage source if it is set to measure current (amps).

Exp. 17 Determination of the sound velocity in air by standing wave

17.1 Objectives

(1) Measure the velocity of sound in air by means of standing wave.

(2) Understand the working principle of piezoelectric transducers.

(3) Learn to deal with data by successive difference method.

17.2 Apparatus

Piezoelectric transducer, oscilloscope, signal generator, etc.

17.3 Principles

Sound is a mechanical disturbance in gaseous, liquid, and solid material. Because its propagation direction is consistent with the vibration direction, the sound is a longitudinal wave. Vibration frequency between 20Hz to 20kHz sound can be heard by people, more than 20kHz frequency called ultrasonic sound waves.

17.3.1 Measurement of sound velocity

The velocity of sound in air can be measured by two very different methods; you can make a direct measurement by measuring the time Δt, it takes a sound pulse to travel a known distance ΔL

(so $v = \Delta L/\Delta t$), or you can measure the wavelength λ of sound of a known frequency f and determine the velocity from the expression $v = \lambda f$.

You will determine the speed of sound in air by measuring the wavelength of a standing wave for a sound of known frequency. A standing wave is what you get when two or more traveling waves combine in such a way that there are some places where there is no motion at all, and those places are called nodes. For any wave with wavelength λ (in m) and frequency f (in vibrations/s, or 1/s, or Hz), the speed of the wave (v, in m/s) is:

$$v = \lambda f \tag{S17-1}$$

A sound wave is a traveling variation in air pressure (the air itself is not transported from one side of the room to the other). The speed it travels depends on the pressure, humidity and temperature of the air. High humidity, high temperature and high pressure all of them will lead to a higher speed v.

From our college physics text, we have a general expression for the speed of sound in a gas, from which we can derive the expression:

$$v = \sqrt{\frac{\gamma RT}{\mu}} \tag{S17-2}$$

where, $\gamma = C_p/C_v = 1.4$, $R = 8314$ J/(kmol · K), T = temperature of the room during the experiment in K, and $\mu = 28.8$ kg/kmol. Thus, by measuring the temperature of the room in "℃" and adding 273 to convert it to "K", you can make an independent estimate of what the speed of sound should be. We can get the sound wave velocity $v_0 = 331.5$ m/s, the room temperature in t(℃), p_s is The saturated vapor pressure of the air in t(℃), $p = 1.013 \times 10^5$ Pa, the formula is given

$$v = v_0\sqrt{1 + \frac{t}{T_0}\left(1 + 0.31\frac{rp_s}{p}\right)} \tag{S17-3}$$

Where $T_0 = 273.15$ K.

17.3.2 Ultrasonic piezoelectric transducer

The transducer is a very important part of the ultrasonic instrumentation system. The transducer incorporates a piezoelectric element, which converts electrical signals into mechanical vibrations (transmit mode) and mechanical vibrations into electrical signals (receive mode). Many factors, including material, mechanical and electrical construction, and the external mechanical and electrical load conditions, influence the behavior of a transducer. Mechanical construction includes parameters such as the radiation surface area, mechanical damping, housing, connector type and other variables of physical construction. As of this writing, transducer manufacturers are hard pressed when constructing two transducers that have identical performance characteristics.

A cut away of a typical contact transducer is shown in Fig. S17-1. It was previously learned that the piezoelectric element is cut to 1/2 the desired wavelength. To get as much energy out of the transducer as possible, an impedance matching is placed between the active element and the face

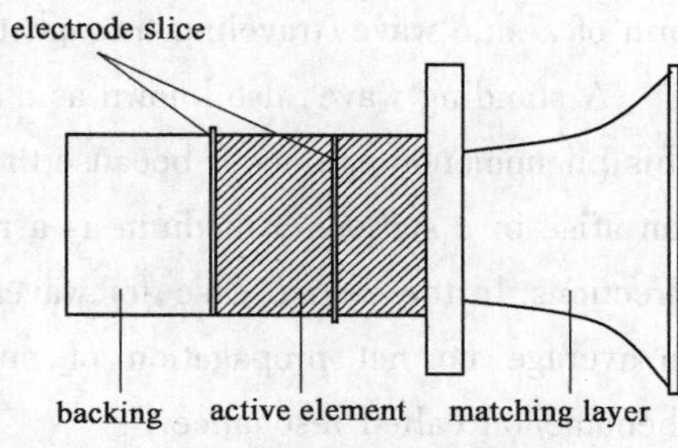

Fig. S17-1 Transducer configuration

of the transducer. Optimal impedance matching is achieved by sizing the matching layer so that its thickness is 1/4 of the desired wavelength. This keeps waves that were reflected within the matching layer in phase when they exit the layer. For contact transducers, the matching layer is made from a material that has an acoustical impedance between the active element and steel. Immersion transducers have a matching layer with an acoustical impedance between the active element and water. Contact transducers also incorporate a wear plate to protect the matching layer and active element from scratching.

The backing material supporting the crystal has a great influence on the damping characteristics of a transducer. Using a backing material with an impedance similar to that of the active element will produce the most effective damping. Such a transducer will have a wider bandwidth resulting in higher sensitivity. As the mismatch in impedance between the active element and the backing material increases, material penetration increases but transducer sensitivity is reduced. It is also important to understand the concept of bandwidth, or range of frequencies, associated with a transducer. The frequency noted on a transducer is the central or center frequency and depends primarily on the backing material. Highly damped transducers will respond to frequencies above and below the central frequency. The broad frequency range provides a transducer with high resolving power. Less damped transducers will exhibit a narrower frequency range and poorer resolving power, but greater penetration. The central frequency will also define the capabilities of a transducer. Lower frequencies (0.5 ~ 2.25MHz) provide greater energy and penetration in a material, while high frequency crystals (15.0 ~ 25.0MHz) provide reduced penetration but greater sensitivity to small discontinuities. High frequency transducers, when used with the proper instrumentation, can improve flaw resolution and thickness measurement capabilities dramatically. Broadband transducers with frequencies up to 150 MHz are commercially available.

17.3.3 Standing wave

A mechanical wave is a disturbance which is created by a vibrating object and subsequently travels through a medium from one location to another, transporting energy as it moves. The mechanism by which a mechanical wave propagates itself through a medium involves particle interaction; one particle applies a push or pull on its adjacent neighbor, causing a displacement of that neighbor from the equilibrium or rest position. As a wave is observed traveling through a medium, a crest is seen moving along from particle to particle. This crest is followed by a trough which is in turn followed by the next crest. In fact, one would observe a distinct wave pattern (in the form of a sine wave) traveling through the medium.

A standing wave, also known as a stationary wave, is a wave that remains in a constant position. This phenomenon can occur because the medium is moving in the opposite direction to the wave, or it can arise in a stationary medium as a result of interference between two waves traveling in opposite directions. In the second case, for waves of equal amplitude traveling in opposing directions, there is on average no net propagation of energy. Standing waves in resonators are one cause of the phenomenon called resonance.

When the plane wave which sound source sent reflected, incident wave piles up reflect wave, as them satisfied certain condition, which would generate standing wave. Supposed two plane wave with same frequency, amplitude and vibration direction, transit on x axes with opposing directions. The vibration mathematic equation is given:

Incident wave

$$y_1 = A\cos\left(\omega t - \frac{2\pi}{\lambda}x\right) \tag{S17-4}$$

Reflect wave

$$y_1 = A\cos\left(\omega t + \frac{2\pi}{\lambda}x\right) \tag{S17-5}$$

Interference wave

$$y = y_1 + y_2 = A\cos\left(\omega t - \frac{2\pi}{\lambda}x\right) + A\cos\left(\omega t + \frac{2\pi}{\lambda}x\right) = 2A\cos\frac{2\pi}{\lambda}x\cos\omega t \tag{S17-6}$$

This describes a wave that oscillates in time, but has a spatial dependence that is stationary: $2A\cos(kx)$. At locations $x = 0, \lambda/2, \lambda, 3\lambda/2, \cdots$ called the nodes the amplitude is always zero, whereas at locations $x = \lambda/4, 3\lambda/4, 5\lambda/4, \cdots$ called the anti-nodes, the amplitude is maximum. The distance between two conjugative nodes or anti-nodes is $\lambda/2$.

In this experiment, we will make use of piezoelectric transducers (PZT) to generate and detect sound waves in several kinds of metal rods. A PZT is a special kind of ceramic that compresses or extends when a voltage is applied, or produces a voltage when it is compressed or extended. It's called a "transducer". because it "transduces" electrical signals into mechanical ones. When we change one transducer position, can observe it nodes or anti-nodes through oscilloscope. The distance between two conjugative nodes or anti-nodes is $\lambda/2$. We will use this method to measure the length of wave.

In this experiment, we will use ultrasonic instrument, its configure is showed in Fig. S17-2.

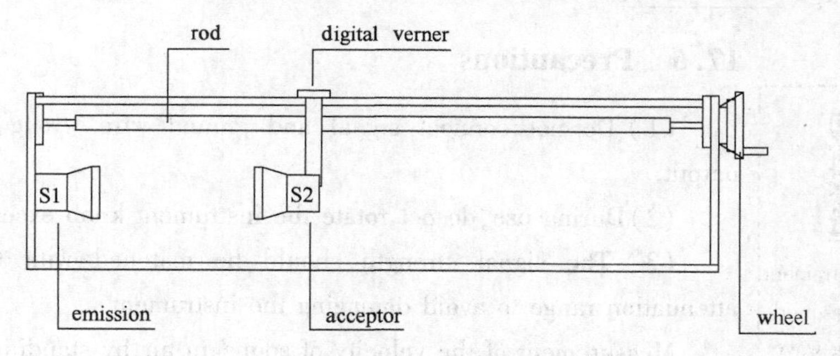

Fig. S17-2 Ultrasonic instrument

17.4 Procedures

(1) Adjust the screw that fixed piezoelectric transducers, make it vertical to main vernier, next to make the face of acceptor transducer parallel to emission transducer's face, then move acceptor transducer near distance 5cm from emission transducer.

(2) Switch on the signal generator, connect output to oscilloscope's input, adjust knob get a amplitude 5V, frequency 30 ~ 40kHz signal.

(3) Connect acceptor transducer to oscilloscope's input, adjust signal frequency and make voltage maximal, this phenomenon called resonance.

(4) keeping signal frequency is unchangeable, move auxiliary Vernier to increase the distance

between two transducer. Record the Vernier value every time ,when the maximum voltage appears.

17.5 Data and Calculations

Record the data in Tab. S17 - 1.

Tab. S17 - 1 Measurement data of wave length

$T=$______℃, $f=$______Hz

No.	1	2	3	4	5	6	7	8
Vernier values								
No.	9	10	11	12	13	14	15	16
Vernier values								
$L_{i+8}-L_i$								

$$\bar{\lambda}=\frac{2\sum_{i=1}^{8}(L_{i+8}-L_i)}{8\times 8},\quad \bar{v}=\bar{\lambda}\times f,\quad \frac{u_v}{v}=\left(\frac{1}{\lambda^2}u_\lambda^2+\frac{1}{f^2}u_f^2\right)^{\frac{1}{2}},\quad u_f=\frac{\Delta f}{\sqrt{3}},\quad \Delta f=2\times 10^{-5}\text{Hz}$$

$$u_\lambda=\frac{2}{8\times 8}\sqrt{\frac{\sum_{i=1}^{8}[(L_{i+8}-L_i)-(\overline{L_{i+8}-L_i})]^2}{8-1}},\ v=v_0\sqrt{1+\frac{t}{T_0}}\ (v_0=331.5\text{m/s},T_0=273.15\text{K}).$$

Standard error: $u_v=\left(\frac{u_v}{v}\right)\bar{v}$.

Relative error: $E=\frac{|\bar{v}-v_t|}{v_t}\times 100\%$.

Result: $\bar{v}\pm u_v=$______.

17.6 Precautions

Video S17 Measurement of the velocity of sound in air by standing wave

(1) Do not connect signal and ground wire wrong or short signal output.

(2) During use, do not rotate the instrument knob excessively.

(3) The signal strength should be matched with the oscilloscope attenuation range to avoid damaging the instrument.

Measurement of the velocity of sound in air by standing wave is shown in Video S17.

Exp. 18 Measurement of low resistance with Kelvin bridge

18.1 Objectives

(1) Understand the composition and principle of Kelvin bridge.

(2) Learn how to measure low-value resistance with Kelvin bridge.

18.2 Apparatus

QJ19 double bridge, mirror galvanometer, ampere-meter, standard resistance, slide rheostat, DC power supply, unknown resistor, reversing switch, wire, etc.

18.3 Principles

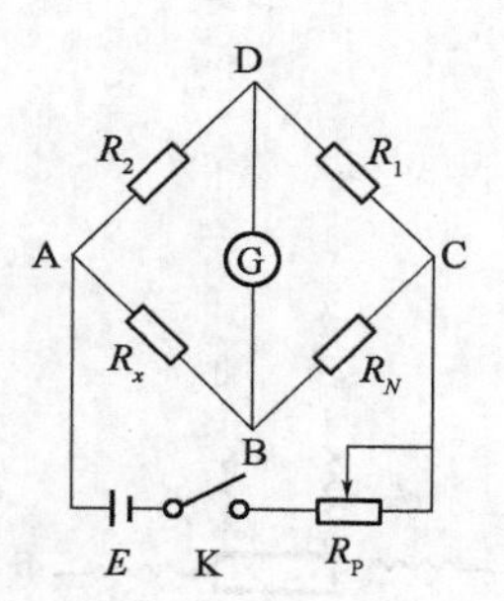

Fig. S18 – 1 Principle of single bridge

As shown in Fig. S18 – 1, we used a Wheatstone bridge (single-arm bridge) to measure a median resistance. In this circuit, there are several wires and four contact points A, B, C and D. The resistance to be measured is $R_x = \frac{R_2}{R_1} R_N$ when the bridge is balanced. The resistance of the wires from points A and C to the power supply can be included in the internal resistance of the power supply, and the resistance of the wires from points D and B to the galvanometer is incorporated into the internal resistance of the galvanometer, all of which have no effect on the measurement result. The proportional arm R_1 and R_2 generally choose a resistor with a larger resistance, so the resistance of the wire connected to these two resistors and the contact resistance at point D can be ignored. When measuring the median resistance, the resistance of the wires connected to R_x and R_N and the contact resistance are also negligible and will not affect the measurement. But if the measured resistance R_x is a low-value resistance, the wire resistance and contact resistance connected to R_x and R_N cannot be ignored. The additional resistance value has exceeded or greatly exceeded the resistance of the measured resistance, which will cause a large error, or even complete no measurement results can be obtained. Therefore, it is impossible to measure low-value resistance accurately with a single-arm bridge. Measures must be taken on the measurement circuit to avoid the influence of additional resistance on the measurement of low-value resistance.

The Kelvin bridge is also called the double-arm bridge. It is developed on the basis of the Wheatstone bridge. It uses the "four-terminal button access method", as shown in Fig. S18 – 2. The resistance to be measured R_x is the resistance between c and d, c and d are called voltage terminals, and a and b are called current terminals. As shown in Fig. S18 – 3, connect the resistance to be measured R_x to the Wheatstone bridge circuit. The values of the additional resistances r_1、r_2、r_1'、r_2' in the figure are relatively small. r_1 is connected with in series with the internal resistance of the power supply and r_1' is connected with in series with the resistance R_2 of the bridge arm, because R_2 and the internal resistance of the power supply is much larger than the additional resistance in series with them, the influence of the additional resistance r_1 and r_1' can be ignored. r_2' is connected in series with galvanometer G, and r_2 is connected in series with standard resistance R_N, the resistance value R_N of is small, so the influence of additional resistor r_2 cannot be ignored. So, how to make the influence of additional resistance r_2 and r_2' can be ignored? Similarly, the standard resistance R_N also adopts the "four-terminal button connection method", as shown in Fig. S18 – 4. The additional resistances r_{N1}、r_{N2}、r_{N1}'、r_{N2}' are both ends of the standard resistance, r_{N1}' are connected in series with the bridge's arm resistance R_1 and r_{N1} is connected with the internal resistance of the power supply after the "four-terminal button access method" is adopted. In order to eliminate the influence of the r_2' and r_{N2}', the resistances with much larger resistance are connected in series with R_3 and R_4, form a new pair of bridge arms. And r_{N2} is in series connected with r_2, set the total resistance value as r, and its influence can be eliminated by adjusting the resistance value of bridge arm resistance R_1、R_2、R_3、R_4.

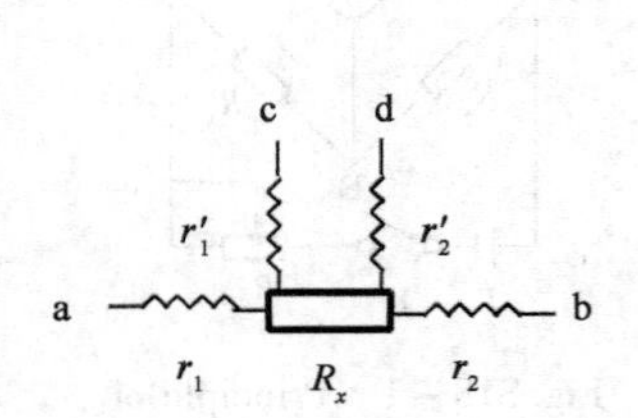

Fig. S18 – 2 Diagram of four terminal resistor

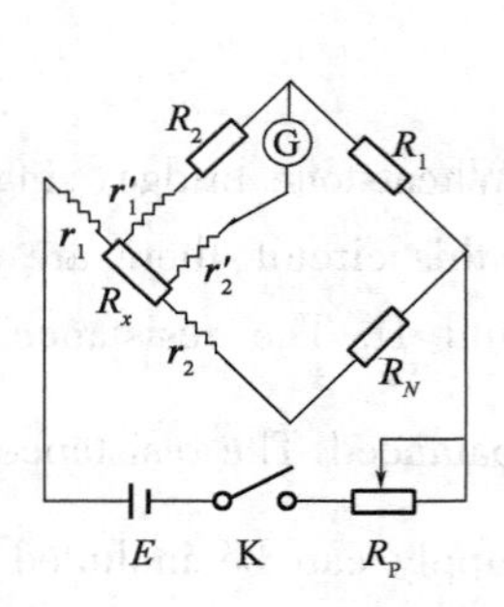

Fig. S18 – 3 Four terminal resistor Connecting method

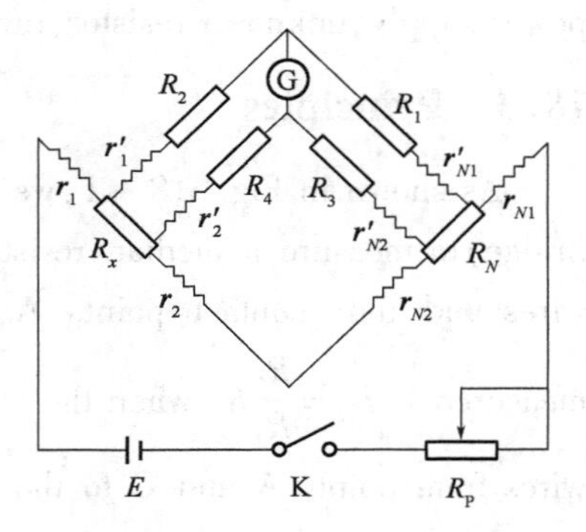

Fig. S18 – 4 Diagram of double arm bridge principle

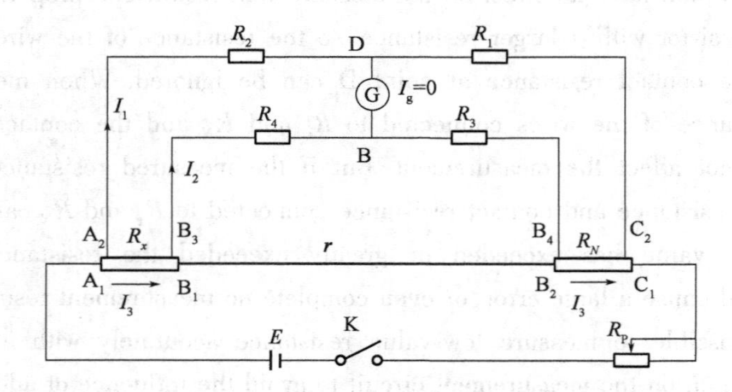

Fig. S18 – 5 Diagram of connecting circuit of double arm bridge

The circuit of the double-arm bridge is shown in Fig. S18 – 5. Since the double-arm bridge has the conditions of $r_2' \ll R_4$, $r_{N1}' \ll R_1$, $r_{N2}' \ll R_3$, adjust R_1、R_2、R_3、R_4, so that when the bridge reaches equilibrium (galvanometer refers to zero), the current flowing through R_1 and R_2 is set to be equal to I_1, and the current on R_3 and R_4 is set to be equal to I_2, The current flowing through R_x and R_N is also set equal to I_3, and there is $I_1 \ll I_3$, $I_2 \ll I_3$, so according to Kirchhoff's Second Law:

$$I_1 R_1 = I_3 R_x + I_2 R_2 \tag{S18 – 1}$$

$$I_1 R_2 = I_3 R_N + I_2 R_4 \tag{S18 – 2}$$

$$I_2 (R_3 + R_4) = (I_3 - I_2) r \tag{S18 – 3}$$

From formula (S18 – 1), formula (S18 – 2), formula (S18 – 3),

$$R_x = \frac{R_2}{R_1} R_N + \frac{r R_3}{R_3 + R_4 + r}\left(\frac{R_2}{R_1} - \frac{R_4}{R_3}\right) \tag{S18 – 4}$$

If $\frac{R_2}{R_1} = \frac{R_4}{R_3}$ or $R_1 = R_3$、$R_2 = R_4$, then the second term on the right side of formula (S18 – 4) is zero, so

$$R_x = \frac{R_2}{R_1} R_N \tag{S18 – 5}$$

The influence of r on the measurement can be eliminated. The $\frac{R_2}{R_1}=\frac{R_4}{R_3}$ or $R_1=R_3$、$R_2=R_4$, is called the auxiliary condition of the double-arm bridge balance.

During the experiment, always maintain $\frac{R_2}{R_1}=\frac{R_4}{R_3}$ or $R_1=R_3$、$R_2=R_4$, and the resistance of the four resistance arms R_1、R_2、R_3、R_4 should be much greater than the wiring resistance and contact resistance of R_x and R_N. These requirements can be ensured by the structure of the electric bridge. Usually the electric bridge is made into a special structure. The two contrast ratio arms and adopt double-decimal resistance boxes with the same ratio, and their rotating arms are connected to the same rotating shaft. Above, in this way, the auxiliary conditions of the balanced bridge can be met at any position of the arm, and the four resistance arms can also get enough resistance. Of course, it is impossible to make, R_1、R_2、R_3、R_4 absolutely equal when manufacturing the instrument. In order to eliminate the influence of r on the resistance R_x to be measured, the wire connecting the R_x and R_N two current terminals is required to be as short and thick as possible to make r as small as possible.

The internal structure principle of the double-arm electric bridge is consistent with the essence of the ideological principle design, but in order to be consistent with the symbols on the QJ19 type electric bridge panel diagram, a slight modification has been made to exchange the numbers with R_2 and R_3. In this way, the auxiliary condition that the double-arm bridge must meet becomes:

$$\frac{R_3}{R_1}=\frac{R_4}{R_2} \quad \text{or} \quad R_3=R_4, R_1=R_2 \tag{S18-6}$$

When the bridge reaches equilibrium,

$$R_x=\frac{R_3}{R_1}R_N=\frac{R_4}{R_2}R_N \tag{S18-7}$$

Usually $R_3=R_4=R$ is specified in the double-arm bridge, then

$$R_x=\frac{R}{R_1}R_N=\frac{R}{R_2}R_N \tag{S18-8}$$

This is the calculation formula for measuring the low-value resistance with the double-arm bridge, where R is the indicated value on the panel when the bridge is balanced.

When measuring the double-arm bridge, since R_x is very small, the voltage across R_x is also very small, and the working current is not very small. At this time, the existence of thermoelectric potential makes the voltage across R_x increase or decrease, which will destroy the balance condition of the bridge, the galvanometer points to zero in the presence of thermoelectric potential, and the bridge arms do not strictly meet the relationship of formula (S18 - 8). Because the thermoelectric potential is related to the direction of the current, it can be eliminated by changing the polarity of the power supply (direction of the current). In the experiment, the reversing switch is used to achieve this goal. Measure different R under different polarities and find the average value. Then formula (S18 - 8) can be written as

$$R_x=\frac{1}{2}(R_+ + R_-)\frac{R_N}{R_1} \tag{S18-9}$$

When the bridge reaches equilibrium, the current passing through the galvanometer $I_g=0$, but whether the galvanometer really points to zero is based on human judgment. Due to the existence of parallax, it will inevitably bring certain errors to the measurement results, and the size of the error depends on the relative sensitivity of the bridge.

The relative sensitivity of the electric bridge is defined as: when the electric bridge reaches equilibrium, the indication value of the electric bridge panel, then adjust the measuring knob to change the indication value of the panel to a certain value ΔR, if the galvanometer deflects from the equilibrium position Δn, the relative sensitivity of the electric bridge is defined as

$$S = \frac{\Delta n}{\frac{\Delta R}{R}} \tag{S18-10}$$

The physical meaning of S is the number of deflection divisions of the galvanometer caused by the relative change in the unit of the bridge arm resistance. The larger the S, the higher the sensitivity of the bridge, and the smaller the measurement error introduced.

18.4 Procedures

18.4.1 Measure the unknown resistance R_x

(1) Connect the circuit as shown in Fig. S18 - 6. Based on the estimates value of R_x and the standard resistance R_N, taken to ensure that the measured value R_x has five significant digits, reasonable choosing $R_1 = R_2$, and estimate the value of R, then put the R in the estimated value.

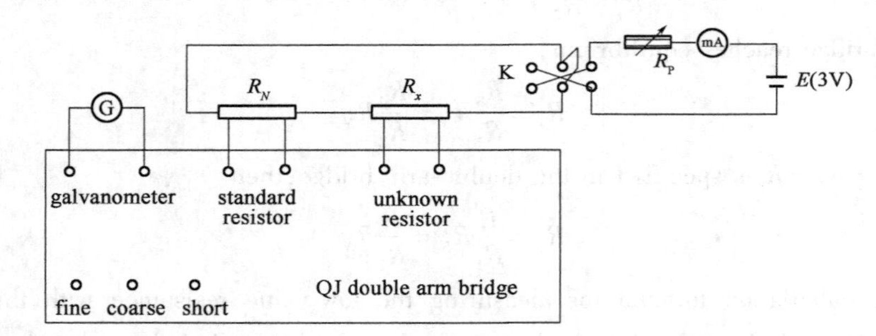

Fig. S18 - 6 Experimental circuit for measuring small resistance with double-arm bridge

(2) The sensitive galvanometer connected to 220V AC power, place the gear knob to "direct" position, adjust the zero point.

(3) The galvanometer gear knob was placed 0.01 position, and then close the switch K to either side, turn adjustable resistor R and make the galvanometer close to zero gradually.

(4) The gear knob were placed in 0.1、1 and direct position, adjusted resistor R in accordance with step (3), and record the value of R_+.

(5) Change the reversing switch to another side and check whether the galvanometer points to zero, if not, adjust R to let the galvanometer to zero, record the value of R_- and calculate the value of the resistance to be measured according to the formula (S18 - 9).

18.4.2 Measure the relative sensitivity S of the bridge

On the basis of balanced bridge, let R changes a value ΔR, record the number of the galvanometer deviation from the zero position Δn, calculated the relative sensitivity of the bridge S according to the following formula:

$$S = \frac{\Delta n}{\frac{\Delta R}{R}}$$

18.5 Data and calculations

18.5.1 Data records (Tab. S18 - 1, Tab. S18 - 2)

Tab. S18 - 1 The data of measuring low-value resistance

Estimated value of R_x = ________, R_N = ________, $R_1 = R_2$ = ________

Readings of R/Ω		Measurements/Ω
R_+	R_-	$R_x = \frac{1}{2}(R_+ + R_-)\frac{R_N}{R_1}$

Tab. S18 - 2 The data of measuring the relative sensitivity of the bridge

Balanced bridge R/Ω	Amount of change $\Delta R/\Omega$	Number of deviation grids Δn	Relative sensitivity $S = \frac{\Delta n}{\frac{\Delta R}{R}}$

18.5.2 Calculations

Calculate of uncertainty of resistance R_x.

18.6 Precautions

(1) Keep the desktop stable during the experiment to prevent the galvanometer from not working properly.

(2) Don't press and lock the "coarse" or "fine" adjustment button for a long time.

(3) Don't turn the knobs excessively avoiding damage the instrument.

(4) The galvanometer should be zeroed before the beginning the experiment.

(5) The voltage terminal and current terminal of the resistor cannot be mixed when using the four-terminal button connection method.

Measurement of low resistance with Kelvin bridge is shown in Video S18.

Video S18 Measurement of low resistance with Kelvin bridge

Exp. 19 Current-voltage characteristic of ohmic and non-ohmic devices

19.1 Objectives

(1) Determine the current-voltage characteristic curve of a resistor and silicon diode.

(2) Determine the resistance of the resistor and diode.

(3) Learn deal with experiment data by plotting graph.

19.2 Apparatus

Power supply, voltmeter, ampere-meter, resistor box, potential resistor, resistor, diode, connecting wire, etc.

19.3 Principles

To make a current flow through a material, three requirements must be met: (1) An electric field must exist; (2) charge carriers must be present in the material; and (3) the charge carriers must be mobile. To establish an electric field, a voltage is applied to the circuit. The charge carriers are the valence electrons in a conductor, or the electrons in the conduction band and the holes in the valence band of a semiconductor or insulator. The mobility is dependent on the crystal structure and the temperature.

19.3.1 Conductor

For a conductor, such as a metal, the valence electrons occupy partially filled energy levels to form a valence band. The crystal structure of a metal allows the valence electrons in the valence band to move freely through the crystal. However, as the temperature increases, the atoms vibrate with greater amplitude, and move far enough from their equilibrium positions to interfere with the travel of the electrons. Only near absolute zero is the mobility at its maximum value.

19.3.2 Semiconductor

For a semiconductor or insulator, the valence electrons occupy a filled valence band. Electrons must move from the valence band to the conduction band (leaving holes, vacancies, in the valence band). Both the electrons in the conduction band and the holes in the valence band are considered charge carriers. The number of these charge carriers is dependent on the temperature and the material. As the temperature increases, more electrons have the energy needed to "jump" to the conduction band. (Important: The electrons do not move from a place in the crystal, called the valence band, to another place, called the conduction band. The electrons have the energy associated with the valence band, and acquire enough energy to have the energy associated with the conduction band. An energy change occurs, not a position change.)

19.3.3 Doping

Doping of a semiconductor material, by adding atoms with one more or one less valence electron than the base material, is one method of increasing the number of charge carriers (such as adding Ga, with three valence electrons, or As, with five valence electrons, to Ge or Si which has four valence electrons). Addition of a Group V element, such as As, forms an N-type material, which provides new "donor" energy levels. Addition of a Group Ⅲ element, such as Ga, forms a P-type material, which provides new "acceptor" energy levels. The energy needed for an electron to move from the valence band to the acceptor level as with Ga (forming a hole), or from the donor level to the conduction band as with As(yielding a conducting electron) is less than the energy needed to make the original "jump" from the valence band to the conduction band of the pure semiconductor material. Thus, for a doped semiconductor material as compared to a pure semiconductor material (at the same temperature), the doped semiconductor would have more electrons in the conduction band (N-type), or more holes in the valence band (P-type). For and N-type material, the carrier of electricity is a negative electron. For a P-type material, the carrier is a positive hole. As the temperature increases, the atoms do vibrate with greater amplitude. However, the increase in number of charge carriers has a greater effect on increasing the material's conductivity than the reduction caused by the vibrating atoms.

19.3.4 Resistor

When a voltage is applied across a resistor, an electric field is established. This electric field "pushes" the charge carriers through the resistor. This "push" gives the charge carriers a "drift velocity" in the direction from high potential energy to low potential energy. As the voltage increases, the drift velocity increases. Since the amount of current flowing through a resistor is directly proportional to the drift velocity, the current is directly proportional to the voltage, which produces the electric field, which produces the drift velocity. This is the origin of Ohm's Law.

19.3.5 Diode

However, in a diode, the number of charge carriers is dependent on the number of electrons that have enough energy to move up an energy hill and across the P-N junction, producing current flow through the diode. The size of this hill, or energy barrier, is dependent on the amount and type of dopants in the semiconductor material of which the diode is made. As a voltage is applied (in the forward bias), the size of the hill is decreased, so more electrons have the energy needed to cross the P-N junction producing current flow. The number of electrons with the energy needed to move up the hill and across the junction increases exponentially as the voltage increases. Thus, the current increases exponentially as the voltage increases.

19.3.6 Applications

The behavior of components in a circuit is a very important aspect of circuit design. Diodes are found in many semiconductor circuits. Their non-linear $I-U$ behavior makes them quite useful for a variety of applications. Resistors are often used in series with another circuit component to reduce the voltage across that component or in parallel to reduce the current through a component.

19.4 Procedures

19.4.1 Measure the characteristic of resistor

(1) Build a circuit as shown in Fig. S19 – 1. Do not turn on the power supply.

(2) Check to make sure the output voltage adjustment knob have been to the minimum.

(3) Connect 250 ohms to R_x position, turn on switch to A port, adjust potential resistor make a values, turn on switch to B port, record the voltmeter and ampere-meter values.

(4) Connect 30 ohms to R_x position , repeat the step(3).

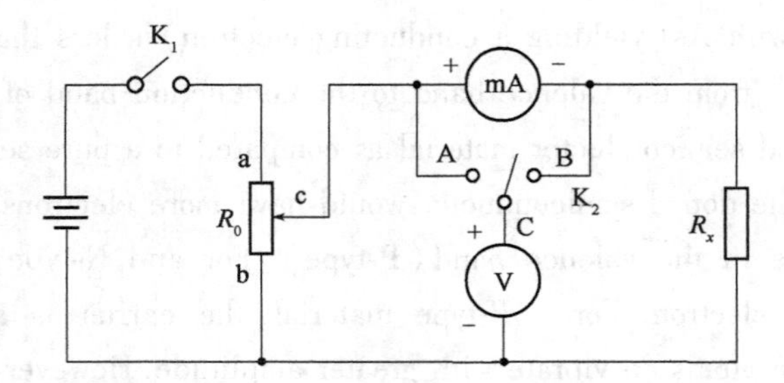

Fig. S19 - 1 Circuit diagram of measuring resistor

19.4.2 Measure the characteristic of diode

(1) Build a circuit as shown in Fig. S19 - 2.

(2) Turn on the switch K, and adjust potential resistor, make the voltmeter increase slowly from zero to 1 V, record the current and voltage.

(3) Build a circuit as shown in Fig. S19 - 3.

(4) Turn on the switch K, and adjust potential resistor, make the voltmeter increase slowly from zero to 1 V, record the current and voltage.

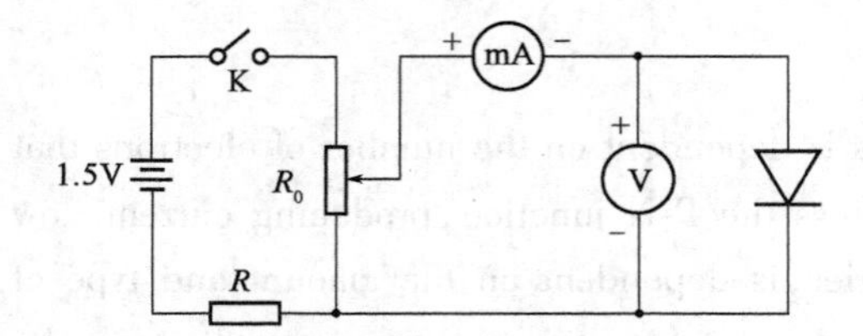

Fig. S19 - 2 Circuit diagram of measuring diode positive resistance

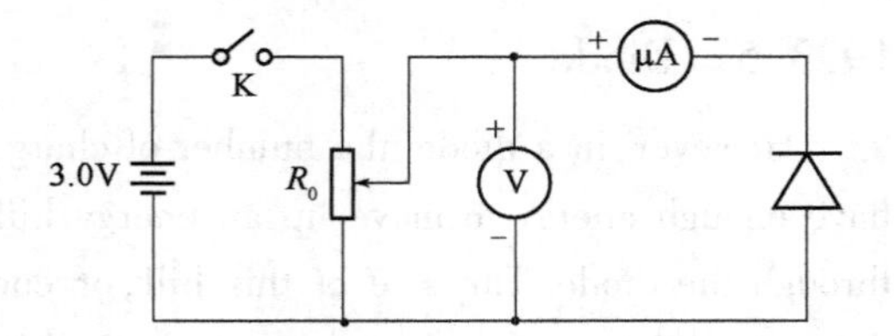

Fig. S19 - 3 Circuit diagram of measuring diode negative resistance

19.5 Data and calculations

Record the data in Tab. S19 - 1 ~ Tab. S19 - 3.

Tab. S19 - 1 The data of measuring resistor

Resistor/ohms	Position	Voltage/V	Current/mA	Calculation/Ω
250	A			
	B			
30	A			
	B			

Tab. S19 - 2 The data of measuring diode (positive)

U/V	0.00	0.10	0.20	0.30	0.40	0.45	0.50	0.55	0.60	0.65	0.70	0.75
I/mA												

Tab. S19 - 3 The data of measuring diode (negative)

U/V	0.00	0.10	1.00	2.00	2.50	3.00	3.50	4.00	4.50
I/mA									

19.6 Precautions

(1) Power supply never be short connected.

(2) The positive and negative electrodes of the electricity meter cannot be connected wrong.

(3) Pay attention to the correct selection of meter range.

(4) When measuring the volt-ampere characteristics of a diode, pay attention to the maximum limited current and the maximum reverse limited voltage.

Exp. 20 Measuring laser wavelength by Michelson interferometer

20.1 Objectives

(1) Know the principle and construction of Michelson interferometer.

(2) Master the method of adjusting Michelson interferometer.

(3) Learn how to measure laser wavelength by Michelson interferometer.

20.2 Apparatus

Michelson interferometer, He-Ne laser generator, beam expanding lens, etc.

20.3 Principles

The Michelson interferometer is the best example of what is called an amplitude-splitting interferometer. It was invented in 1893 by Albert Michelson, to measure a standard meter in units of the wavelength of the red line of the cadmium spectrum. With an optical interferometer, one can measure small distances directly in terms of wavelength of light used, by counting the interference fringes. In the Michelson interferometer, coherent beams are obtained by splitting a beam of light that originates from a monochromatic source with a partially reflecting mirror called a beam splitter. The resulting reflected and transmitted waves are then redirected by ordinary mirrors to a screen where they superimpose to create fringes. This is known as interference by division of amplitude. This interferometer, used in 1817 in the famous Michelson-Morley experiment, demonstrated the non-existence of an electromagnetic-wave-carrying ether, thus paving the way for the Special theory of Relativity.

20.3.1 Brief introduction of Michelson interferometer

The frame diagram of Michelson interferometer is showed as Fig. S20 – 1. M_1 and M_2 are fine polish plane mirror, which are fixed on two uprightness arms of interferometer. Both M_1 and M_2 have three screw of adjusting lean on back, which can change mirror's orientation.

Mirror M_2 can move 100mm on worm stick, accuracy is 0.0001mm. Turning large wheel make coarse move, turning small wheel make fine move. The move distance can be read from interferometer directly, the values include three parts: main ruler, large wheel, and small wheel. The main ruler is number mm, reading window(large wheel) is 0.01mm, small wheel reading is 0.0001mm, we need estimate or guesstimate one number, so get 0.00001mm.

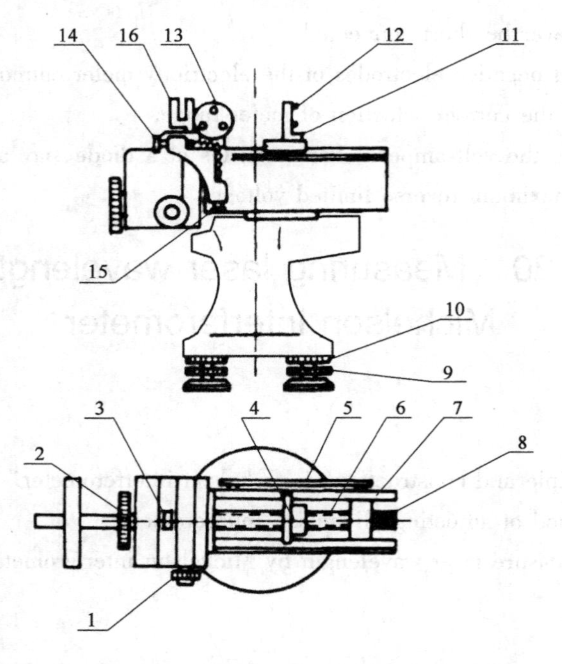

Fig. S20 – 1 Frame diagram of interferometer

1—fine adjusting wheel; 2—coarse adjusting wheel; 3—reading window; 4—worm gear nut; 5—main ruler; 6—worm stick; 7—slid way; 8—head nut; 9—level screw; 10—lock ring; 11—movable mirror M_1; 12—coarse adjusting screw; 13—fixed mirror M_2; 14—horizontal tension spring screw; 15—perpendicular tension spring screw; 16—splitter and compensating plate

20.3.2 The interference principle of Michelson interferometer

A simplified schematic diagram of Michelson interferometer is shown in Fig. S20 – 2, and the interference pattern of Michelson interferometer is shown in Fig. S20 – 3.

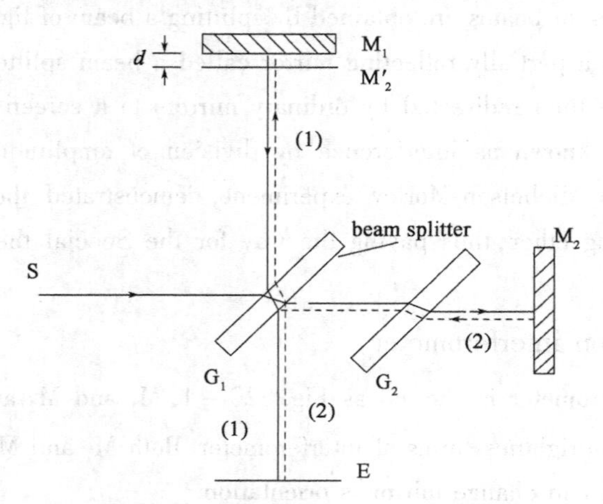

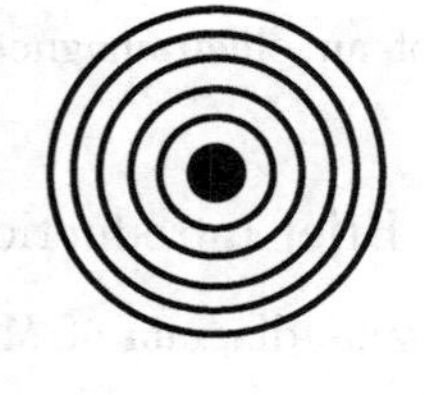

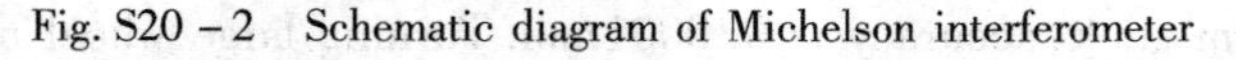

Fig. S20 – 2 Schematic diagram of Michelson interferometer

Fig. S20 – 3 Interference pattern of Michelson interferometer

Light from a monochromatic source S is divided by a Beam Splitter (BS), which is oriented at an angle 45° to the beam, producing two perpendicular beams of equal intensity. About 50% of the light is reflected from the surface and 50% is transmitted. The Reflected beam (R) travels to mirror M_1 and is reflected back to BS. The transmitted beam (T) is reflected back towards the BS by a fix mirror M_2, each reflected beam passes straight through Beam Splitter and reaches the screen for observation.

Since the reflector of the BS is the surface on its right surface, the transmitted beam starting from the source S and undergoing reflection at the mirror M_2 passes through the beam splitter three times, while the Reflected beam reflected at M_1 travels through BS only once. The optical path length through the glass plate depends on its index of refraction, which causes an optical path difference between the two beams. To compensate for this, a glass compensated plate (CP) of the same thickness and index of refraction as that of BS is introduced between M_1 and BS. The recombined beams interfere and produce fringes at the screen E. The relative phase of the two beams determines whether the interference will be constructive or destructive. By adjusting the inclination of M_1 and M_2, one can produce circular fringes, straight-line fringes, or curved fringes. This experiment uses circular fringes, shown in Fig. S20 – 3.

From the screen, an observer sees M_2 directly and the virtual image M_1' of the mirror M_1, formed by reflection in the beam splitter, as shown in Fig. S20 – 4. This means that one of the interfering beams comes from M_2 and the other beam appears to come from the virtual image M_1'. If the two arms of the interferometer are equal in length, M_1' coincides with M_2. If they do not coincide, let the distance between them be d, and consider a beam from a point S. It will be reflected by both M_1' and M_2, and the observer will see two virtual images, S_1 due to reflection at M_1', and S_2 due to reflection at M_2. These virtual images will be separated by a distance $2d$. If θ is the angle with which the observer looks into the system, the path difference between the two beams is $2d\cos\theta$. When the beam that comes from M_1 undergoes reflection at BS, a phase change of π occurs, which corresponds to a path difference of $\lambda/2$.

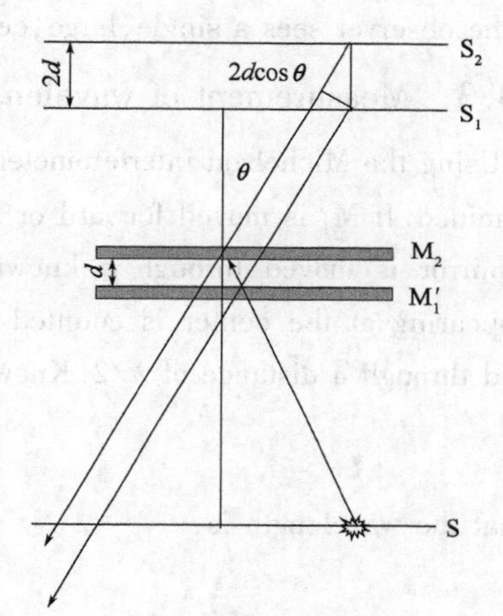

Fig. S20 – 4 Explanation of circular fringes

Therefore, the total path difference between the two beams is

$$\Delta = 2d\cos\theta + \frac{\lambda}{2} \qquad (S20-1)$$

The condition for constructive interference is then,

$$\Delta = 2d\cos\theta + \frac{\lambda}{2} = m\lambda \quad (m = 0, 1, 2, \cdots) \qquad (S20-2)$$

For a given mirror separation d, a given wavelength λ, and order m, the angle of inclination θ is a constant, and the fringes are circular. They are called *fringes of equal inclination*, or *Haidinger fringes*. If M_1' coincides with M_2, $d = 0$, and the path difference between the interfering beams will be $\lambda/2$. This corresponds to destructive interference, so the center of the field will be dark.

If one of the mirrors is moved through a distance $\lambda/4$, the path difference changes by $\lambda/2$ and a maximum is obtained. If the mirror is moved through another $\lambda/4$, a minimum is obtained; moving it by another $\lambda/4$, again a maximum is obtained and so on. Because d is multiplied by $\cos\theta$, as d increases, new rings appear in the center faster than the rings already present at the periphery disappear, and the field becomes more crowded with thinner rings toward the outside. If d decreases, the rings contract, become wider and more sparsely distributed, and disappear at the center.

For destructive interference, the total path difference must be an integer number of wavelengths plus a half wavelength,

$$\Delta_{\text{destr}} = 2d\cos\theta + \frac{\lambda}{2} = \left(m + \frac{1}{2}\right)\lambda \quad (m = 0, 1, 2, \cdots) \qquad \text{(S20-3)}$$

If the images S_1 and S_2 from the two mirrors are exactly the same distance away, $d = 0$ and there is no dependence on θ. This means that only one fringe is visible, the zero order destructive interference fringe, where

$$\Delta_{\text{destr}} = \frac{\lambda}{2} = \left(m + \frac{1}{2}\right)\lambda \quad (m = 0) \qquad \text{(S20-4)}$$

and the observer sees a single, large, central dark spot with no surrounding rings.

20.3.3 Measurement of wavelength

Using the Michelson interferometer, the wavelength of light from a monochromatic source can be determined. If M_1 is moved forward or backward, circular fringes appear or disappear at the center. The mirror is moved through a known distance Δd and the number N of fringes appearing or disappearing at the center is counted. For one fringe to appear or disappear, the mirror must be moved through a distance of $\lambda/2$. Knowing this, we can write,

$$\Delta d = \frac{N\lambda}{2} \qquad \text{(S20-5)}$$

So that the wavelength is,

$$\lambda = \frac{2\Delta d}{N} \qquad \text{(S20-6)}$$

20.4 Procedures

(1) Turn on the laser generator, adjust altitude of laser so that the laser beam just incident to the center of the beam splitter.

(2) Adjust coarse adjusting screws of two plane mirror until they are middle, then adjust the level screw of interferometer until reflect light near the aperture of laser generator by two mirror.

(3) Shade M_2, turn the adjusting screw of M_1 until the reflect light enters the aperture of laser generator.

(4) Shade M_1, turn the adjusting screw of M_2 until the reflect light enters the aperture of laser generator.

(5) Several laser spots can be observed on the screen. Adjust the two screws on the back of M_1/M_2 so as to superimpose the two brightest light spots.

(6) Insert a Beam expander lens after the laser to spread out the beam (ideally the laser beam

should be pass through the center of the lens to preserve alignment). After the beams traverse through the system, the image of the interfering rays will be a circular pattern projected onto a screen. The two beam reflected off the mirrors should be aligned as parallel as possible to give you a circular pattern.

Note the reading on the micrometer. Focus on a particular fringe (the center is a good place). Begin turning the micrometer so that the fringes move (for example, from bright to dark to bright again is the movement of 1 fringe). After counting a total of about 50 fringes and please record the reading on the micrometer, and fill in the Tab. S20 – 1. Calculate the wavelength from equation (S20 – 6).

A final mean value of λ and its uncertainty should be generated. Compare your value with the accepted value (given by the instructor).

20.5 Data and calculations

20.5.1 Data records (Tab. S20 – 1)

Tab. S20 – 1 Measuring interference fringes data

$\lambda_{theory} = 632.8\text{nm}$

Fringes numbers N_1	0	50	100	150	200
d_i/mm					
Fringes numbers N_2	250	300	350	400	450
d_{i+5}/mm					
$\Delta N = N_2 - N_1$	250	250	250	250	250
$\Delta d_i = \frac{d_{i+5} - d_i}{5}$/mm					
$\Delta(\Delta d_i) = \Delta d_i - \overline{\Delta d}$/mm					
$\overline{\lambda} = \frac{2\overline{\Delta d}}{N}$/mm ($N = 50$)					

20.5.2 Calculations

$$u_{\Delta d_A} = 2.78\sqrt{\frac{\sum_{i=1}^{5}[\Delta(\Delta d_i)]^2}{5\times(5-1)}} = \quad, \quad u_{\Delta d_B} = \frac{0.00005}{\sqrt{3}}, \quad u_{\Delta d} = \sqrt{u_A^2 + u_B^2},$$

$$u_\lambda = \frac{2}{N}u_{\Delta d}, \quad \lambda = \overline{\lambda} \pm 2u_\lambda.$$

20.6 Precautions

(1) Never look directly at the laser beam. Align the laser so that it is not at eye level.

(2) In order to make the measurement results correct, avoid the introduction of backlash error, the hand adjusting wheel should be rotated in one direction after the fringe starts to move when recording the data.

Video S20 Measuring laser wavelength by Michelson interferometer

Measuring laser wavelength by Michelson interferometer is shown in Video S20.

Exp. 21 Use of the potentiometer

21.1 Objectives

(1) Understand the working principle of potentiometer.

(2) Learn how to use a portable potentiometer.

21.2 Apparatus

UJ33a DC potentiometer, unknown voltage, several wires.

21.3 Principles

If we want to measure the electromotive force (EMF) of a battery, we cannot simply attach a voltmeter to the battery because the battery has internal resistance, which will affect the voltage measurement when the battery sends current through the closed loop. Therefore, in order to measure accurately the EMF of a battery, we need a measurement method that does not draw current through the battery. The potentiometer is such an instrument designed for this purpose, and it was proposed by Johann Christian Poggendorff around 1841.

Fig. S21 - 1: is the schematic circuit diagram of the compensation measuring method. The galvanometer G is a sensitive device capable of indicating the presence of very small current. Its function is to accurately indicate a condition of zero current, rather than to indicate any specific quantity as a normal ammeter would. The unknown EMF(E_x) is connected with the resistance AB. By adjusting the sliding endpoint, a and b, a status can be found, where there is no deflection of the galvanometer is observed. It means the unknown EMF(E_x) is equal to the voltage between a and b. If the resistor has uniform cross section throughout its length, the electrical resistance per unit length of the resistor is also uniform throughout its length. Suppose the electric current through the resistor is I and resistance per unit length of the resistor is R. Then the voltage appears per unit length across the resistor would be "IR". So we can get the unknown EMF $E_x = IR$.

Fig. S21 - 2 displays the working principle diagram of the DC potentiometer. There are three main parts: (1) the operating current regulation circuit: E, r, S_0, R_N and R_0. (2) the correction circuit, comprised of S(N-terminal), R_N, and E_N. (3) the measurement circuit include E_x, R and S (x-terminal).

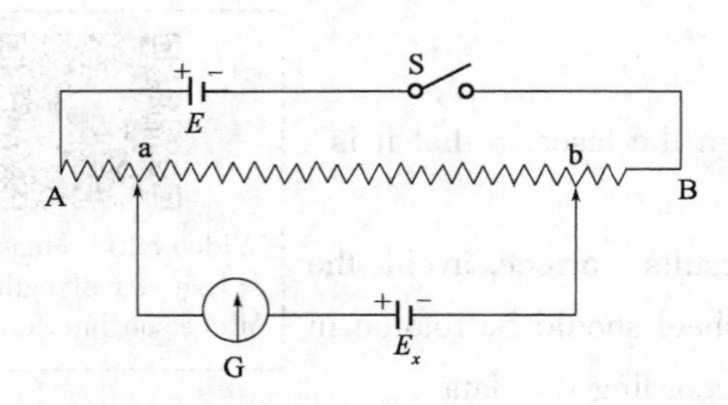

Fig. S21 - 1 Principle circuit of the compensation measuring method

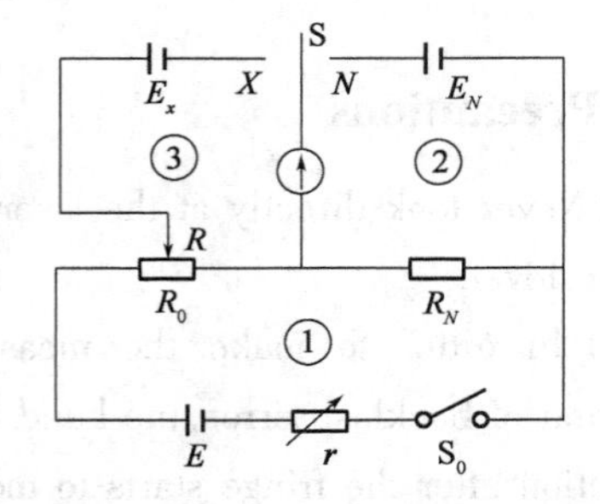

Fig. S21 - 2 Schematic circuit diagram of the potentiometer

Before measuring, we should standardize the operating current (for a given type of potentiometer, the operating current is pre-designed). When turn S to N-terminal and S_0 is closed, the resistor r can be adjusted until the galvanometer shows zero. Then, the standard EMF E_N is equal to the voltage drop on the standard resistor R_N.

$$E_N = I_0 R_N \tag{S21-1}$$

When measuring, switch S to x-terminal and adjust R until the galvanometer shows zero again, then the unknown EMF E_x.

$$E_x = I_0 R \tag{S21-2}$$

Dividing one equation by formula (S21-1), we get

$$E_x = \frac{E_N}{R_N} R \tag{S21-3}$$

It can clearly be seen, when the potentiometer is used to measure an EMF, the result is depended on the precision of the standard EMF E_N, standard resistance R_N and compensation resistance R, as well as the sensitivity of the galvanometer.

The operating current is an important parameter of potentiometer. The scales marked on the instrument panel are determined by $I_0 R$. Therefore, the current must be standardized before the measurement, and unchanged in the measurement.

21.4 Procedures

(1) Preparation: According to the instructions, you can connect the circuit and adjust the potentiometer in use. Take UJ33a type potentiometer as an example: the override switch from "Off" to the desired multiple, then the internal power supply is switched on. Adjust the "Zero" knob below the galvanometer, so that the galvanometer pointer shows zero. Connect the unknown EMF to "Unknown" port with the same polarity; "Measuring / Output" switch to "Measuring" position.

(2) Calibration: According to the instrument supplement, you can calibrate the potentiometer operating current. Take UJ33a type potentiometer as an example: turn switch S_2 to "Standard", then adjust "Coarse/Fine" knob until the galvanometer shows zero. At this time, the operating current of the potentiometer can has been standardized.

(3) Measurement: Measure an unknown EMF. Take UJ33a type potentiometer as an example: turn the switch S_2 to "Unknown." Regulate Ⅰ, Ⅱ, Ⅲ measuring dial until galvanometer shows zero, the unknown EMF equal to the product of multiple K and the sum of Ⅰ, Ⅱ, Ⅲ dial readings.

21.5 Data and calculations

Record the data in Tab. S21-1.

Tab. S21-1 The results of the unknown EMF E_x

Multiple K	×10	×1	×0.1	U_x
×0.1				
×1				
×5				

21.6 Precautions

(1) Measure voltage not exceeding the maximum range of the potentiometer.

Video S21 Use of the potentiometer

(2) Pay attention to whether the operating current changes and calibrate it in time during measurement.

(3) Be sure to turn off the power supply of the potentiometer and the switch "S_2" in the middle position after the measurement is completed.

Use of the potentiometer is shown in Video S21.

Exp. 22 Modification and calibration of electric meters

22.1 Objectives

(1) Master the method of modification and calibrating electric meters.

(2) Learn how to determine the accuracy class of electric meters.

22.2 Apparatus

Microampere meter, ammeter, voltmeter, DC power supply, resistance box, slide-wire rheostat, switch, and several wires.

22.3 Principles

In order to meet the different needs of production and experimentation, ammeter and voltmeter are generally modified from microampere meters with different circuits and components. There are two important parameters of the micro-ammeter, which is full-scale current I_g and internal resistance R_g. A microampere meter can only measure small currents.

22.3.1 Converting microampere meter into ammeter

Modifying ammeter, in other words, is to expand the measuring range of microampere meter. I_g is the maximum current which can be measured by the micro-ammeter. If want to measure higher currents, you must expand the range of microampere meter. That is modifying a microampere meter to a milliampere meter or an ammeter. To measure higher currents, we can connect a shunt resistance R_s in parallel with micro-ammeter. Let majority of current flow through the shunt resistance R_s, and the current flow through is still in the measuring range of the microampere meter (Fig. S22 - 1).

The value of the shunt resistance is determined by the measuring range of the modifying ammeter. Suppose $I = n \times I_g$ is the full scale deflection current of the modified ammeter。 The equivalent voltage across the parallel circuit:

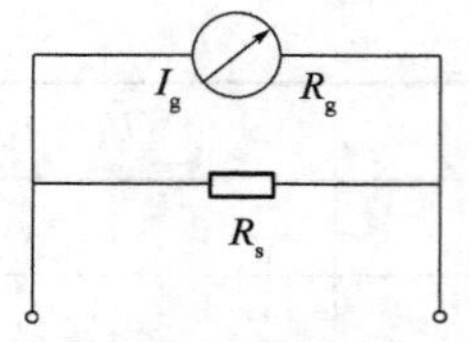

Fig. S22 - 1 Principle circuit of modifying ammeter

$$I_g R_g = (I - I_g) R_s = (n - 1) I_g R_s$$

Then the required value of shunt resistance is given by,

$$R_s = \frac{R_g}{n-1} \quad (S22-1)$$

The above formula shows that, once the microampere meter is connected in parallel with a shunt resistance $R_s = R_g/n-1$, its measuring range can be expanded to n times. For example, refit a microampere meter with $I_g = 1\text{mA}$ and $R_g = 500\Omega$ to a milliammeter with its range $I = 50.0\text{mA}$, the calculating method of shunt resistance is: According to $I = n \times I_g$, we can obtain the value of $n = I/I_g = 50$. Then we can calculate R_s use formula (S22 − 1): $R_s = R_g/(50-1) \approx 10.20\Omega$.

22.3.2 Modify microampere meter to voltmeter

When current I_g flow through a micro-ammeter with resistance R_g, the voltage upon it can be calculated by $U_g = I_g R_g$, according to Ohm's Law. So although its full scale current is very small, a microampere meter can also be used as a voltmeter. If want to measure higher voltage, it must be modified with a series resistance R_p. Let the majority measured voltage landed on the series resistance and the voltage on the micro-ammeter will not exceed the measuring range. The principle circuit is shown in Fig. S22 − 2.

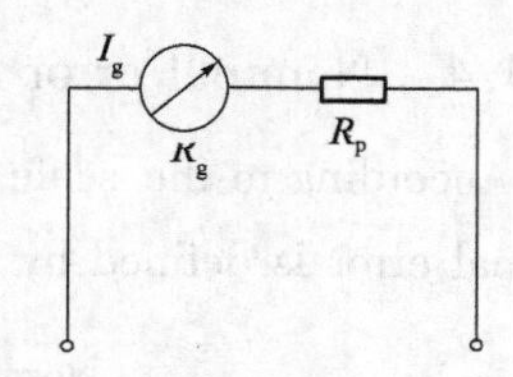

Fig. S22 − 2 Principle circuit of modifying voltmeter

The value of the series resistance is determined by the full scale deflect voltage of the modified voltmeter. Suppose the voltage range of the refitted voltmeter $U = n \times U_g = n \times I_g \times R_g$. According to Ohm's Law, we have

$$I_g(R_g + R_p) = U = nU_g$$

Then we can obtain,

$$R_p = \frac{nU_g}{I_g} - R_g = (n-1)R_g \quad (S22-2)$$

The above formula shows that, the microampere meter can be used to be a voltmeter, as long as a resistance $R_p = (n-1)R_g$ is connected in series with. For example, modify a microampere meter with $I_g = 50\mu\text{A}$ and $R_g = 5\text{k}\Omega$ to a voltmeter with its range $U = 5.0\text{V}$, the calculating method of the series resistance is: According to $U = nI_g R_g$, we can obtain the value of $n = U/I_g R_g = 20$. Then we can calculate R_s use formula (S22 − 1): $R_p = (n-1)R_g = 95\text{k}\Omega$.

22.3.3 Calibrate the modified ammeter and voltmeter

Based on the calculated value of the shunt resistor R_s and the division resistor R_p, we can build up the modified ammeter and voltmeter. Due to various reasons, the values that the modified meters show cannot be guaranteed to consistent with the required measurement standards. Therefore, the refitted meters must be calibrated by the standard meters. The adjustable errors, such as the incorrect range, must be adjusted, and the uncorrectable errors can be reflected in the calibration curve.

The calibration circuit of ammeter and voltmeter are shown in Fig. S22 − 3 and Fig. S22 − 4, respectively. In which, A_{st} and V_{st} are standard ammeter and voltmeter; the modified meters are in the dashed box; R is adjustable resistance, which can be used to change the current flow through ammeter or the voltage land on the voltmeter. Compare the difference between the standard meters and the modified ones, the difference in the measuring range can be corrected by adjusting R_s and

R_p, according to the standard meters' values. For other scale values, the calibration curves $\Delta I - I_M$ and $\Delta U - U_M$ should be plotted. ΔI and ΔU is defined as $I_s - I_M$ and $U_s - U_M$, respectively.

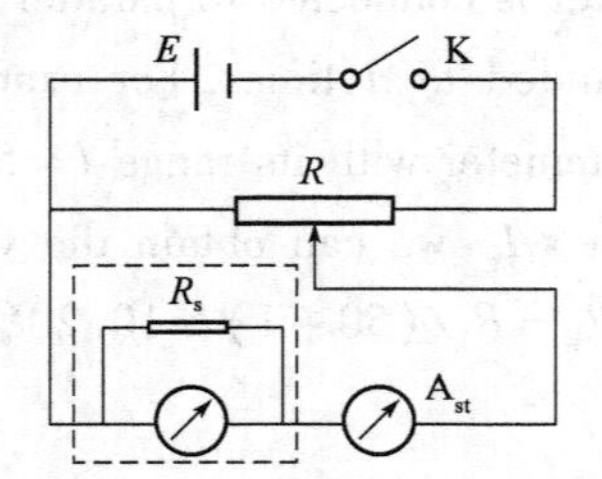

Fig. S22 - 3 Calibration circuit of ammeter

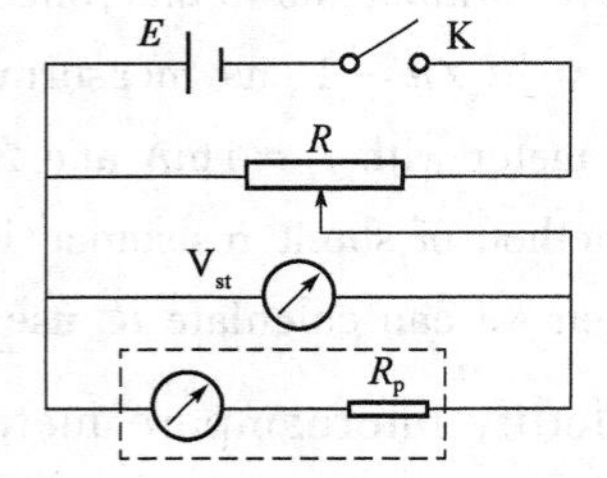

Fig. S22 - 4 Calibration circuit of voltmeter

22.3.4 Nominal error and accuracy level of refitted meters

According to the scale values of standard meters, we can find a series of absolute error. The nominal error is defined by the following formula,

$$\text{Nominal error} = \left| \frac{\text{Maximum absolute error}}{\text{Measuring range}} \right| \times 100\% \qquad (S22-3)$$

The accuracy class of the modified meters is decided by the nominal error. For example, the modified ammeter, $\Delta I_{max} = 0.05\text{mA}$, the measuring range is 5mA, so we have

$$\text{Nominal error} = \left| \frac{\text{Maximum absolute error}}{\text{Measuring range}} \right| \times 100\% = \left| \frac{0.05}{5} \right| \times 100\% = 1.0\% \qquad (S22-4)$$

So the modified ammeter is at class 1.0. According to the national standard, there are seven classes for meters' accuracy: 0.1, 0.2, 0.5, 1.0, 1.5, 2.5 and 5.0.

22.4 Procedures

(1) Modify a 50μA microampere to 5mA milliampere.

①According to the experimental content requirements, you can calculate the multiple index n for modifying ammeter.

②According to the formula (S22 - 1), calculate the value of the parallel-resistor R_s. Connect the resistance-box showing R_s with microampere meter in parallel, then the modified ammeter is formed.

③In accordance with the circuit diagram, connect the calibration circuit. (The ammeter's range should equal to or larger than the modified ammeter's range.)

④Calibrate range: adjust slide-wire rheostat R until the pointer of standard ammeter point to 5mA. At the same time, the modified ammeter pointer must point to full scale 5mA. If not, adjust slide-wire rheostat R and the parallel resistor R_s.

⑤Calibrate other scale: adjust slide-wire rheostat R, allow the current flow through the ammeter descending. When the modified ammeter shows 4.00mA, 3.00mA, 2.00mA, 1.00mA, 0.00mA, record the values of standard ammeter. Then let the current increasing and record the corresponding values. At last, plot the $\Delta I - I$ calibration curve of modified ammeter.

(2) According to the experimental principle, based on the experimental procedures of modifying ammeter, refit voltmeter and measure the corresponding data.

22.5 Data and calculations

(1) Modify a 50μA microampere to 5mA milliampere.

Full scale deflection current of standard ammeter: ________ mA.

Accuracy class of standard ammeter: ________.

Full scale deflection current of microampere meter: ________ μA.

Resistance of microampere meter: ________.

Accuracy class of microampere meter: ________.

Multiple index n: ________.

Calculated value of R_s: ________.

Measured value of R_s: ________.

Record the modified ammeter data in Tab. S22 – 1.

Tab. S22 – 1 Modified ammeter data

Modified Ammeter		Standard Ammeter			$\Delta I = \bar{I}_{st} - I_{re}$
Reading number	I_{re}/mA	descending	increasing	$\bar{I}_{st}$	mA
0	0.00				
10	1.00				
20	2.00				
30	3.00				
40	4.00				
50	5.00				

①Plot the calibration curve $\Delta I - I_M$ of modified ammeter;

②Calculate the nominal error of the modified meters and identify their accuracy classes.

(2) Modify a 50μA microampere meter to 5V voltmeter.

Full scale deflection current of standard voltmeter: ________ mA.

Accuracy class of standard voltmeter: ________.

Full scale deflection current of microampere meter: ________ μA.

Resistance of microampere meter: ________.

Accuracy class of microampere meter: ________.

Multiple index n: ________

Calculated value of R_p: ________.

Measured value of R_p: ________.

Record the modified voltmeter data in Tab. S22 – 2.

Tab. S22 – 2 Modified voltmeter data

Modified voltmeter		Standard voltmeter U_{st}/V			$\Delta U = \bar{U}_{st} - U_{re}$
Reading numbers	U_{re}/V	descending	increasing	$\bar{U}_{st}$	V
0	0.00				
10	1.00				
20	2.00				
30	3.00				
40	4.00				
50	5.00				

①Plot the calibration curve $\Delta U - U_M$ of modified voltmeter;

②Calculate the nominal error of the modified meters and identify their accuracy classes.

22.6 Precautions

Video S22 Modification and calibration of electric meters

(1) Before connecting the circuit, adjust the zero point of the modified meter and the standard meter, and make the slider rheostat set the minimum value.

(2) When plot calibrating the curve, the line between the two points should be a line.

(3) Don't exceed the maximum range of meter when measuring.

Modification and calibration of electric meters is shown in Video S22.

Exp. 23 Frank-hertz experiment

23.1 Objectives

(1) Understand the principle of Frank-Hertz experiment;

(2) Measure the first excitation potential of argon atom and prove the existence of the atomic energy level with experimental method.

23.2 Apparatus

ZKY-FH-2 intelligent Frank-Hertz experiment instrument, an Oscilloscope.

23.3 Principles

On the basis of Rutherford's atomic model, Bohr atomic theory points out that: (1) Electrons in an atom can move in some specific circular orbit, but not radiating electromagnetic wave, and the atoms is in a stable state, and has a certain energy; (2) Atoms need to emission photons that atomic frequency ν when they jump from the high energy to low energy, and

$$h\nu = E_2 - E_1 \qquad \text{(S23-1)}$$

Where h is Planck constant, $h = 6.63 \times 10^{-34}$ J · s.

The electrons of the atom in the normal state move in the first pathway, atomic energy minimum, and that is at the lowest level, this state is called the ground state. Atomic transitions from the ground state to a higher energy state called excited state. Energy required is called the critical energy when atomic transitions from the ground state to the first excited state. To change the state of atomic, the method is: (1) Atoms itself absorb or emit electromagnetic radiation; (2) Atoms collide with other particles and energy exchange. Franck-Hertz used the latter.

Franck-Hertz experiment apparatus principle shown in Fig. S23 - 1. In Franck-Hertz tube tilled argon, electrons emitted by the thermal cathode, tend space G_1G_2 under the action of U_{G_1K}, and then accelerated movement to the grid electrode G_2 under the action of U_{G_2K}. There is reverse voltage U_{G_2A} between plate electrode A and grid electrode $R_x = \dfrac{R_2}{R_1}R + \dfrac{rR_1'}{R_1' + R_2' + r}\left(\dfrac{R_2}{R_1} - \dfrac{R_2'}{R_1'}\right)$. As long as the electron energy is sufficient to overcome the rejection voltage U_{G_2K}, electron can reach plate electrode A, the electron flow is formed. Micro-current electron flow meter can display the intensity. Experiment maintained U_{G_1K}, U_{G_2A} unchanged, U_{G_2K} increases, electron energy increases gradually, and also increases. In this process, electronics and argon atoms in tube elastic collisions, and electron energy cannot be obtained by argon atoms. With the increase of U_{G_2K}, when the electron energy is equal to or greater than the critical energy of argon atoms, all or most of the electron energy is passed to the argon atoms. The energy of the electrons decreases sharply, so that the electrons cannot overcome the effects of rejection voltage, resulting in a sharp reduction of electrons that reaches plate electrode, then I_A decreased sharply. In this process, electronic and atom inelastic collision, argon atoms jump from the ground state to the excited state after getting energy from the electron. Continuing to increase U_{G_2K}, electronic access to accelerate, and its energy increase, then electrons that overcome the rejection voltage and reached the plate electrode continue to increase, I_A increased. When the electron energy is equal to or greater than the critical energy of argon atoms, all or most of the energy of electrons passed to the argon atoms. Then electron energy reduced again, so that most of the electron cannot reach plate electrode, it means that I_A decreased again. So, with U_{G_2K} continued to increase, the electron gets accelerated to reach plate electrode, I_A increases. As long as the electron energy is sufficient to overcome the rejection voltage U_{G_2A} again, argon atoms can obtain energy from the electron to achieve at excited state. Electrons lost all or most energy, so they can't reach plate electrode, and then I_A decreased sharply. With the increase of U_{G_2K}, I_A changes in the performance characteristics of a distinct peaks and valleys. It shows in Fig. S23 - 2.

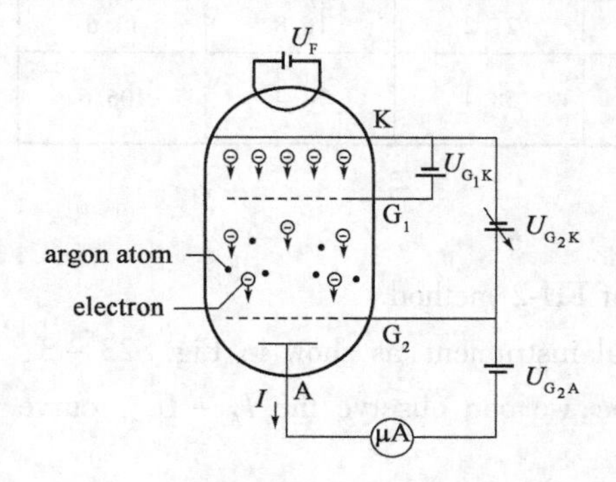

ig. S23 - 1 Schematic diagram of $F-H$ tube

Fig. S23 - 2 $I_A - U_{G_2K}$ relation map

This suggests that atomic absorption energy quantization characteristics. It found by curve $I_A - U_{G_2K}$: potential difference between two adjacent peaks are equal, and it is 11.6V. When A is integer times of 11.6V, I_A decreased sharply. When the electron energy is 11.6eV, Argon atoms will absorb the energy of an electron and inspire. It can be considered that 11.6eV is the needed energy of exciting argon atoms from the ground state to the first excited state. It can be seen, the potential difference between adjacent two peaks curve $I_A - U_{G_2K}$ is the first excitation potential of argon atom, and $U_0 = 11.6$ V. The critical energy value is obtained by argon atoms is:

$$eU_0 = E_2 - E_1 \qquad \text{(S23 - 2)}$$

The time that the atom is in the excited state is usually not very long. Excited atoms will automatically transition to the ground state and simultaneously release the energy they have obtained.

Energy is radiated in the form of light, and the light frequency is

$$\nu = eU_0/h \tag{S23-3}$$

Presumably, in the experiment radiation should be able to be seen that argon atoms transition from the first excited state back to the ground state. Its energy eU_0, in the form of photons emitted wavelength

$$\lambda = hc/(eU_0) \tag{S23-4}$$

Where, Planck constant $h = 6.63 \times 10^{-34}$ J·s; speed of light $c = 3.00 \times 10^8$ m/s; electron charge $e = 1.6 \times 10^{-19}$ C; the first excitation potential of argon atoms $U_0 = 11.6$V. Combine with equation (S23-4):

$$\lambda = 6.63 \times 10^{-34} \times 3.00 \times 10^8 / (1.6 \times 10^{-19} \times 11.6) = 1.07 \times 10^2 \text{ nm}$$

Franck-Hertz observed spectral lines of mercury whose $\lambda = 2.53 \times 10^2$ nm. This is in agreement with the results, $\lambda = 2.53 \times 10^2$ nm calculated by $U_0 = 11.6$V. This fully confirms the atomic level does exist. For exciting atoms to an excited state, atoms must absorb the energy certain magnitude, and the energy is not continuous.

If the Franck-Hertz tube is filled with other elements, the first excitation potential of the elements can be measured. Tab. S23-1 lists several first excited potential of elements and radiation wavelength from the first excited state to the ground state.

Tab. S23-1 Several first excited potential of elements and radiation wavelength from the first excited state to the ground state

Elements	Sodium (Na)	Potassium (K)	Lithium (Li)	Magnesium (Mg)	Helium (He)	Neon (Ne)	Argon (Ar)
U_0/V	2.12	1.63	1.84	2.71	21.2	16.8	11.6
λ/nm	589.0 589.6	766.4 769.9	690.8	457.1	58.4	74.4	106.6

23.4 Procedures

(1) Be familiar with using the experimental apparatus of FH-2 method.

(2) Check the panel wiring of Frank-Hertz experimental instrument, as show in Fig. S23-3, confirm and start. Connect the instrument to the oscilloscope, we can observe the $I_A - U_{G_2K}$ curve show on the screen visually.

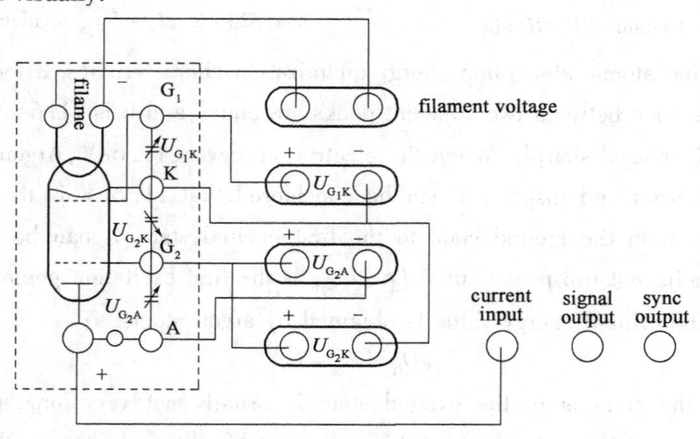

Fig. S23-3 Frank-Hertz experimental instrument panel wiring diagram

(3) The First excitation voltage of argon element are measured with manual mode.

①Set the instrument to "manual" work status, Press the "manual/automatic" key, and "manual" indicator light.

②Set current range (Current range may refer to data provided by the chassis cover), press the appropriate current range key, the corresponding range indicator light.

③Set the power supply voltage (Settings can refer to data provided by the chassis cover), and complete with the key of ↓, ↑, ←, →. Parameters have to be set: filament voltage U_F、the first acceleration voltage U_{G_1K}, rejection voltage U_{G_2A}.

④Press the "Start" button to start the experiment. Adjusting the voltage U_{G_2K} with ↑, ↓, ←, →. Start from 0.0 V. Step length is 1V or 0.5V. At the same time, use an oscilloscope to observe the change of the plate current I_A with U_{G_2K}. (Remember that in order to ensure the uniqueness of the experimental data, U_{G_2K} must be adjusted unidirectionally from small to large, and cannot be repeated in the process. After recording the last set of data, immediately reset U_{G_2K} to zero.)

⑤Restart. During the manual test, press the start button, U_{G_2K} will be set to zero, the internally stored test data will be cleared, and the curve displayed on the oscilloscope will be cleared, but the initial setting parameters U_F, U_{G_1K}, U_{G_2A} and the current gear, etc. will not change. At this time, you can perform the test again in this state, or perform the test after modifying the state.

Suggestion: manually test $I_A - U_{G_2K}$, do it once or do it again after modifying the U_F value.

(4) The first excitation voltage of argon element are measured with testing automatic mode.

In addition to manual testing, the intelligent Frank-Hertz experimental instrument can also perform automatic testing.

During the automatic test, the experimental instrument will automatically generate the U_{G_2K} scanning voltage to complete the entire test process; connect the oscilloscope to the experimental instrument, and the curve of the Frank-Hertz tube plate current changing with the voltage of the U_{G_2K} can be seen on the oscilloscope.

①Set the auto test status: All the operation processes, i. e. the parameter setting of U_F, U_{G_1K}, U_{G_2A} and current gear, the connection of Frank-Hertz tube, are the same as the manual test.

②Set the end value of scan voltage U_{G_2K}

During automatic testing, the experimental instrument will automatically generate U_{G_2K}. The initial value of U_{G_2K} is zero by default, and it is incremented by 0.2 volts approximately every 0.4 seconds until the end voltage.

③Select U_{G_2K} as the voltage source, and then press the "Start" button on the panel to start the automatic test.

23.5 Data and calculations

(1) Draw the corresponding curve of each group of $I_A - U_{G_2K}$ data (Tab. S23-2).

(2) Calculate the difference ΔU_{G_2K} of U_{G_2K} corresponding to every two adjacent peaks or valleys, and find the average value $\overline{U}_0$. Compare the experimental value $\overline{U}_0$ with the first excitation potential $U_0 = 11.61\text{V}$ of argon to calculate the relative error. And give the final result.

Tab. S23 – 2 Data record form

	F_1	G_1	F_2	G_2	F_3	G_3	F_4	G_4
U_{G_2K}/V								
I_A/μA								

23.6 Precautions

(1) Before open the power supply, please check the attachment, make sure it is connected correctly, if not, please report the teacher.

(2) Frank-Hertz tube is easily damaged due to improper voltage setting. Please set parameter refer to data provided by the chassis cover.

(3) In order to guarantee the uniqueness of the experimental data, the voltage U_{G_2K} must be one-way from small to large, the process cannot be repeated. After recording the final data, the voltage U_{G_2K} rapidly return to zero immediately.

Exp. 24 Photoelectric effect experiment

24.1 Objectives

(1) Be familiar with the conditions of the photoelectric effect.

(2) Understand the basic laws of photoelectric effect and the quantum nature of light.

(3) Verify the Einstein photoelectric equation and measure the Planck ' s constant.

24.2 Apparatus

ZKY-GD-4 photoelectric effect experiment instrument, mercury lamp, color filter, aperture, phototube, tester, the structure of which is shown in Fig. S24 – 1.

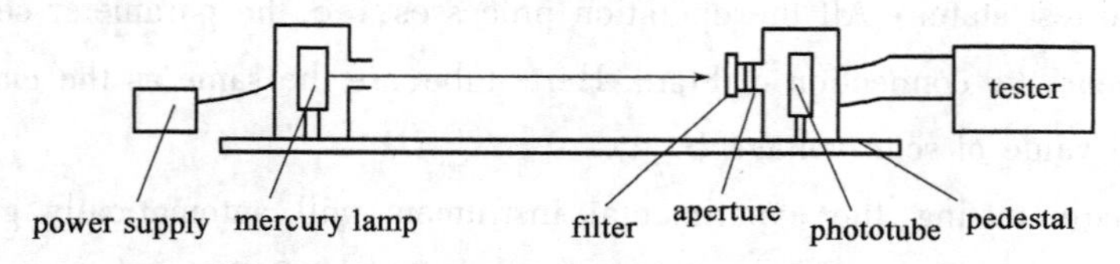

Fig. S24 – 1 Structure of photoelectric effect instrument

24.3 Principles

Planck constant is related with the ubiquitous wave-particle duality and quantized energy exchange phenomenon in the microscopic world, and plays an important role in modern physics. Measuring Planck constant through photoelectric effect experiment can help students understand the quantum of light and universal constant h.

When light with a certain frequency irradiates the metal surface, there are electrons escaping from its surface, and this phenomenon is known as the photoelectric effect. Its basic experimental facts are:

(1) Photoelectron emission rate (photocurrent) is proportional to the light intensity [Fig. S24 - 2 (a)、(b)].

(2) There is a threshold frequency (or cutoff frequency) of the photoelectric effect. When the frequency of the incident light is below a certain threshold value ν_0, regardless of the intensity of light, no photoelectron will generate [Fig. S24 - 2(c)].

(3) The initial kinetic energy of photoelectron is independent of light intensity but proportional to the frequency of the incident light [Fig. S24 - 2(d)].

(4) Photoelectric effect is an instantaneous effect, i. e. once light irradiates the metal surface, photoelectrons are immediately produced. However, the experimental facts above cannot be explained by Maxwell's classical electromagnetic theory.

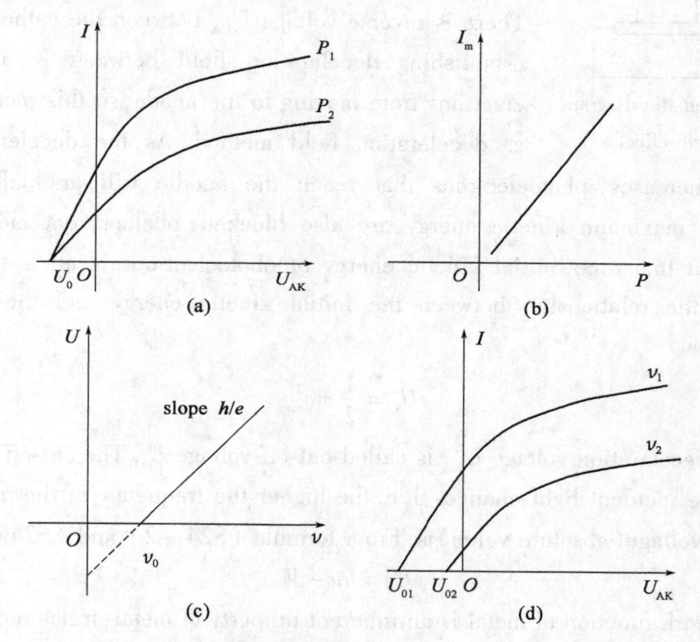

Fig. S24 - 2 Several characteristics of the photoelectric effect

Einstein believed that the light emitted from a point doesn't follow Maxwell electromagnetic theory, i. e. not in the form of a continuous distribution of the energy spreading into the space, but the light, whose frequency is ν, radiates outwards one by one with the energy unit of $h\nu$.

According to this theory, in the photoelectric effect, when a free electron in the metal absorbs the energy of $h\nu$ from a photon of the incident light, if not losing energy due to collision on the way, a part is used for the work function W_s, and the rest corresponds to the largest kinetic energy after the electron escaping from the metal surface, i. e.

$$\frac{1}{2}mv_{\max}^2 = h\nu - W \qquad \text{(S24 - 1)}$$

This is the famous Einstein photoelectric equation. In this formula, h is Planck's constant, the known value is $6.6260755 \times 10^{-34}$ J · s. The formula (S24 - 1) successfully explains the law of the photoelectric effect:

(1) When photon energy $h\nu < W$, the photoelectric effect cannot be produced.

(2) Only when the frequency is greater than the threshold frequency of the incident light $\nu_0 = W_s/h$, the photoelectric effect can be produced. The higher the frequency of the incident light is, the larger the initial kinetic energy of the escaping photoelectron is;

(3) The size of the light intensity means the size of the photon flux density, i. e. light intensity only affects the size of photocurrent. The size of saturated photocurrent is proportional to the size of light intensity of the incident light.

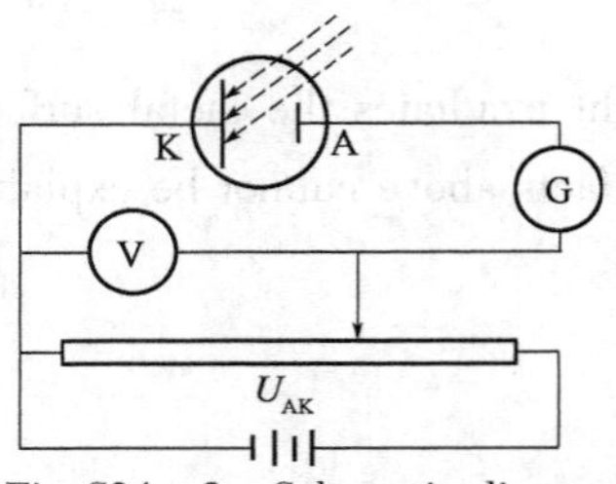

Fig. S24 - 3 Schematic diagram of photoelectric effect

Fig. S24 - 3 is the experimental device of this experiment: the light, whose frequency is ν and light intensity is I, irradiates on the cathode K of phototube, photoelectrons emitted from K move to the anode A, and form photocurrent along the outer loop. There is reverse voltage U_{KA} between the cathode and the anode, establishing deceleration field between K and A to prevent electrons from moving to the anode, so this method is also known as deceleration field method. As the deceleration voltage U_x. (absolute value) increases, photoelectrons that reach the anodic will gradually decrease. Until photoelectrons with maximum kinetic energy are also blocked; photocurrent along the outer loop decreases to zero; at this time, initial kinetic energy of photoelectrons is all used to overcome the deceleration field, the relationship between the initial kinetic energy and the deceleration field satisfies this formula:

$$eU_s = \frac{1}{2}mv_{max}^2 \qquad (S24-2)$$

At this time, deceleration voltage U_{KA} is called cut-off voltage U_s. The cut-off voltage changes as the frequency of the incident light changes, i. e. the higher the frequency of the incident light is, the greater the cut-off voltage (absolute value) is. From formula (S24 - 2) and (S24 - 1), and we get:

$$eU_s = h\nu - W \qquad (S24-3)$$

Because the work function of metal is an inherent property of metal, it has nothing to do with the frequency of the incident light. According to the formula (S24 - 3), for the same kind of cathode, the relationship between cut-off voltage U_s and the frequency of the incident light ν is linear, and the slope of the line is h/e. Thus, as long as measuring cut-off voltage U_s of the different frequencies of light, making a curve of $U_s - \nu$, and finding the slope of this curve, you can find the value of Planck constant h. Electron charge is $e = 1.60 \times 10^{-19}$ C.

Fig. S24 - 4 represents the change of photocurrent-voltage curve that is a theoretical case. In practical measurement, there are some unfavorable factors that affect the measurement results. If not making reasonable treatments to these unfavorable factors, it will bring great errors to experiment results. These unfavorable factors mainly refer to:

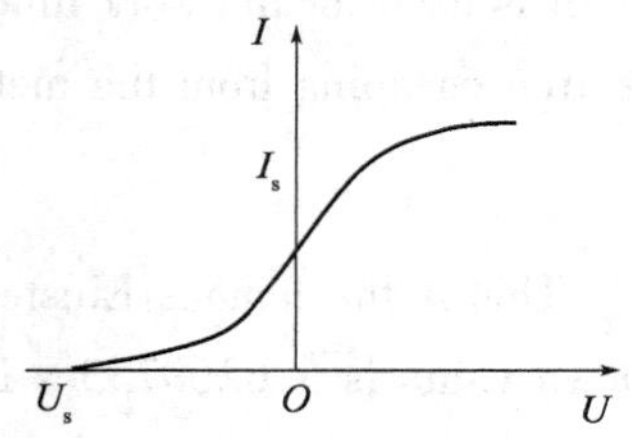

Fig. S24 - 4 Volt-ampere characteristic of phototube

(1) Dark current. When there is no light in phototube, there is weak current in the phototube under the effect of deceleration voltage, and because there is thermion emission at normal temperature and insulation resistance between cathode and anode is

not enough high, etc. The relationship between the volt-ampere characteristic of phototube dark current and external voltage is basically linear.

(2) Anode emission current. The anode of phototube is made of high work function materials such as platinum, tungsten, etc. Because of depositing cathode materials, it will emit photoelectrons when meeting visible light. The electric field which has deceleration effect for electrons emitting from cathode is accelerating electric field for electrons emitting from anode, and it will make reverse saturation current in phototube. Avoiding the beam pointing at the anode directly is required when using the device, but the scattering light is inevitable from the cathode; therefore, reverse saturation current exists.

(3) The cathode of phototube is made of low work function materials. This material is still easy to be oxidized in high vacuum, so the work function of the cathode surface varies. With the increase of the reverse voltage, the photocurrent is not cut-off abruptly, but decreasing faster and gently reaching zero points; therefore, a high sensitivity galvanometer is needed.

Due to the above reasons, the $I-U$ relationship curve of phototube is shown in Fig. S24 – 5.

Actually each current value on the measured curve includes three parts, i. e. two curves mentioned above and forward current produced by cathode photoelectric effect, so the voltage-current curve is not tangent to the U-axis. Because the value of dark current is small compared with cathode forward current, their effects on the cut-off voltage can be ignored. The current emitting from anode is significant in practice, but it is subject to a certain rule. Through the analysis of these unfavorable factors, the reasonable design of phototube structure and the use of rational data processing method can reduce or even eliminate these disturbances. Two methods are usually adopted when determining the value of cut-off voltage: ①Intersection point method. The anode of phototube is made of high work function materials, so prevent possible evaporation of the cathode materials in producing process. Energize the anode of phototube before experiment, reduce its sputtered cathode materials and avoid incident light irradiating the anode. In this way, reverse current can greatly be reduced, and its volt-ampere characteristic curve is very close to Fig. S24 – 4. Therefore, the potential difference at the intersection point between the measured curve and the U-axis is approximately equal to the cut-off voltage U_s, which is known as the intersection point method. ②Inflection point method. Photocurrent emitting from phototube anode is comparatively great, but in structural design, if anode current can saturate as quickly as possible, volt-ampere characteristic curve will have an obvious inflection point after the saturation of the cathode current. As shown in Fig. S24 – 4, the potential difference at this inflection point is the cut-off voltage U_s.

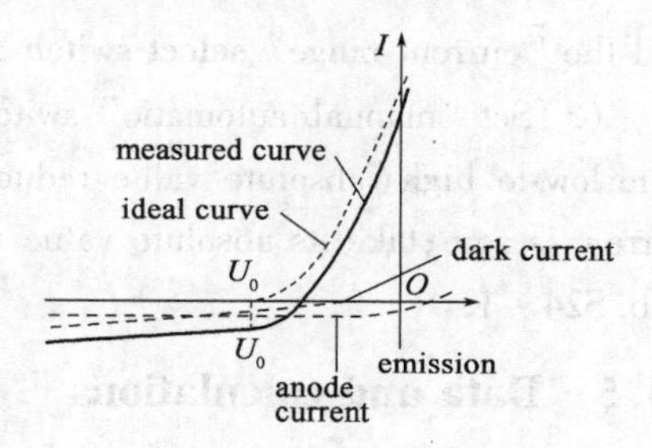

Fig. S24 – 5 $I-U$ characteristic curve of phototube

24.4 Procedures

24.4.1 The preparation work before experiment

(1) Cover the dark box baffle of phototube and port of mercury lamp.

(2) Turn on the tester and mercury lamp and preheat them for 20 minutes.

(3) Adjust the distance between phototube and mercury lamp (about 40cm) and remain it unchanged on the instrument base, and make the port of mercury lamp point at the input port of the dark box of phototube.

(4) Connect voltage input terminal of phototube and voltage output terminal at the tester back panel with special wires (attention: red to red, and blue to blue).

(5) Rotate the selector button of current range on the tester panel to desirable gear (10A). After fully preheating, adjust current to zero before test. That is, rotate current zero switch and the current indication will be 000.0. After the adjustment, press "zero adjustment confirmation/system reset" switch, and the system comes into the test status.

24.4.2 The measurement of cut-off voltage

(1) Install aperture of 4mm and filter of 365.0nm at input port of phototube box (attention: mercury lamp cover can't be taken away during this process to avoid the damage of phototube).

(2) Set "volt-ampere characteristic test/ cut-off voltage test" switch at "cut-off voltage test", and the "current range" select switch at 10^{-13} A position.

(3) Set "manual/automatic" switch at manual mode, and open the lamp cover. Adjust voltage from low to high (absolute value reduces); observe the changes of current value; search U_s when current is zero; take its absolute value as the value of U_s of this wavelength, and record the data in Tab. S24 - 1.

24.5 Data and calculations

Record the data in Tab. S24 - 1.

Tab. S24 - 1 $U_0 - \nu$ data

Aperture ϕ = ______ mm

Wavelength λ_i/nm		365.0	404.7	435.8	546.1	577.0
Frequency ν_i/ $\times 10^{14}$ Hz		8.214	7.408	6.879	5.490	5.196
Cut-off voltage U_{0i}/V	manual					
	auto					

The experimental data in Tab. S24 - 1 can be used to draw the $U_0 - \nu$ curve, and then the Planck constant can be calculated from the slope k of the $U_0 - \nu$ curve. Compare the result with the recognized standard value to calculate the relative error.

24.6 Precautions

(1) During the use of the instrument, the mercury lamp should not directly irradiate the photocell, nor should it be continuously irradiated for a long time, otherwise the service life of the photocell will be reduced.

(2) Please make sure not to ground the output terminal of the power supply to avoid burning the power supply.

(3) After the experiment is completed, please cover the photocell cassette with a light shield to protect the photocell.

Exp. 25 Hall effect experiment

25.1 Objectives

(1) Measure the voltage of Hall device in different areas of the solenoid.

(2) Describe the distribution curve of magnetic induction in the solenoid.

25.2 Apparatus

ZKY-HS Hall effect tester, Hall effect measurement instrument, wires.

25.3 Principles

25.3.1 Hall effect

The Hall effect is caused by the movement of the electric charge in the magnetic field coming in for the Lorentz force. The current in the semiconductor is formed by the directional motion of the charge carriers (electron or hole). Put a Hall device (thickness d, width b, length l) made of semiconductor material in the magnetic field B. The direction of magnetic field is along the z axis.

Assuming that the control current I_s flows forward through the semiconductor along the r axis. If the carrier in the semiconductor is an electron, the charge is e, the average migration velocity is v, and the carrier in the magnetic field for the Lorentz force can be expressed as

$$F_B = evB \tag{S25-1}$$

Under the effect of F_B, the electron flow is deflected. Some of the electrons gather on the transverse face N of the slice, so that some positive charges appear on the transverse face M. The agglomeration of electric charges caused a transverse electric field E_H, the direction is from the M to N. the electron experiences an opposite direction force F_E compared with F_B. the size can be expressed as

$$F_E = eE_H \tag{S25-2}$$

Because of the electrostatic field force F_E, the further accumulation of charge is hindered, at the beginning, F_E is small, electrons can continue gather, the electric field E_H gradually increases. Until the role of the carrier on the Lorentz force and electric field force achieves dynamic equilibrium state: $F_B = F_E$. so

$$evB = eE_H = eU_H/b \tag{S25-3}$$

At this time, the electric potential difference between MN is the Hall voltage

$$U_H = vbB \tag{S25-4}$$

The relationship formula of I_s, e, n, v and b, d is

$$I_s = nevbd \tag{S25-5}$$

and

$$U_H = \frac{I_s B}{ned} = R_H \frac{I_s B}{d} = K_H I_s B \tag{S25-6}$$

If we know the sensitivity of Hall components, using the instrument to measure I_S and U_H, we can calculate the size of the magnetic induction intensity B, which is the principle of magnetic field measurement of the Hall Effect.

25.3.2 Eliminate the influence of Hall component secondary effect on the measurement results

Measuring the Hall voltage U_H will inevitably produce some side effects, because of these secondary effects with potential superimposed on the Hall voltage, system errors exist in the measurement.

(1) Unequal potential difference U_0.

In production, it is difficult to weld the potential electrode on the same equipotential surface, so when the current flows through the Hall piece, even without the magnetic field, there will be a potential difference U_0 between the two potential electrodes, $U_0 = I_s R_x$. This potential difference is called the unequal potential difference. It is only related to the current I_S, independent of the magnetic field B.

(2) Ettingshausen effect U_E.

Because of the different carrier migration velocities, the Lorentz force is not equal, so the circular motion orbit radius is not equal. The carrier with higher rate will move along the larger radius of circular orbit, causing one Hall side to appear more fast-carrier with high temperature, the other side appeared more slow-carrier and low temperature. Because of the different temperature between two surfaces of temperature-different electromotive forces, this generated effect is called the Ettingshausen effect. U_E is proportional to the product of $I_C B$. The direction is changed with B and I_s.

(3) Nernst effect U_N.

Because the Hall element current with lead-out solder joint contact resistance is different, with current after the different degree of the Joule heat of fever caused by difference, generating the electromotive force between the two poles, the electromotive force generated by the electromotive current (known as the diffusion current) Q, the thermal current will deflect under the action of the magnetic field, resulting in the additional potential difference U_N in the r direction. The direction of U_N is related to the direction of the magnetic field B, independent of the direction of current I_s.

(4) Righi-leduc effect U_R.

The carrier diffusion rates of the above thermal diffusion currents are different. Under the action of the magnetic field, in addition to producing the potential difference in the r direction, it will also cause the temperature difference on both sides of the Hall piece in the r direction. This temperature difference also produces an additional thermo-electromotive force U_R in the z direction, which is related to the direction of the magnetic field B, independent of direction of current I_s.

The sum of the potential differences caused by the above-mentioned side effects, sometimes much larger than the Hall voltage, result in systematic errors in the measurement, the true result of which is the comprehensive effect. In order to reduce and eliminate the additional voltage caused by these effects, we use these additional voltages and the relationship between the control current I_s and the magnetic field B of the Hall element.

If $+I_M$ and $+I_s$ $\quad U_{CD1} = +U_H + U_o + U_E + U_N + U_R$

If $+I_M$ and $-I_s$ $\quad U_{CD2} = -U_H - U_o - U_E + U_N + U_R$

If $-I_M$ and $-I_s$ $\quad U_{CD3} = +U_H - U_o + U_E - U_N - U_R$

If $-I_M$ and $+I_s$ $\quad U_{CD4} = -U_H + U_o - U_E - U_N - U_R$

Then we get

$$\frac{1}{4}(U_{CD1} - U_{CD2} + U_{CD3} - U_{CD4}) = U_H + U_E \tag{S25-7}$$

By changing the direction of I_s and B, U_0, U_N, and U_R disappear from the calculation, and direction of U_E is always consistent with the U_N direction; it can't be eliminated in the experiment. But the U_E is usually much smaller than the U_H, so the error it brings in is negligible. So the Hall voltage is

$$U_H \approx U_H + U_E = \frac{1}{4}(U_{CD1} - U_{CD2} + U_{CD3} - U_{CD4}) \tag{S25-8}$$

25.4 Procedures

(1) Connect the wires according to the experimental instrument manual.

(2) Turn on the excitation current, double-pole, double-throw switch of Hall current, adjust the solenoid's excitation current $I_M = 300$ mA.

(3) Adjust the working current $I_s = 1.00$mA, 2.00mA, ···, 10.00mA (step Length 1.00mA), measure the corresponding voltage $U_i (i = 1,2,3,4)$ between C and D, under the four conditions of $(+I_M, +I_s)$, $(+I_M, -I_s)$, $(-I_M, -I_s)$ and $(-I_M, +I_s)$, respectively. Fill in the data obtained in Tab. S25 – 1.

(4) Place the Hall piece in the position of the solenoid $x = 0$, adjust the excitation current $I_M =$ 600mA, the working current $I_s = 5.00$mA. Then, record the voltage value $U_i (i = 1,2,3,4)$ at the position $x = 0$, 2mm, ···, 12mm (step Length 2mm), then 15mm, 20mm, ···, 50mm (step Length 5mm). Fill in the data obtained in Tab. S25 – 2.

25.5 Data and calculations

Record the data in Tab. S25 – 1、Tab. S25 – 2.

Tab. S25 – 1 Hall effect measurement data (Current)

I_s/mA	U_1/mV	U_2/mV	U_3/mV	U_4/mV	U_H/mV
	$+I_M$, $+I_s$	$+I_M$, $-I_s$	$-I_M$, $-I_s$	$-I_M$, $+I_s$	
1.00					
2.00					
3.00					
4.00					
5.00					
6.00					
7.00					
8.00					
9.00					
10.00					

Tab. S25 - 2 Hall effect measurement data (Position)

X/mm	U_1/mV	U_2/mV	U_3/mV	U_4/mV	U_H/mV
	$+I_M$, $+I_s$	$+I_M$, $-I_s$	$-I_M$, $-I_s$	$-I_M$, $+I_s$	
0					
2					
4					
6					
8					
10					
12					
15					
20					
25					
30					
35					
40					
45					
50					

25.6 Precautions

(1) As the excitation current I_s is large, do not connect I_M and I_s incorrectly, otherwise the excitation current will burn out the Hall device.

(2) Hall device and two-dimensional moving ruler are easy to break, attention should be paid to avoid extrusion, collision and so on. Before the experiment, the electromagnet should be checked for looseness and displacement.

(3) Disconnect the excitation current switch except for reading the relevant data in a short time and using the excitation current I_M.

(4) The instrument should not be operated and stored in the environment of strong light, high temperature, strong magnetic field and corrosive gases.

中文部分

(Chinese part)

第1章 绪　论

物理学是研究物质的基本结构、基本运动形式、相互作用及其转化规律的学科。物理学是自然科学的基础,是产生新技术的源泉。高新技术无一例外地与物理学的基本概念、基本理论、基本实验方法有着密切的联系。诸如从激光技术、半导体技术、真空技术、计量技术等,大到宇航技术,小到微电子技术无不与物理实验有着千丝万缕的联系。

物理实验是新兴科学技术的生长点,在推进科学技术的进步和国民经济的发展中起着重要的作用。由于实验是科学前沿最为活跃的领域,当新的自然现象被发现和新的实验技术产生后,人们往往利用这些新成果,进一步跟踪创新,扩大成果,创造出一个又一个崭新的科学技术。

科学实验,首先是物理实验,它和生产技术紧密相关。它可能凭借实验室的优越条件,超越生产实践的某些局限,步入生产实践的前列,为生产技术的发展开辟出新的道路。

科学史表明,自然科学的重大突破,一般不是直接来自生产实践,往往要通过科学实验。例如,电磁感应定律的确立、狭义相对论的发现、量子理论的兴起、基因学说的形成,都不是直接来源于生产实践,而是源于实验研究的结果。

爱因斯坦有句名言:“发展独立思考和独立判断的一般能力应该始终放在首位,而不应该把获得专业知识放在首位。”诺贝尔物理学奖获得者李政道也说过,动手多了会让人聪明。物理实验就是通过学生的动手动脑,让学生学会最基本的实验知识、最基本的物理思想和最基本的科学实验方法。

1.1　物理实验课程的目的和任务

物理实验既是对学生进行科学实验的基本训练,又是学生进入大学后接受系统实验方法和实验技能训练的一门实验课程。它是一门独立设课、学生必修的实验基础课,为学生学习后续课程的实验和进行工程实验打下必要的基础。物理实验课的任务是:

(1)通过对实验现象的观测、分析和对物理量的测量,学习并掌握物理实验的基本知识、基本方法和基本技能。

(2)使学生学会常用物理仪器的调整方法及正确的使用方法。

(3)使学生初步具备处理数据、分析实验结果和撰写实验报告的能力。

(4)培养学生对待科学实验一丝不苟的严谨态度和实事求是的工作作风。

1.2 物理实验课程的教学环节

1.2.1 实验前的预习

课前认真阅读教材和有关资料,弄清实验的目的、原理、仪器的使用和测量方法,了解实验的主要步骤及注意事项等,在此基础上写出明晰的预习报告,预习报告应简明扼要,它包括实验名称、目的、原理、仪器规格和实验需要记录的数据表格。

1.2.2 实验操作

学生进入实验室,首先要遵守实验室的规章制度,认真领悟实验的注意事项、仪器的操作规程及实验方法。在实验操作中,注意观察各过程中的物理现象,仔细分析实验现象所揭示的物理规律是否正确、实验条件是否满足等。这是实验的基础,只有打好这样的基础,才能进行下一步的测量,否则,测量过程就会成为只动手不动脑的机械操作,毫无意义。在测量过程中,要不断对实验数据进行分析,通过分析,随时改进实验方法,以求取最满意的实验结果。通过实验的亲身实践,要得到更多的启发,例如,本实验还有无其他的设计方案?有无其他更好的测量方法?要大胆提出自己的创新思想。

实验完毕,学生记录的数据应提交指导教师进行审查验收,并将仪器复位归整好后,方能离开实验室。

1.2.3 撰写实验报告

实验报告是实验工作的最后总结。通过撰写实验报告,可以培养学生对科学实验报告或科学技术报告的写作能力和对研究工作进行总结的能力。实验报告的字迹要清楚、文理要通顺、内容简明扼要、图表明晰正确。实验报告的内容应包括:

(1)实验名称。

(2)实验班级、姓名、学号、实验日期。

(3)实验目的:要用自己的语言简明扼要地写明为什么要做这个实验。

(4)实验原理:要用自己的语言简明扼要地写明实验的基本理论依据,包括:物理模型和数学模型;画出实验装置示意图;明确测量方法以及实验条件等。

(5)实验仪器:简要写明主要仪器的型号、规格、精度或最小分度值。

(6)实验操作步骤:按实际操作情况简明扼要地列出。

(7)实验数据:应将实验过程记录下的原始数据仔细记录到报告中所列的数据表格中。

(8)数据处理:简要写明数据处理的方法,突出处理方法的计算过程或所作图线的拟合,以最终测量结果表达的形式将本实验结果报告出来。

(9)不确定度分析:简要明晰地分析出影响测量结果不确定度的主要因素,并指出解决的途径。

(10)问题讨论:对本实验所观测到的现象、涉及的实验方法及测量技术进行优劣分析和讨论,回答所布置的问题。

(11)结论:针对测量结果,简要明晰地写出你所得出的结论。

(12)创新:通过本实验的学习,完成教材提出的知识拓展题目。

1.3 物理实验室守则

(1)进入实验室,要树立安全第一的观念,实验过程中确保人身及设备安全。

(2)课前做好预习,进入实验室,应对号入座,首先从外观检查和核实实验仪器、用具、材料等,发现有短缺,应及时报告指导教师,切勿搬运他人的仪器和器材。

(3)实验过程中,应保持严肃认真和实事求是的科学态度,严格遵守实验室的纪律,保持实验室的安静环境。

(4)爱护仪器设备,如有损坏、丢失,应立即报告指导教师。由于粗心大意或违反操作规程而损坏仪器者,除按规定赔偿外,还应做出书面检讨,并按情节轻重,做出处理。

(5)凡使用电源的实验,必须经指导教师检查电路并同意后,才能开启电源。

(6)实验完成后,测量数据要经教师审查核对签字。离开实验室前,应将仪器整理还原,桌面收拾整洁,凳子摆放整齐。负责值日的学生做好实验室的卫生。

(7)按时完成和上交实验报告。

第 2 章　测量误差和不确定度分析

科学实验离不开测量，凡测量就必然存在测量的误差。本章将介绍误差的基本概念、误差的分类以及测量误差的分析与评价方法等内容。

2.1　测量与测量误差

2.1.1　被测量及其单位

在科学实验中，一切物理量都是通过测量得到的。所谓测量，就是将被测物理量与规定作为标准单位的同类物理量（或称为标准量）通过一定方法进行比较，得出被测物理量大小的过程，得到的大小即为被测量以此标准量为单位的测量值。测量值是由数值和单位两部分组合而成。

根据国际标准化组织（ISO）编辑出版的 ISO 标准手册和我国的国家标准 GB 3100—1993《国际单位制及应用》、GB 3101—1993《有关量、单位和符号的一般原则》，物理量可划分为 11 类：

（1）空间和时间量，如长度、角度、时间、速度等（GB 3102.1—1993）；

（2）周期及其有关量，如频率、波长、振幅、阻尼系数等（GB 3102.2—1993）；

（3）力学量，如质量、密度、力、功、能、流量等（GB 3102.3—1993）；

（4）热学量，如热力学温度、热量、热容、热导率等（GB 3102.4—1993）；

（5）电学和磁学量，如电流、电位、磁通量、磁导率等（GB 3102.5—1993）；

（6）光及有关电磁辐射的量，如发光强度、光通量、照度、辐射强度等（GB 3102.6—1993）；

（7）声学量，如声压、声速、声强等（GB 3102.7—1993）；

（8）物理化学和分子物理学量，如阿伏加德罗常数、摩尔质量、玻耳兹曼常数等（GB 3102.8—1993）；

（9）原子物理学和核物理学的量，如原子质量、电子质量、普朗克常数、里德伯常数等（GB 3102.9—1993）；

（10）核反应和电离辐射的量，如离子通量密度、能通量密度、活动、吸收剂量等（GB 3102.10—1993）；

（11）固体物理学量，如霍尔系数、汤姆逊系数、理查逊常数等（GB 3102.13—1993）。

工程上除上述物理量外，还有许多其他的量，如物质的硬度、弹性模量、感光度等。

然而，就量的种类而言，可以毫不夸张地说，工程上的绝大部分的测量都是对物理量的测量，这些量有些是表征物性的，有些则是反映物理学规律的。

测量必须注意其测量值与单位的组合结果的完整表述，也必须注意被测量单位的国际单位制（SI）的表述。

2.1.2　测量分类与测量条件

1. 测量分类

测量离不开测量人员从测量仪器获取测量值的方法,测量需要在确定的条件下进行。

(1)按测量方法的不同,可将测量分为直接测量和间接测量两类。

①直接测量:指借助测量仪器能直接读出被测量量值的测量。例如,用米尺测量长度、用天平称衡物体质量、用秒表测量时间等都属于直接测量,而相应的被测量——长度、质量、时间等称为直接被测量。

②间接测量:指需要先获取直接被测量的值,然后按照已知的函数关系经过一定运算才能求得被测量的量值的测量。例如,测量圆柱形物体的体积 V 时,首先用测长仪器对圆柱形物体的直径 d 和高度 h 进行直接测量,然后将测量值 d 和 h 代入公式 $V=\pi d^2h/4$ 计算出圆柱形物体的体积 V 的值,此例中,d 和 h 是直接被测量,V 是间接被测量。

一般地说,大多数测量都是间接测量。但随着科学技术的发展,很多原来只能以间接测量方式来获得的物理量,现在也可以以直接的方式测量了。例如,电功率的测量,现在可用功率表直接测量、速度可用速率表直接测量等。

(2)按测量条件是否相同,可将测量分为等精度测量和非等精度测量两类。

①等精度测量:在相同条件下进行一系列的测量。例如,同一人员在相同的环境下,在同一台仪器上,采用相同的测量方法,对同一被测量进行多次测量。

②非等精度测量:在测量过程中,若测量人员、测量仪器、测量方法、测量参数及测量环境等测量条件发生了改变,则这样的测量就是非等精度测量。

2. 测量条件

测量条件是指一切能够影响测量结果,而本质上又可控制的全部因素。测量条件包括进行测量的人员、测量方法、测量仪器与调整方法、环境条件等。环境条件是指测量过程中的环境温度、湿度、大气压、气流、振动、辐射等。

2.1.3　测量误差

任何物理量,在一定条件下,都存在一个客观真实的数值,这个值称为该物理量的真值。测量的目的就是要力求得到这些物理量的真值。但测量总是要依赖于测量人员,使用一定仪器,依据一定的理论和方法,在一定的环境和实验条件下进行的。在测量过程中,由于受到测量人员、测量仪器、测量方法和测量条件的限制或受一些不确定因素的影响,使测量结果与客观存在的真值之间总是存在一定的差异,也就是说,被测值只能是该真值的近似值。因此,任何一种测量值与其真值之间总会或多或少地存在一定的差值,这种差值称为该测量值的测量误差(或称被测值的绝对误差),即

$$\Delta N = N - N_0 \qquad (2-1-1)$$

由于测量条件、环境等因素具有随机性,因此,误差也是随机的,它既没有确定的大小,也没有确定的方向,但可以用特征值来表示其统计意义上的大小,这样的特征值主要有误差期望值、误差的标准差、误差极限值等。

误差存在一切测量之中,而且贯穿测量过程的始末,因此,测量者的根本任务就在承认误

差存在这一基本事实的前提下，通过科学的测量方法，将测量误差限制到最低限度，这就是一个优秀的实验工作者应遵循的最基本的原则，也是一切科技工作者进行科学实验应有的基本素质。

式(2-1-1)所表示的是绝对误差，其值可正可负，它是随机变量，没有确定的大小。绝对误差并非"误差"的"绝对值"，它表示被测量的测量值偏离真值的大小和方向。当两被测量相同时，可以用绝对误差的大小评定测量的精密度，但当两被测量不相同时，只从绝对误差的大小，是难以评定各测量结果谁优谁劣的。例如，用同一把米尺测量两物体长度，分别测得 $L=2.5(\mathrm{mm})\pm0.5$ 和 $L'=250.0(\mathrm{mm})\pm0.5$，两者的绝对误差都是 0.5mm，哪一个测量的精密度高呢？为了比较，只能用相对误差的概念，它定义为：

$$E=\frac{\Delta N}{N_0} \tag{2-1-2}$$

由于式(2-1-1)和式(2-1-2)中的真值 N_0 不能确定，因此，两式所表示的绝对误差和相对误差也无法确定。在实际测量中，常用被测量的实际值 N 取代真值 N_0。

引入相对误差的概念后，上述的用同一把米尺测量的两物体长度代入式(2-1-2)中，并分别以百分数表示为：

$$E_L=\Delta L/L\times100\%=0.5/2.5\times100\%=20\%$$

$$E_L'=\Delta L'/L'\times100\%=0.5/250.0\times100\%=0.2\%$$

通过比较，显然，L'的相对误差小于 L 的相对误差，表明 L'的精密度要比 L 高得多。

2.2 误差分类及相关知识

从数学的观点来看，自然界存在着两种类型的规律——确定性规律与统计性规律，这两种规律分别与两类现象相对应。

确定性规律：在一定的条件下，某个结果一定会出现，例如同性电荷相斥。这里，电荷同性为条件，相斥是结果，可用库仑定律描述这一条件与结果的函数关系。显然，由这种现象的前因引起的后果是可以预知的。

统计性规律：在一定的条件下，某个结果可能出现，也可能不出现。例如，用同样的力和同样的方式上抛硬币，下落后正面向上的结果是不确定的，既不是一定的，也不是不可能的。正面向上这一结果完全是随机的，每次抛掷的结果事先都是不可预言的，然而实践证明，抛掷次数足够多时，正面向上的机会约为 1/2，这就表明，在特定条件下，这类事先不可预知结果的现象也存在着规律性，寻求这种规律的方法是对大量的数据进行统计分析，这正是将其称为统计规律的原因。完善地描述统计规律的数学方法是说明分布函数或分布密度函数。

例如，气体分子按速率的分布可用气体分子按速率(比例)的分布函数描述。数学上已总结出来的分布有多种，物理实验将提到的有正态分布和均匀分布等。用分布函数描述统计规律并没有改变被描述现象的随机性。本例中"比例数"一词换用"概率"更为恰当，因为反映比例数的分布函数值是概率。

从测量学的观点来看，确定性规律与统计性规律也分别与以下两类误差相对应。

2.2.1 系统误差

1. 系统误差的概念及分类

测量中，由于测量所依据的理论的近似性、测量仪器、测量方法、测量环境、测量人员等因素所引入的，具有确定性的测量误差称为系统误差。

在同一实验条件下，多次测量同一物理量时，误差的绝对值和正负号保持不变，或按一定规律变化，这种保持不变或事先可预知其变化并服从确定性规律的误差，就是系统误差。由于系统误差具有确定性，所以一般情况下都可以找到产生系统误差的原因并可事先修正。系统误差主要来源于以下几个方面：

(1)理论(方法)误差。理论(方法)误差归因于测量所依据的理论公式本身的近似性，或实验条件不能满足理论公式的要求，或测量方法的不完善等引起的误差。例如，单摆周期公式 $T=2\pi\sqrt{l/g}$ 的成立条件是摆角趋于零，这实际上是不可能达到的，所以测量值必然引入误差。

(2)仪器误差。测量仪器本身的误差或缺陷所引起的测量误差。例如，用未经校准零位的千分尺测量某工件长度，必将引入误差。

(3)环境条件误差。由于温度、气压、空气湿度等环境的变化引起的误差。

(4)人员误差。由于测量者的生理或心理因素决定着观测的反应速度和观测习惯引起的误差。例如，不是正视刻线，而总是斜视刻线等引起的误差。

2. 系统误差的发现与消除

测量者的根本任务是将测量误差限制到最小限度。对于系统误差，其规律及来源可能是测量者已知的，称为可定系统误差；也可能是未知的，称为未定系统误差。对于前者，一般可在测量中采取一定措施予以减小、消除，或在测量结果中予以修正，而对于后者一般难以做出修正，只能估计它的取值范围。

1)可定系统误差的发现

为了发现系统误差，测量者应全面仔细考察和分析实验的理论模型建立、测量方法、仪器及其状态调整、环境条件等。由于系统误差的存在，对于不同实验来说，缺少共性。下面介绍几种分析和发现系统误差的方法。

(1)实验对比法。

①用不同的方法测量同一被测量，比较测量结果是否一致。例如，用单摆法、复摆法、自由落体法分别测量重力加速度 g，各自所得测量结果在随机误差允许范围内不重合，说明至少其中两个存在系统误差。

②用两个电流表串联于电路中测量流过同一电路的电流强度，比较读数是否一致，若读数不一致，说明至少有一个电流表不准。如果，其中一个表为标准表，则另一个表有待校准。

③通过交换测量方法，比较交换前后过程的测量结果是否重合，如天平的称衡换位、金属丝拉伸时的增减砝码。若交换前后所得测量结果不一致，说明存在系统误差。

④通过更换测量人员，比较各自测量结果是否一致。若两人用同一仪器、采用同一测量方法测量同一被测量，所测得结果不一致，说明至少有一人的测量存在系统误差。

(2)理论分析法。

①分析理论公式所要求的条件与实际情况有无差异。如在单摆实验中，公式 $T=2\pi\sqrt{l/g}$ 作了摆角 $\theta\approx0$ 的近似、摆球视为质点、忽略摆线质量、空气浮力和阻力等。这些都是被理想化了的，实际测量单摆周期 T 时，存在系统误差。

②分析仪器所要求的工作条件是否满足要求。例如，物理天平的刀承面是否处于水平、立柱是否处于铅直、三线悬摆的摆盘是否处于水平等。如不能满足仪器使用条件，在实际称衡物体质量和测量扭摆周期时，存在系统误差。

③分析对多次测量的数据,如明显不服从统计规律时,且偏差又呈现单一方向变化时,说明存在系统误差,如偏差的正负号有规律地交替变化,说明存在周期性系统误差。

2)可定系统误差的消除

可定系统误差是容易被发现的,首先是立足于避免,如保证仪器使用满足规定条件等,如无法避免的,可针对不同的实验方法按以下不同的方法予以消除。

(1)直接校准法。

测量前要先检查零位,如各类电表在未通电时,指针不指零位,可通过调整"机械调零器",将指针校准到零位,对于一些自身就带有"短路"调零旋钮的平衡指示仪,可随时监测并调零,消除由于零位不准引起的系统误差。

(2)引入修正值对测量结果修正。

①量具或测量仪器的零值误差的修正。许多测量仪器的零位不能直接校准,只能在测量前将零位偏离值读出,按偏离的正负对测量结果进行修正。

②通过标准仪器与一般仪器的校准值或校准曲线对长期使用过的仪器进行校准。

③根据理论分析导出不满足测量条件的近似公式的修正式。消除这类系统误差的方法是将公式略去部分作为修正值,对测量结果进行修正。例如,用无阻尼单摆测量重力加速度 g,若不考虑摆角 θ 的影响时,单摆周期公式为 $T=2\pi\sqrt{l/g}$,若考虑摆角 θ 的影响时,则公式为

$$T=2\pi\sqrt{l/g}\left(1+\frac{1}{4}\sin^2\frac{\theta}{2}+\cdots\right)$$

相应可得

$$g=4\pi^2\frac{l}{T^2}\left(1+\frac{1}{4}\sin^2\frac{\theta}{2}+\cdots\right)^2$$

第一项之后的高级小量就是修正量。经这样的修正,就能部分地消除摆角对测量结果的影响。

(3)通过仪器的设计和测量方法的选择抵消系统误差的影响。

常通过仪器的设计和选择适当的测量方法,使系统误差能够抵消或补偿。常用的方法有:

①利用仪器读数装置的对称设计抵消系统误差的影响。例如,分光计读数盘设计为相隔180°的左右角游标,以消除偏心差引起的系统误差。

②将测量中的某些因素(如被测物)交换位置。例如,用天平称衡物体质量时,常采用将被测物和砝码在盘中的位置互换,分别测出质量 m 和 m',按公式 $m=\sqrt{m\times m'}\approx\frac{1}{2}(m+m')$ 计算后,即可消除天平不等臂引起的系统误差。其实质是使产生系统误差的原因对测量结果起相互反作用。

③在一定条件下,用一已知量替代被测量来消除系统误差的影响。例如,先取大小不同、质量不等的替代物与被测物平衡,然后在不改变替代物质量的情况下,取下被测物换上砝码并使天平再次平衡,则砝码的质量即为被测物的质量,从而消除了天平结构缺陷带来的系统误差。

④通过改变测量的方向使系统误差产生等值异号,然后求其平均作抵消。例如,用霍尔元件测磁场时,将霍尔元件的工作电流和励磁电流的方向换向四次,不等电位出现了两次正值、两次负值、四次平均,因此不等电位就被抵消。

⑤利用某种系统误差在任何相差半个周期的两对应点处的绝对值相等而符号相反的情况,可通过半周期的偶数法予以消除这种系统误差,即每次都在相差半个周期处测量两次,取

其平均值作为测量结果，便可消除此类系统误差。

3）未定系统误差的处理

未定系统误差大都包含在仪器和量具的仪器误差（限）Δ_{ins} 中，它是指在满足仪器规定使用条件下正确使用仪器时，仪器的示值与被测量值的真值之间可能产生的最大误差的绝对值。它是既无法消除，也不能做修正，只能作为一种分量加以合成。

导致仪器产生仪器误差的因素是多方面的，现以电表为例说明组成电表仪器误差的因素。

（1）轴尖轴套间的摩擦；

（2）磁场不均匀；

（3）游丝弹性系数不均匀及游丝老化；

（4）分度刻线不均匀；

（5）调整仪表指针到要求的示值所引起的起伏；

（6）外界条件的变动对仪表读数的影响。

若仪器的可定系统误差为 ε_e，未定系统误差为 ε_s，随机误差为 ε_R，则仪器误差可表示为

$$\Delta_{ins} = |\varepsilon_e| + c\sqrt{\varepsilon_R^2 + \varepsilon_s^2} = |\varepsilon_e| + c\varepsilon \tag{2-2-1}$$

式中，c 为包含因子，其值取决于 ε 所遵从的规律及 $c\varepsilon$ 值的包含水平。从上式也可看出，仪器误差既包含系统误差，也包含随机误差。对级别不高的仪表，仪器误差则主要是系统误差。为了简化计算，在物理实验教学中，可约定仪器误差直接作为不确定度中的非统计方法估算的分量来处理。常用物理实验仪器的仪器误差见表 2-2-1。

表 2-2-1　常用物理实验仪器的仪器误差表

仪器名称	仪器误差 $\Delta_{仪}$	说明
毫米尺	0.5mm	最小分度值的 1/2
游标卡尺	0.05mm（1/20 分度） 0.02mm（1/50 分度）	最小分度值
千分尺	0.004mm（或 0.005mm）	（最小分度值的 1/2）
读数显微镜	0.004mm（或 0.005mm）	（最小分度值的 1/2）
测微目镜	0.004mm（或 0.005mm）	（最小分度值的 1/2）
水银温度计（最小分度值 1℃）	0.5℃（或 1℃）	最小分度值的 1/2（或最小分度值）
计时仪器	1s，0.1s，0.01s（各类机械表）	最小分度值
	$(5.8\times10^{-6}t+0.01\text{s})$（电子表）	t 为时间的测量值
物理天平	0.05g（感量 0.1g） 0.01g（感量 0.02g）	物理天平标尺最小分度值的 1/2
分光计	1′	最小分度值
电桥	$k\cdot\%\left(R+\frac{R_0}{10}\right)$	k 为电桥的精度级别； R 为测量值；R_0 为基准值
电位差计	$k\cdot\%\left(U+\frac{U_0}{10}\right)$	k 为电位差计的精度级别； U 为测量值；U_0 为基准值
电阻箱	$k\cdot\%\cdot R$	k 为电阻箱的精度级别； R 为测量值

续表

仪器名称	仪器误差 $\Delta_{仪}$	说明
指针式电表 (电流表,电压表)	$k \cdot \% \cdot N_m$	k 为电表的精度级别; N_m 为电表的满量程值
各类数字仪表	$k \cdot \% \cdot N_x + \xi \cdot \% \cdot N_m$ 或 $k \cdot \% \cdot N_x + n$ 或仪表最小读数单位	k 为仪表的精度级别;N_x 为测量值; N_m 为仪表的满量程值; ξ 为误差绝对项系数; n 为仪器固定项误差,是最小量化单位的 n 倍

2.2.2 随机误差

1. 随机误差的概念

测量中,由于随机因素和不确定因素所引入的误差称为随机误差,其特征是随机性。若系统误差已减小到可忽略的程度,则在等精度条件下对该被测量进行多次重复测量时,测量值时大时小,符号时正时负,无确定的变化规律,完全表现为随机性;但当测量次数无限增加时,它们一般服从统计规律。随机误差主要来自以下几个方面:

(1)主观方面:实验者的感官灵敏度和仪器分辨力有限,以及实验者对仪器操作不熟练、对仪器的示值估读不准等。

(2)客观方面:外界温度的涨落、气流的扰动、电磁场的干扰等,造成对实验者的干扰,在有限的测量中,既无法排除,又无法估量其影响的大小。

(3)其他不可预测的因素的影响。

2. 随机误差的规律与特征

随机误差的出现,就某一次测量值来说,毫无规律可循,其大小和方向都是事先无法预知的,但对于同一被测量在相同条件下进行足够多次的测量,就会发现随机误差是按一定的统计规律分布的。理论和实践都证明,大部分测量的随机误差都服从正态分布,如图 2-2-1,横坐标表示测量误差 $\varepsilon_i = X_i - X_0$($X_i$ 表示只含有随机误差的第 i 次测量值,X_0 为被测量的真值),纵坐标表示与误差出现的概率有关的概率密度分布函数 $f(\varepsilon)$。

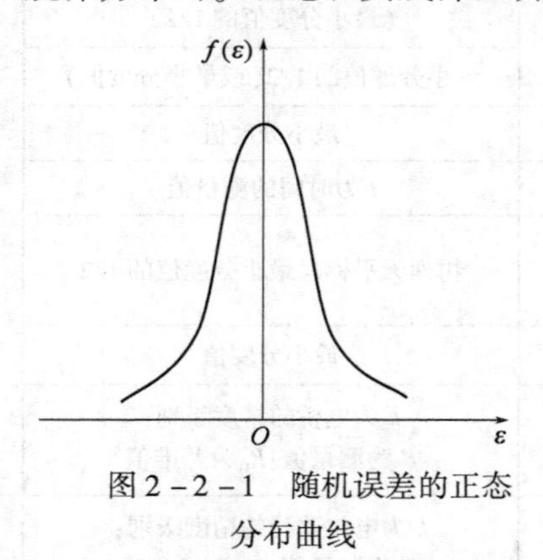

图 2-2-1 随机误差的正态分布曲线

由概率论的数学方法可得

$$f(\varepsilon) = \frac{1}{\sigma\sqrt{2\pi}} e^{-\frac{\varepsilon^2}{2\sigma^2}} \qquad (2-2-2)$$

式中,特征量

$$\sigma = \sqrt{\frac{\sum \varepsilon^2}{n}} \qquad (n \to \infty)$$

称为测量列的标准误差,它具有明确的统计意义。由式(2-2-2)表示的正态分布函数和概率论的数学方法可得

$$\int_{-\infty}^{\infty} f(\varepsilon)\,\mathrm{d}\varepsilon = 1 \qquad (2-2-3)$$

$$\int_{-\sigma}^{+\sigma} f(\varepsilon)\mathrm{d}\varepsilon = P(\sigma) = 0.683 \tag{2-2-4}$$

$$\int_{-2\sigma}^{+2\sigma} f(\varepsilon)\mathrm{d}\varepsilon = P(2\sigma) = 0.954 \tag{2-2-5}$$

$$\int_{-3\sigma}^{+3\sigma} f(\varepsilon)\mathrm{d}\varepsilon = P(3\sigma) = 0.997 \tag{2-2-6}$$

式(2-2-3)表明:当$n\to\infty$时,任何一次测量值与真值之差出现在区间$(-\infty,\infty)$内的概率为1,即满足归一化条件。而其他各式则表明:出现在区间$(-\sigma,\sigma)$内的概率为0.683,即包含概率$P=0.683$;出现在区间$(-2\sigma,2\sigma)$内的概率为0.954,即包含概率$P=0.954$;出现在区间$(-3\sigma,3\sigma)$内的概率为0.997,即包含概率$P=0.997$。

由此看出统计特征量σ表明在等精度条件下测量列的随机误差的概率分布情况,特别看出,当测量次数无限多时,测量误差的绝对值大于3σ的概率仅为0.3%,这对于有限次测量来说,这种可能性几乎是微乎其微的,也只能认为这种测量是失误的,不可信的,应予剔除。这就是判别过失误差的3σ准则。

由上所述,标准误差σ是随机误差离散情况的量度。它在正态分布曲线中具有如下特征:

(1)单峰性。绝对值小的误差出现的概率比绝对值大的误差出现的概率大,且绝对值越小的误差出现的概率越大。

(2)对称性。大小相等、符号相反的误差出现的概率相同,因此,当测量次数增加时,它们可能正负抵消。

(3)有界性。非常大的正负误差出现的概率很小,甚至趋近于零。

(4)抵偿性。当测量次数$n\to\infty$时,由于正负误差相互抵消,所以各个误差的代数和趋于零。

2.2.3 系统误差与随机误差的关系

在任何一次测量中,误差既不会是单纯的系统误差,也不会是单纯的随机误差,而是两者兼而有之,各自所占的比例与具体的测量有关,不能一概而论。在以后的讨论中,需要单独谈一种误差时,必须意识到,另一种误差并非没有,而是假定已消除或是减小到忽略的程度。

系统误差与随机误差之间有一定的联系,它们在一定条件下还会相互转化。例如,千分尺不调零位也不作修正时,测量中反映了系统误差的性质,一次性调零并作修正,其剩余分量中一部分依然反映出未定系统误差的性质,多次测量前均调零,相当于将未定系统误差值随机化了,因而增加了测量值的随机误差的分布范围。由此可见,尽管把误差分为系统误差与随机误差。但这种分法还不够完善。

2.2.4 测量的最佳估计值——算术平均值

假设系统误差已被消除或被减小到可忽略的程度,在等精度测量条件下,对某一被测量进行了n次测量,其测量值分别为$X_1,X_2,X_3,\cdots,X_n$(又称为测量列),其算术平均值$\bar{X}$为

$$\bar{X} = \frac{1}{n}\sum_{i=1}^{n} X_i \tag{2-2-7}$$

容易证明,测量值的算术平均值最接近被测量的真值。由最小二乘法原理,一列等精度测量的最佳估计值(近真值)是能够使各次测量值与该值之差的平方和为最小的那个值。设被

测量的真值的最佳估计值为 X,可写出差值平方和如下:

$$f(X) = \sum_{i=1}^{n} (X_i - X)^2 \tag{2-2-8}$$

令 $\frac{df(X)}{dX}=0$,求极值:

$$\frac{df(X)}{dX} = -2\sum_{i=1}^{n}(X_i - X) = 0 \tag{2-2-9}$$

则
$$X = \frac{1}{n}\sum_{i=1}^{n} X_i = \bar{X} \tag{2-2-10}$$

因此,可用算术平均值表示近真值,每次测量值 X_i 与算术平均值 $\bar{X}$ 之差称为偏差,即

$$\Delta X_i = X_i - \bar{X} \tag{2-2-11}$$

显然,这些偏差也有正有负,有大有小,它们反映了测量结果的离散性。

2.2.5 随机误差的估算——用标准偏差作为标准误差的估算值

无穷多次等精度测量的标准误差 σ_X 为

$$\sigma_X = \sqrt{\frac{1}{n}\sum_{i=1}^{n} (X_i - X_0)^2} \quad (n \to \infty) \tag{2-2-12}$$

式中,σ_X 是无穷多次测量这一"总体"的特征参数。在实际测量中,测量次数 n 总是有限的,且真值 X_0 也是一个未知数,因此,标准误差也未知的,实际应用中,只能对它进行估算。这样,可用 $\bar{X}$ 取代式(2-1-12)中的真值 X_0,其结果定义 S_X 为有限次测量标准偏差,即

$$S_X = \sqrt{\frac{1}{n-1}\sum_{i=1}^{n} (X_i - \bar{X})^2} \tag{2-2-13}$$

式中,S_X 是 σ_X 的估算值,式(2-2-13)称为贝塞尔公式。其统计意义为,当测量次数足够多时,测量列中任一测量值与平均值的偏离出现在区间($-S_X$, $+S_X$)内的概率为68.3%。如果测量中只含有随机误差,当测量次数 $n \to \infty$ 时,$S_X \to \sigma_X$。实验结果的最佳估计值是测量列的算术平均值 $\bar{X}$,人们也更加关注它的标准误差的估算。根据数理统计理论,算术平均值 $\bar{X}$ 的标准差(简称为平均值的标准差)为

$$S_{\bar{X}} = \frac{S_X}{\sqrt{n}} = \sqrt{\frac{\sum_{i=1}^{n} (X_i - \bar{X})^2}{n(n-1)}} \tag{2-2-14}$$

$S_{\bar{X}}$ 同样具有十分明确的意义,即在一组测量次数 n 足够大的测量中,测量值的算术平均值 $\bar{X}$ 出现在区间($\bar{X} \pm S_{\bar{X}}$)内的概率为68.3%,如果测量中只含有随机误差,当测量次数 $n \to \infty$ 时,真值 X_0 出现在区间($\bar{X} \pm S_{\bar{X}}$)的概率为68.3%,包含概率为68.3%。

同理,经过理论分析表明,若将包含区间放大为($\bar{X} \pm 2S_{\bar{X}}$),则包含概率为95%;若再将包含区间放大为($\bar{X} \pm 3S_{\bar{X}}$),则包含概率为99.7%。

由以上分析可以看出,若将 $S_{\bar{X}}$ 前倍乘以一个不同的用以反映包含区间大小的覆盖因子 k_P

(下标 P 为包含概率)，即可得到不同的包含概率 P。

然而，实际测量中，测量次数 n 也只能为有限的，因而，测量值 X_i 将偏离正态分布而服从一种 t 分布。当测量结果在已确定的包含概率的情况下，覆盖因子 k_P 的大小与测量次数 n 密切相关。表 2－2－2 给出了 t 分布，可从中了解到包含概率 P、测量次数 n 及 t 分布因子 $t_P(n)$ 的关系。覆盖因子 $k_P = t_P(n)$ 在不会引起误解时，$t_P(n)$ 可简写为 t_P。

表 2－2－2　t 分布[$t_P(n)$]

P \ n	2	3	4	5	6	7	8	9	10	15	20	∞
0.997	235.8	19.21	9.22	6.62	5.51	4.90	4.53	4.28	4.09	3.64	3.45	3.00
0.950	12.70	4.30	3.18	2.78	2.57	2.45	2.36	2.31	2.26	2.14	2.09	1.96
0.900	6.31	2.92	2.35	2.13	2.02	1.94	1.90	1.86	1.83	1.76	1.72	1.65
0.683	1.84	1.32	1.20	1.14	1.11	1.09	1.08	1.07	1.06	1.04	1.03	1.00

从表 2－2－2 中看出：当测量次数 $n=5$，要求包含概率 $P=0.95$，则 $t_P=2.78$，测量中只含有随机误差时，被测量的真值 X_0 出现在区间($\bar{X} \pm 2.78S_{\bar{X}}$)的包含概率 P 为95%。通观 t 分布[$t_P(n)$]的全局表明：t_P随测量次数 n 的增加而减小，$n>10$ 以后，t_P 减小很缓慢，因而一般测量中 n 很少大于 10。

这一结果似乎与前面所推证出的增加测量次数对于提高算术平均值的可信度是有利的结果相矛盾。如果增加测量次数，则必将导致延长测量时间，这就难以保证测量条件在测量时间内的稳定不变，测量条件的变化，必将加大测量误差。况且增加测量次数对减小系统误差毫无作用。因此，实际进行的有限次测量时，随机误差的分布是由正态分布退化为 t 分布，因此，以上关于测量次数 n 的多少的讨论中，仅仅是从实际需要出发，两者并不矛盾。

长期以来，在一般测量中，采用扣除已知系统误差的最佳估计值表示测量结果的大小，采用平均值的标准差表示测量误差。这样一来，将一些未知分布的系统误差忽视，或者无法归属于哪一种分布，甚至于将一些未知的系统误差值随机化，从而增加了随机误差的分布范围。也就是说，前面所述的系统误差和随机误差的分类本身具有不完善性，也就无法用统计方法处理那些误差分量在测量结果中的具体表现了，因此这种处理方法具有相当大的局限性。随着误差理论研究的深入及科学技术的发展，人们普遍认识到，用"测量不确定度"的概念，能对测量结果作出更为合理的评价。

2.3　测量不确定度及其评定

1981 年，国际计量委员会(CIPM)批准发表了关于测量不确定度的正式文件——《测量不确定度工作组织建议书》INC－1(1980)。1991 年 8 月 5 日我国国家技术监督局批准颁布了 JJG 1027—1991《测量误差及数据处理(试行)》计量技术规范。该规范明确规定，在报告最后测量结果的表示形式中使用总不确定度。INC－1(1980)只是一份十分简单的纲要性文件，不便实施，所以国际标准化组织(ISO)在国际计量局(BIPM)等七个国际组织的支持下，于 1993 年制定了《测量不确定度表示指南 ISO 1993 (E)》(简称 GUM93)，并于 1995 年作了一些修改，修改后简称 GUM95。为了与国际接轨，我国颁布了新的国家计量技术规范 JJF 1059—1999《测量不确定度评定与表示》用以取代 JJG 1027—1991 中的测量误差部分。根据 ISO/IEC 指南 98 - 3:2008,又颁布了新的国家计量技术规范 JJF 1059.1—2012《测量不确定度评定与表示》和国家标准 GB/T 27418—2017《测量不确定度评定和表示》,用以取代原来的技术规范和标准。在新的国际标准和我国新的计量技术规范中,用“包含区间”和“包含概率”替代国际标准中的“置信区间”和“置信水平”。

2.3.1 测量不确定度的含义

测量不确定度是指由于测量误差的存在而使被测量值不能确定的程度。从此意义上讲,测量不确定度是评定被测量的真值所处范围的一个参数。用不确定度来评定实验结果,可反映出各种来源不同的误差对结果的影响,而对它们的计算又反映出这些误差所服从的分布规律。

2.3.2 不确定度的分类

测量结果的不误确定度一般包含有几个分量,按其数值的评定方法,这些分量可归为两大类——A 类分量(或称为 A 类评定)和 B 类分量(或称为 B 类评定)。

(1)A 类分量:多次等精度重复测量时,可用统计方法求出的分量。

(2)B 类分量:不能用统计方法处理,而需要用其他方法估算的分量。

2.3.3 测量不确定度的评定

测量不确定度的评定方法不是唯一的。按国际计量局的建议,测量不确定度可用算术平均值的标准差$S_{\bar{X}}$、标准误差 σ 和自由度 ν 等来评定。考虑到实验教学需要,为了便于操作,本书作统一的简化处理,即省略有关自由度的计算。

1. 直接测量量不确定度的评定

1) 多次直接测量量不确定度的评定

A 类不确定度分量用统计学方法估算得出。可用平均值 $\bar{X}$ 的标准差$S_{\bar{X}}$与 t_P 因子的乘积来估算,即

$$u_A = t_P \sqrt{\frac{\sum_{i=1}^{n} (X_i - \bar{X})^2}{n(n-1)}} \qquad (2-3-1)$$

式中,t_P 因子与测量次数 n 和对应的包含概率 P 有关,当包含概率 $P=0.95$、测量次数 $n=6$ 时,从表 2-2-2 中可查阅计算得到 $t_{0.95}/\sqrt{6} \approx 1$,则有

$$u_A = S_X = \sqrt{\frac{1}{n-1} \sum_{i=1}^{n} (X_i - \bar{X})^2} \qquad (2-3-2)$$

在物理实验教学中,测量次数 n 取 $5<n<10$,$\frac{t_{0.95}}{\sqrt{n}} \approx 1$,则有 $u_A \approx S_X$。为了简化和统一,本书约定包含概率 P 取 95%。

B 类不确定度分量用非统计学方法估计得出。要做到估计适当,对于初学者来说是很困难的,因为需要实验者确定出这类误差的分布规律、估算误差限等。本书对 B 类不确定度分量的估计也作简化处理,约定 B 类不确定度分量仅涉及仪器的最大允差(仪器误差)Δ_{ins}或 $\Delta_{仪}$。

多次直接测量的总不确定度是 A 类不确定度分量 u_A 与 B 类不确定度分量 u_B 的方差合成，即

$$u_X = \sqrt{u_A^2 + u_B^2} \tag{2-3-3}$$

在物理实验教学中，考虑到 Δ_{ins} 一般服从均匀分布，因而 B 类标准不确定度分量可取为

$$u_B = \frac{\Delta_{ins}}{\sqrt{3}} \tag{2-3-4}$$

则总合成标准不确定度由下式进行计算：

$$u_X = \sqrt{\left(\frac{t_p}{\sqrt{n}}\right)^2 S_X{}^2 + \left(\frac{\Delta_{ins}}{\sqrt{3}}\right)^2} \tag{2-3-5}$$

2）单次直接测量的测量不确定度的评定

物理实验中，如果符合下列两种情况，则可以考虑进行单次测量。两种情况为：

（1）多次测量的 A 类不确定度分量对实验的最后结果的总不确定度影响甚小。

（2）因测量条件的限制，不可能进行多次测量。

在这两种情况下，单次直接测量的不确定度可取

$$u_X = u_B = \frac{\Delta_{ins}}{\sqrt{3}} \tag{2-3-6}$$

应当指出，式（2-3-6）并不说明单次测量的总不确定度 u_X 比多次测量的 u_X 值小，只能说明这种估算比式（2-3-5）更为粗糙。

3）直接测量的不确定度计算

对于被测量 X 的直接测量结果，可按如下程序计算：

$$\bar{X} = \frac{1}{n}\sum_{i=1}^{n} X_i \quad （单位）$$

$$u_A = t_P \cdot \sqrt{\frac{\sum_{i=1}^{n}(X_i - \bar{X})^2}{n(n-1)}} \quad (P = \quad)；\quad u_B = \frac{\Delta_{ins}}{\sqrt{3}}$$

$$u_X = \sqrt{u_A^2 + u_B^2} \ （单位） \quad (P = \quad)$$

$$E_r = \frac{u_X}{\bar{X}} \times 100\% \quad (P = \quad)$$

括号中的 P 是包含概率。

若测量中存在可修正的系统误差（可定系统误差）Δ，则应对测量值进行修正，这时的最佳值应为

$$\bar{X} = \frac{1}{n}\sum_{i=1}^{n} X_i - \Delta \tag{2-3-7}$$

4）直接测量结果的表示

对于被测量 X 的直接测量结果最终可表示为

$$X = \bar{X} \pm u_X \quad （单位） \quad (P = \quad)$$

$$E_r = \frac{u_X}{\bar{X}} \times 100\% \quad (P = \quad)$$

括号中的 P 是包含概率。上述的表示结果既反映了多次直接测量的结果,也反映了单次直接测量结果。实验者都应学会怎样正确、规范、科学地表述测量结果的最终报告形式,下一节还要作进一步的介绍。

2. 间接测量量的不确定度的评定

在物理实验中,一些物理量的测量值是要通过它与直接测量量的某些函数关系计算出来的。由于每个直接测量量都存在误差,则这种误差必将通过函数关系传递给间接测量量,使间接测量量也产生误差,因此间接测量量也就有了自己的不确定度。

1)间接测量量的平均值

设间接测量量 Y 和各直接测量量 $X_1, X_2, \cdots, X_i, \cdots, X_n$ 有下列函数关系:

$$Y = f(X_1, X_2, \cdots, X_i, \cdots, X_n) = f(X_i) \tag{2-3-8}$$

该物理量的平均值为

$$\bar{Y} = f(\bar{X}_1, \bar{X}_2, \cdots, \bar{X}_i, \cdots, \bar{X}_n) \tag{2-3-9}$$

即间接测量量的最佳值由各直接测量量的最佳值代入函数表达式求得。

2)间接测量量的不确定度

对表征间接测量量 Y 的函数关系式(2-3-8)求全微分,得

$$\mathrm{d}Y = \frac{\partial f}{\partial X_1}\mathrm{d}X_1 + \frac{\partial f}{\partial X_2}\mathrm{d}X_2 + \cdots + \frac{\partial f}{\partial X_n}\mathrm{d}X_n \tag{2-3-10}$$

上式表明,当各直接测量量 $X_1, X_2 \cdots, X_i \cdots, X_n$ 有微小改变 $\mathrm{d}X_1, \mathrm{d}X_2 \cdots, \mathrm{d}X_i \cdots, \mathrm{d}X_n$ 时,间接测量量 Y 也将改变 $\mathrm{d}Y$。通常误差远小于测量值,故可以将 $\mathrm{d}X_i$ 和 $\mathrm{d}Y$ 看作误差,式(2-3-10)就是误差传递公式。如果求得了各直接测量量 X_i 的合成不确定度 u_{X_i},则间接测量量 Y 的不确定度即可求得,设 $u_{X_1}, u_{X_2}, \cdots, u_{X_i}, \cdots, u_{X_n}$ 分别为 $X_1, X_2 \cdots, X_i, \cdots, X_n$ 等相互独立的直接测量量的不确定度,则间接测量量的总不确定度为

$$u_Y = \sqrt{\left(\frac{\partial f}{\partial X_1}\right)^2 u_{X_1}^2 + \left(\frac{\partial f}{\partial X_2}\right)^2 u_{X_2}^2 + \cdots + \left(\frac{\partial f}{\partial X_n}\right)^2 u_{X_n}^2} \tag{2-3-11}$$

式中,偏导数 $\frac{\partial f}{\partial X_1}, \frac{\partial f}{\partial X_2}, \cdots, \frac{\partial f}{\partial X_n}$ 称为传递系数,它的大小直接代表了各直接测量量的不确定度对间接测量量不确定度的贡献。间接测量量的相对不确定度可表示为

$$\frac{u_Y}{\bar{Y}} = \sqrt{\left(\frac{\partial \ln f}{\partial X_1}\right)^2 u_{X_1}^2 + \left(\frac{\partial \ln f}{\partial X_2}\right)^2 u_{X_2}^2 + \cdots + \left(\frac{\partial \ln f}{\partial X_n}\right)^2 u_{X_n}^2} \tag{2-3-12}$$

式中,$\ln f$ 表示对函数 f 取自然对数。式(2-3-11)和式(2-3-12)就是不确定度传递的基本公式。实际计算时,传递系数 $\frac{\partial f}{\partial X_i}$ 或 $\frac{\partial \ln f}{\partial X_i}$ 中的各直接量均以平均值代入即可。并且由式(2-3-8)判断间接测量量和各直接测量量存在的函数形式进行简便计算,对于和差形式的函数,用式(2-3-11)计算较为方便,而对于积商、乘方、开方形式的函数,则用式(2-3-12)较为方便。为了方便读者,表2-3-1将常用函数的不确定度传递公式列入其中。

表 2-3-1　常用函数不确定度使用说明

间接测量结果的函数表达式	不确定度的传递公式	说　明
$N = X \pm Y$	$u_N = \sqrt{u_X^2 + u_Y^2}$	直接求 u_N
$N = X \cdot Y$	$E_N = \frac{u_N}{\overline{N}} = \sqrt{\left(\frac{u_X}{\overline{X}}\right)^2 + \left(\frac{u_Y}{\overline{Y}}\right)^2}$	宜先求相对不确定度 E_N
$N = \frac{X}{Y}$	$E_N = \frac{u_N}{\overline{N}} = \sqrt{\left(\frac{u_X}{\overline{X}}\right)^2 + \left(\frac{u_Y}{\overline{Y}}\right)^2}$	宜先求相对不确定度 E_N
$N = \frac{X^a Y^b}{z^c}$	$E_N = \frac{u_N}{\overline{N}} = \sqrt{a^2\left(\frac{u_X}{\overline{X}}\right)^2 + b^2\left(\frac{u_Y}{\overline{Y}}\right)^2 + c^2\left(\frac{u_Z}{\overline{Z}}\right)^2}$	宜先求相对不确定度 E_N
$N = AX$	$u_N = Au_X$；$E_N = \frac{u_N}{\overline{N}} = \frac{u_X}{\overline{X}}$	直接求 u_N
$N = \sqrt[n]{X}$	$E_N = \frac{u_N}{\overline{N}} = \frac{u_X}{nN}$	宜先求相对不确定度 E_N
$N = \sin X$	$u_N = u_X \cos X$	直接求 u_N

3）扩展不确定度 u_P 的评定

将间接测量量的合成标准不确定度 u_Y 乘以一个与所要求的包含概率 P 相关的覆盖因子 k_P，即构成相应的扩展不确定度 u_P，即

$$u_P = k_P u_Y \tag{2-3-13}$$

应该指出，直接测量量的合成标准不确定度 u_X 乘以一个与所要求的包含概率 P 相关的覆盖因子 k_P，也可以构成相应的扩展不确定度。但是，由于在报告中只需要报告间接测量量的扩展不确定度，直接测量量的扩展不确定度便没有必要计算了。另外，如果最终测量结果只由直接测量量构成，实际上是间接测量的特例，即函数关系为 $Y = X$，则被测量 Y 的合成标准不确定度与直接测量量的合成标准不确定度相同。

覆盖因子 k_P 的大小不仅与包含概率 P 有关，还与需要进行扩展的标准不确定度所服从的分布类型及自由度有关。当确定间接测量量的合成标准不确定度的输入参数较多，且自由度又非常高时，可按正态分布处理。而实际测量中自由度的大小通常是有限的，所以 k_P 需要根据 t 分布来确定。由于 B 类不确定度“等效自由度”的确定和合成标准不确定度“等效自由度”的计算过于复杂，因此，在物理实验中，为了简化计算，就约定包含概率为 $P = 0.95$，近似取 $k_{0.95} = 2$，即

$$u_{0.95} = 2u_Y \tag{2-3-14}$$

4）最终测量结果的扩展相对不确定度的计算

在约定包含概率 $P = 0.95$ 的情况下

$$\frac{u_P}{\overline{Y}} \times 100\% = \frac{u_{0.95}}{\overline{Y}} \times 100\% = \frac{2u_Y}{\overline{Y}} \times 100\% \tag{2-3-15}$$

5）间接测量结果的表示

对于间接测量的最终结果可表示为

$$Y = \bar{Y} \pm u_P \quad (\text{单位}) \quad (P = \quad)$$

$$E_Y = \frac{u_P}{\bar{Y}} \times 100\% \quad (P = \quad)$$

在约定包含概率 $P = 0.95$ 的情况下，可表示为

$$Y = \bar{Y} \pm 2u_Y \quad (\text{单位}) \quad (P = 0.95)$$

$$E_Y = \frac{u_P}{\bar{Y}} \times 100\% = \frac{2u_Y}{\bar{Y}} \times 100\% \quad (P = 0.95)$$

其中，($P =$ 　　)是测量结果表达式的格式，需要自己确定包含概率 P。上述的表示结果，作为一种教学规范形式，实验者务必正确、规范、科学地表述测量结果的最终报告形式。

2.4 测量结果

由于任何测量都存在误差，测量不可能得到被测量量的真实值，只能是近似值。测量数据中，有些数位是准确的，有些数位是欠准确的，在获得测量值时，保留哪些测量数据位，能够科学准确地反映待测量的大小，为此，在物理实验中引入了有效数字的概念。

2.4.1 有效数字的概念

为了说明有效数字的概念，可参照用米尺测量某一长度的实例，如图 2-4-1 所示。

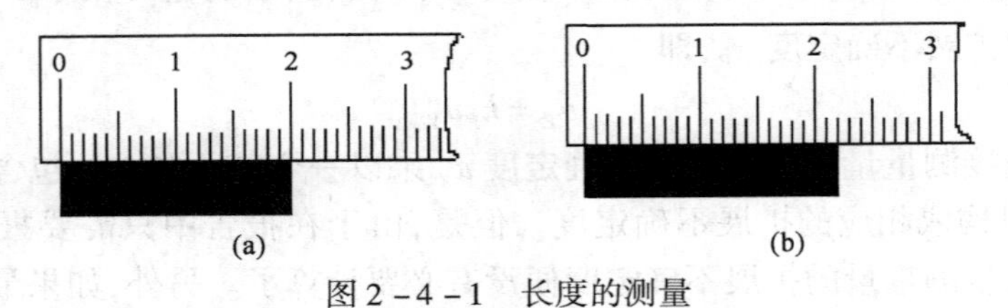

图 2-4-1　长度的测量

可看出图 2-4-1(a)中的物体长度在 2.0~2.1cm 之间，虽然米尺没有毫米以下的刻度，但可凭借眼力估计到$\frac{1}{10}$mm，因而可以读出该物体的长度为 2.01cm、2.02cm 或 2.03cm。前两位数字都相同，但第三位不同，这说明前两位数字是可靠的，第三位数字是观测者估读出来的，由于观测者或环境条件的不同估读有所差异，因此这一位数字是欠准的数字，是含有误差的，它以下的各位数字的估读就毫无意义了。而这三位数字都是有效的，缺少了任何一位都不能正确表示该物体长度的测量值了。

在图 2-4-1(b)中，物体的末端正好与某刻线对齐，估读位数字为“0”，若物体末端与某刻线稍有偏移，则这一估读位数字就不再为“0”了。因此读出物体长度不能记为 2.2cm，而应写为 2.20cm，可靠数字仍然是两位，“0”所在的位仍然是欠准位，则 0 是欠准数字，同样，这三位数字也都是有意义的，缺少了任何一位都不能正确表示该物体长度的测量值。

综述此例，人们将测量结果中的所有可靠数字和最后一位欠准数字合起来，称为测量结果的有效数字，或者说，从发生误差的这一位起算，包含这一位及以上所有位的数字都属于有效数字。本例中物体长度测量结果都是三位有效数字。

应该指出，在表示间接测量结果大小的有效数字中，欠准数字应保留几位，是由测量结果

的不确定度经修约后来决定的。

如果用同一量具测量两个大小不同的同种物理量，量值大的一方测量结果有效数字位数可能会多一些，如本例中物体长度超过10cm，则有效数字位数就会达到或超过四位，测量值的相对不确定度就会变小。

如果用两种量具测量同一物体长度，精度高的量具所测数据的有效数字位数比精度低的量具所测数据的有效数字位数要多。例如，用测量精度较米尺高很多的千分尺（最小分度值为0.01mm）测量本例中的物体长度，欠准数字一般产生在$\frac{1}{1000}$位上（即0.001mm），测量结果的有效数字位数较米尺的测量结果多两位。显然，测量结果的有效数字位数的多少取决于所用测量工具的精度高低以及被测量的大小。

综上所述，测量量的原始数据位数的确定原则为：

（1）不多取得不到的信息，即不应估读到最小分度后二位有效数字位。

（2）不丢失有用的信息，即应该估读到最小分度值的下一位，而不应只读到分度位。

2.4.2 关于有效数字的几点说明

（1）有效数字的位数与小数点的位置无关。也就是说，表示小数点位置的“0”不是有效数字。例如：0.0202m = 2.02cm = 20.2mm都是三位有效数字。

（2）不得随意在测量数据的末尾添加或删减数字“0”。由上例（图2－4－1）可知，有效数字末尾的“0”表示可疑数字的位置，随意增减会人为夸大或亏损测量精度或者是测量误差。

（3）用科学计数法表示太大或太小的有效数字。所谓科学计数法，就是采用乘以10的整数幂的方式（$\times 10^n$）表示数值的大小，既可以缩短较大计量单位下数值的位数，也可以避免在较小计量单位下无法正确书写测量结果。这种书写方式，是实验数据记录和表示的标准方式。例如：$2.02\text{cm} = 2.02 \times 10^{-5}\text{km} = 2.02 \times 10^{-2}\text{m} = 2.02 \times 10^{7}\text{nm}$。计量单位改变，有效数字的位数不能改变。

（4）一些参与函数运算的整数型或非整数型的数学常数，如4、$\sqrt{2}$、π等，可以把它们看作是位数为无穷多的有效数字。在参与运算时，它们的取位原则是要比参与运算的其他因子中有效数字位数最少的多取一位。参与函数运算的物理量的公认值（如电子的电量e等），也应照此处理。

（5）在函数运算过程中，中间结果应多保留1位，以免因舍位过多或修约过早带来过大的附加误差。

2.4.3 数据的修约

实验数据的有效修约是处理实验数据的一个重要环节。

1．测量数据结果的确定

（1）游标类量具，如游标卡尺、带游标的千分尺、分光计中的角游标等，一般应读到游标分度值即可，无须再进行估读。

（2）指针式仪表一般应估读到最小分度值的1/4～1/10，或估读到基本误差限的1/3～1/5。

（3）数显仪表及十进步进式标度盘的仪表，如电阻箱、电桥等，一般应直接读取仪表的示值即可。

2. 测量不确定度的结果表示

1)测量不确定度的有效数字的选取

作为一种教学规范,本教材约定,当直接测量结果是最终结果时,不确定度有效数字可以取1~2位。若其作为间接测量的中间结果时,不确定度的截取应再多取1位,以避免过早舍入造成误差的亏损和积累效应。对于相对不确定度,则一律采用2位有效数字(百分数)表示。

2)测量不确定度的修约

在确定测量不确定度的有效数字时,为保证不确定度的包含概率水平不降低,本教材约定,不确定度的值在截取时采取宁大勿小的原则,即"只入不舍"的原则。例如:$u=0.3214\text{mm}$,取2位时,则$u=0.33\text{mm}$。

3. 测量结果的修约

1)报告测量结果的书写方式

测量结果中的测量值(平均值)的最末位数应与不确定度u的最末位数对齐。在评价测量结果X时的u_A、Δ_A、u、E_X等量时,必须用括号注明包含概率值($P=$　　),测量结果表达式中还应包含该物理量的单位。例如:$R=(4.045\pm0.012)\times10^3\Omega$ ($P=$　　);$R=(1.346\pm0.007)\times10^3\Omega$($P=$　　)等都是正确的表示方法;而$R=(4.04\pm0.012)\times10^3\Omega$、$R=(1.3464\pm0.007)\times10^3\Omega$等都是错误的书写方法。

2)测量结果中的测量值(平均值)有效数字的修约

根据测量不确定度的大小,在对测量结果进行截断时,有效数字的末位需要作进位或舍位处理。处理方法应遵循《数值修约规则与极限数值的表示和判定》(GB/T 8170—2008)中的进舍规则。

(1)进舍规则:①拟舍弃数字的最左一位数字小于5时,则舍去,即保留的其他各位数字不变。例如:若根据测量不确定度的大小需将15.1498修约到一位小数,则经修约后得15.1;若根据测量不确定度的大小需将15.1498修约成两位有效数字,则经修约后得15。②拟舍弃数字的最左一位数字大于5或者等于5时,而其后跟有并非全部为0的数字时,则进1,即保留的末位数字加1。例如:若根据测量不确定度的大小需将1356修约到二位有效数字,则经修约后得1.4×10^3;若根据测量不确定度的大小需将1356修约成三位有效数字,则经修约后得1.36×10^3;若根据测量不确定度的大小需将20.502修约到个位数,则经修约后得21。③拟舍弃数字的最左一位数字为5,且其后有非0数字时进一,即保留数字的末尾数字加1;而其后无数字或皆为0时,若所保留的末位数字为奇数1、3、5、7、9时,则进1,为偶数2、4、6、8、0时,则舍弃。例如:若根据测量不确定度的大小需将3.050修约到一位小数,则经修约后得3.0;若根据测量不确定度的大小需将0.350修约到一位小数,则经修约后得0.4。

(2)不允许连续修约,拟修约数字应在确定修约位数后一次修约获得结果,而不得多次按前述规则连续修约。例如,将14.4546修约到个位,正确的做法为14.4546→14;不正确的做法为14.4546→14.455→14.46→14.5→15。

2.4.4 有效数字运算规则

进行有效数字运算涉及两种情况,一是由直接测量量计算间接测量量的近真值;二是在未估算出测量不确定度时最后结果的有效数字取位,两种情况都需要按照以下运算规则进行。

有效数字运算规则是:(1)可靠数字与可靠数字的运算结果仍为可靠数字;(2)可疑数字与可靠数字或可疑数字与可疑数字的运算结果为可疑数字,但进位为可靠数字;(3)在运算的最后结果中,一般只保留一位可疑数字,其余可疑数字根据位数取舍规则处理。具体分为:

(1)加减运算时,以参与加减的各量的末位数中量级最大的那一位为结果的末位。

例如,$18.45+13.2=31.65$,各量末位数中量级最大的一位是13.2中的0.2,按本规则,该位即为结果的欠准位,所以两者之和应为31.6。

(2)乘除运算时,以参与乘除各量中有效数字位数最少的那一项与结果的有效位数相同。例如,$13.25\times26.2=347.15$,参与运算各量中有效数字位数最少的是26.2,三位有效数字,所以两者之积应为347。

(3)成方、开方、对数、指数等运算结果的有效数字位数不变。

2.5　实验数据处理的基本方法

物理实验是从被观测的物理现象中记录相关的数据,对这些数据进行处理,给出实验结果及不确定度的评价的处理过程,包括记录、整理、计算和分析等处理步骤。用简明而严密的方法将所观测到的物理现象用数据所表现出的内在规律提炼出来就是数据处理,下面主要介绍列表法、图示法、图解法、逐差法及最小二乘法等常用的数据处理方法。

2.5.1　列表法

列表法是把测量数据按一定规律列成表格的数据处理方法。它是记录数据和处理实验数据最常用的方法,又是其他数据处理的基础。数据表格必须简单而明确地表示出有关物理量之间的对应关系,便于检查核实,易于分析和比较,以便及时发现问题,有助于找出与实验现象相关联的规律性,并能求出经验公式等。列表的要求如下:

(1)表格设计要合理、简单明了,能完整地记录原始数据,并反映相关量之间的函数关系。

(2)表格的标题栏中注明物理量的名称和单位,各物理量的名称应使用符号表示,单位不必在数据栏内重复书写。

(3)表格中的数据应能正确反映测量量的有效数字位数,同一列数值的小数点应上下对齐。

(4)表格中还应包括各种所要求的计算平均值和误差。

(5)提供与数据处理有关的说明和参数,包括表格名称、主要测量仪器的规格(如分度值、仪器误差限、准确度等级等)、测量环境参数(如温度、湿度等)及可修正误差的修正值等。

2.5.2　图示法

图示法是将实验测得的数据用各种图线表示的数据处理方法。

1.图示法在数据处理中的作用

(1)图示法形象直观,使人看起来一目了然。它既能简明地显示出物理量之间的相互依赖关系和变化趋势,又能方便地找出函数的极大值、极小值、转折点、周期性、其他性质和规律。

尤其是对于那些很难用一个简单的函数解析式表示清楚的物理量之间的关系（如一天内的气温变化关系、晶体管的伏安特性等），用图示法能比较简单地表示这些关系。

(2)图示法在科学研究中，常常先由实验图线通过解析几何或其他的数学方法找出物理量之间的对应函数关系或经验公式，从中探求物理量之间的变化规律。

(3)在图示的曲线上，可以直接读出未观测的物理量的值（称内插法），也可以从图线的延伸部分读出测量数据以外对应点的值（称外推法）。

(4)由图示的曲线可以判断实验中的个别点的测量是否存在失误和一些系统误差是否存在等。

2. 物理实验中常用到的三种图示曲线

(1)物理量之间的函数关系图线、反映元件的特性图线和传感器的定标曲线等，这类曲线一般表现为连续光滑的曲线或直线。

(2)用以计算取值的图线。这类图线是根据较精密的测量数据经过整理后，精心细致地绘制在标准坐标纸上，以供计算和查对。

(3)仪器仪表的校准图线。这类图线的特点是两个物理量之间并无确定的函数关系，其图线不是连续光滑的曲线，而是无规则的折线。

上述这三种图线虽有各自不同的特点和应用，但它们的基本图示原则是一致的。

3. 作图的步骤及规则

物理实验用作数据处理的图示法的作图，不是示意图，它既要准确表达物理量之间的函数关系，又要反映物理量测量的精确程度，因此必须按照一定要求作图。要做好一张正确、实用、美观的实验图线，可按以下步骤和规则进行。

1)选择合适的坐标图纸

常用的坐标纸有直角坐标纸、对数（单对数和双对数）坐标纸、极坐标纸三种。图纸大小的选择原则是：既不损失实验数据的有效数字，又能包含所有实验数据点作为选取坐标纸大小的最小限度。图纸的分格大小选取应根据坐标的具体要求来确定，在同样分格标度值的情况下，分格大小决定着表现同一函数关系图线的放大或缩小，分格大，估读间隔宽，分格小，估读间隔窄。无论选取分格大的坐标纸还是选取分格小的坐标纸，图上的最小分格至少应与实验数据中的最后一位可靠数字相对应，即测量数据的可靠位在图中也应是可靠数字位。

2)确定坐标轴和坐标分度

坐标轴的确定常常习惯上将以自变量作为横坐标轴，因变量作为纵坐标轴。在两个变化的物理量中，究竟哪个是自变量，哪个是因变量，应根据实验方法和实验数据的特性来确定。例如用伏安法测电阻时，主回路用限流电路，则电流是自变量；主回路用分压电路，则电压是自变量。

坐标轴确定后，应在沿轴方向标明该轴所代表物理量的名称和单位。还应在轴上均匀地标明该物理量在坐标上的分度。坐标分度选取原则是：(1)应使所作出的图线比较对称地充满整个图纸，而不偏于一边或一角，横坐标、纵坐标的起点不一定要从零值开始，两轴的比例也可以不同；(2)应使每一个点的坐标值能迅速方便地读出，一般用一大格（10mm）代表1，2，5，10个单位为宜，而不采用一大格代表3，6，7，9个单位的分度，因为那样标度对于描点和读数都造成不便，而且很容易出错。

3)在坐标系中准确标出各个数据点

在实验前应预先考虑图线的特征,如果图线是直线,则应使测量点大体沿直线均匀分布。如果图线是曲线,则在曲线突变的某些点附近应适当增加测量点。描点时用直尺和削尖的铅笔准确而点点不漏地标出测量的数据点,常用作描点的符号有“+、×、△、⊙”等,同一图上描绘出不同的图线应该用不同的符号表示,即便是相重合的区间也应如此。

4)连接数据点成图线

连线时应根据自变量和因变量的对应关系是否在某一范围内连续、各点的分布趋势、图上的点数多少来确定连线方法。连线方法一般有两种,一种是直接将各点逐一用直线连接起来成为一条折线,这只有在作仪表的校准曲线或测量数据不够充分、连图的点数过少、自变量和因变量对应关系难以确定时才能采用。另一种是根据图上各点分布趋势作出一条光滑的连续曲线或直线。连线时要兼顾大多数点在曲线或直线上,而少量的点分布在曲线或直线附近的两旁,对于个别偏离曲线或直线较远的点,要进行审核,通过分析判断后决定是否属于坏值而剔除掉。只有在确定两个物理量之间的关系是线性的或所有数据点用“斜视法”沿曲线看去都在某一直线附近且均匀分布其两侧时,才能将图线连成一条直线。

5)图注和说明

描完图线后,在图纸的显著位置处注明图名、作者和作图日期,有时还应附上简要的说明,如实验条件、数据来源等。

2.5.3 图解法

由实验所得的图示曲线,运用解析几何的知识进一步求得曲线方程或经验公式的方法称为图解法,特别当图线为直线时,用图解法求经验公式就更加方便了。

1.线性图解的步骤

若作出的实验图线是一条直线,直线方程为 $y = kx + b$ 的形式,实际上就是求直线斜率 k 和截距 b,其步骤如下:

(1)选点。在直线上选取两点 $A(x_1, y_1)$ 和 $B(x_2, y_2)$,选取原则是:①由于利用图线求解相当于取平均值的作用,为减少相对误差,这两点相距尽量较远一些,但不应超出实验范围;②所选取的点不应该是测量的数据点,这两点的标注应与数据标注符号相区别,并在标注符号旁注明其坐标值,如图2-5-1所示。

(2)求斜率 k。将 $A(x_1, y_1)$ 和 $B(x_2, y_2)$ 的坐标值代入直线方程 $y = kx + b$ 中,可求得斜率:

$$k = \frac{y_2 - y_1}{x_2 - x_1} = 10.0\Omega$$

需指出,用图解法求得的斜率 k 不同于直线的几何斜率,它应是一个有单位的物理量,如此例中的 k 为电阻 R。

(3)求截距 b。如果横坐标起点为零,则直线的截距可直接由图中读出,否则,可由下式计算:

$$b = y_3 - \frac{y_2 - y_1}{x_2 - x_1}x_3 = -11.0\text{V}$$

图解法求出的斜率 k 和截距 b 具有较大的任意性,其结果随图线的质量和实验者的经验不同而不同,通常在测量数据离散度不大、线性较好或对曲线拟合精度要求不高时使用。在高

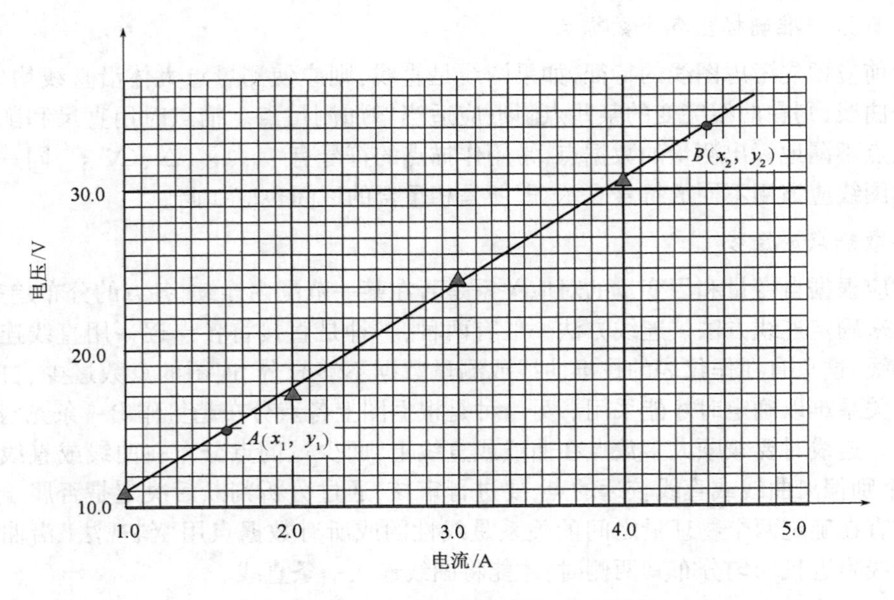

图 2-5-1　线性电阻伏安特性曲线

精度的要求下,就必须作出若干次修正。即将各次测量值(x_i,y_i)代入第一次图解得到的方程$y=k_1x+b_1$中,求出每一次的残差:

$$\Delta y_i = y_i - (k_1x_i + b_1)$$

再作残差图,即 $\Delta y—x$ 图线。若第一次拟合良好,则残差图上的点呈随机分布。若第一次拟合存在系统误差,则残差图上的点呈现有规律的直线,再求出该直线的斜率 k_2 和 b_2,修正后的直线斜率 $k=k_1+k_2$,截距 $b=b_1+b_2$,可得到一条优于第一次图解直线,经反复多次拟合,直到满意为止。

2. 曲线改直

实际上许多物理量之间的关系并非都是线性的,但可以通过适当变换后成为线性关系,这种变换成为曲线改直。

1)自由落体公式

$$h = v_0t + \frac{1}{2}gt^2$$

式中,v_0、g 均为常数。只要对等式两边同除以 t,即可得到:

$$\frac{h}{t} = v_0 + \frac{1}{2}gt$$

作$\frac{h}{t}—t$ 图线,即可得到一直线,其斜率为$\frac{1}{2}g$,截距为 v_0。

2)电容器放电曲线

$$q = q_0\exp\left(-\frac{1}{RC}\right)t$$

式中,q_0、R、C 均为常数。只要对等式两边各取自然对数,即可得到:

$$\ln q = \ln q_0 - \frac{1}{RC}t$$

作 $\ln q—t$ 图线,即可得到一直线,其斜率为$\frac{1}{RC}$,截距为 $\ln q_0$。

2.5.4 逐差法

逐差法是针对自变量等量变化,所测得有序数据进行等间隔项相减后取其逐差平均值得到的结果。其优点是充分利用了测量数据,具有对数据取平均的效果。它也是物理实验中处理数据常用的一种方法。

逐差法处理数据适应的条件如下:

(1)自变量 x 等间隔变化。

(2)被测的物理量之间的函数形式可写成 x 的多项式,即

$$y = \sum a_m x^m = a_m x^m + a_{m-1}x^{m-1} + \cdots + a_2 x^2 + a_1 x + a_0 \qquad (2-5-1)$$

下面以拉伸法测量弹簧的劲度系数为例,介绍逐差法的应用。

设实验中等间隔在弹簧下加砝码,共加 9 次(每次加 1kg),分别记录弹簧下端点对应位置 $l_0, l_1, \cdots, l_9$,则用逐差法处理如下:

①验证函数形式是线性关系。将所测得数据逐项相减,即 $\Delta l_1 = l_1 - l_0, \Delta l_2 = l_2 - l_1, \cdots, \Delta l_9 = l_9 - l_8$,若 Δl_i 均基本相等,则验证了外力与弹簧伸长量之间的函数关系为线性关系,即 $F = k\Delta l$。

②求弹簧伸长量的平均值。若要求出每加 1kg 砝码时,求弹簧的平均伸长量,应有

$$\overline{\Delta l} = \frac{1}{9}\sum_{i=1}^{9} \Delta l_i = \frac{l_9 - l_0}{9}$$

从上式可以看出,中间的测量值全部抵消了,只有始末两次测量起作用,这与一次加 9kg 砝码的测量结果完全等价。

通常是等间隔地将测量量分成前后两组,前一组为 $l_0, l_1, \cdots, l_4$,后一组为 $l_5, l_6, \cdots, l_9$,将前后两组的对应项相减为

$$\Delta l'_1 = l_5 - l_0;\ \Delta l'_2 = l_6 - l_1;\ \cdots;\ \Delta l'_5 = l_9 - l_4$$

取其平均值

$$\overline{\Delta l'} = \frac{1}{5}\sum_{i=0}^{4} (l_{5+i} - l_i)$$

式中,$\overline{\Delta l'}$是增加 5kg 砝码时弹簧的平均伸长量。由此可见,每一测量数据都对平均值有贡献,对应项逐差可以充分利用测量数据,具有对数据取平均值和减小误差的效果。

2.5.5 实验数据的直线拟合

1. 用最小二乘法进行直线拟合

最小二乘法是一种比较精确的曲线拟合方法。它的主要原理是:若能找到一条最佳的拟合曲线,则各测量值与这条拟合曲线上对应点之差的平方和最小。

现假设两物理量之间可满足线性关系,其函数形式为 $y = mx + b$,并等精度地测得一组数据(x_i、y_i, $i = 1, 2, 3, \cdots, k$)。因为误差总是伴随测量而存在,因此 x_i 和 y_i 中均含有误差,相对来说,x_i 的误差远比 y_i 的误差小。为了简便起见,可认为 x_i 值是准确的,而误差只与 y_i 相关联。假如对于一组($x_i, y_i, i = 1, 2, 3, \cdots, k$)数据点,$y = mx + b$ 是最佳拟合方程,则每一次测量

值与按方程 $y=mx+b$ 计算出的 y 值之间偏差为

$$v_i = y_i - (mx_i + b)$$

根据最小二乘法原理,所有偏差平方和为最小,即

$$S(m,b) = \sum_{i=1}^{k} v_i^2 = \sum_{i=1}^{k} [y_i - (mx_i + b)]^2 = \text{最小} \qquad (2-5-2)$$

式中,x_i、y_i 是已经测定出的数据点,不再是变量,要使所有偏差平方和为最小,只能变动 m 和 b,如果设法确定这两个参数,则该直线也就随之确定了。由求解极小值的条件,式(2-5-2)对 m 和 b 的一阶导数分别为零,即

$$\frac{\partial S}{\partial m} = -2\sum_{i=1}^{k} x_i(y_i - mx_i - b) = 0 \qquad (2-5-3)$$

$$\frac{\partial S}{\partial b} = -2\sum_{i=1}^{k} (y_i - mx_i - b) = 0 \qquad (2-5-4)$$

(1)求解 m 和 b,联立求解式(2-5-3)和式(2-5-4),得

$$m = \frac{\bar{x} \cdot \bar{y} - \overline{xy}}{(\bar{x})^2 - \overline{x^2}} \qquad (2-5-5)$$

$$b = \bar{y} - m\bar{x} \qquad (2-5-6)$$

其中 $\bar{x} = \frac{1}{k}\sum_{i=1}^{k} x_i$; $\bar{y} = \frac{1}{k}\sum_{i=1}^{k} y_i$; $\overline{x^2} = \frac{1}{k}\sum_{i=1}^{k} x_i^2$; $\overline{xy} = \frac{1}{k}\sum_{i=1}^{k} x_i y_i$

要验证式(2-5-2)表示的极值最小,还需证明二阶偏导数大于零,这里不再证明了。实际上由式(2-5-5)和式(2-5-6)给出的 m 和 b 对应的 $\sum_{i=1}^{k} v_i^2$ 就是最小值。

(2)各参量的标准误差,测量值 y 的标准误差为

$$\sigma_y = \sqrt{\frac{\sum_{i=1}^{k} (y_i - mx_i - b)^2}{k-2}} \qquad (2-5-7)$$

式(2-5-7)中分母是 $k-2$,是因为确定两个未知数要用到两个方程,多余的方程数为 $k-2$。

斜率 m 值的标准误差为

$$\sigma_m = \frac{\sigma_y}{\sqrt{k[\overline{x^2} - (\bar{x})^2]}} \qquad (2-5-8)$$

截距 b 值的标准误差为

$$\sigma_b = \frac{\sqrt{\overline{x^2}}}{\sqrt{k[\overline{x^2} - (\bar{x})^2]}}\sigma_y \qquad (2-5-9)$$

(3)拟合直线的检验,在待定参量确定之后,还要检验拟合直线是否成功。为此,引入一个称为线性相关系数的参量,记为 r。它定义为

$$r = \frac{\overline{xy} - \bar{x} \cdot \bar{y}}{\sqrt{[\overline{x^2} - (\bar{x})^2][\overline{y^2} - (\bar{y})^2]}} \qquad (2-5-10)$$

线性相关系数 r 表征了两个物理量之间关于线性关系的符合程度。r 值总是在 0 与 ±1 之间。r 值越接近于 1,表明 y_i 和 x_i 各实验点聚集在一条直线上或者直线附近,越符合所求得的直线,或 y_i 和 x_i 之间线性关系越好。相反,r 值等于零或趋近于零,表明实验点分布很分散,

y_i 和 x_i 相互独立，无线性关系，不能用线性函数拟合，而用其他函数重新试探。$r>0$，拟合直线斜率为正，称为正相关；$r<0$，拟合直线斜率为负，称为负相关。

2. 用 Excel 软件进行直线拟合

具有曲线拟合功能的软件很多，例如 Excel、Origin、Matlab 等。Excel 软件是微软 Office 办公套件的一个组件，一般来说，安装了 Word 软件的计算机，也安装了 Excel，因此 Excel 软件很容易获取。下面简单介绍如何用 Excel 进行直线拟合。

1）LINEST 函数

LINEST 函数是 Excel 软件提供的多元回归分析函数。直线拟合只是多元回归的特例，因此也可以用 LINEST 函数进行直线拟合，其函数句型和相应的参数选择列于表 2－5－1 中。LINEST 函数不仅可以直接给出拟合直线的截距、斜率、相关系数和因变量标准差等参量，还可以直接给出斜率标准差、截距标准差和残差平方和等参量，是非常方便的拟合工具。

在使用 LINEST 函数时，首先在 Excel 表格中输入原始数据，然后在任一空白单元处键入函数，即可得到计算结果。

表 2－5－1　Excel 软件中的 LINEST 函数

参　量	Excel 函数
斜率 b	INDEX(LINEST($y_1:y_n,x_1:x_n$,1,1),1,1)
截距 a	INDEX(LINEST($y_1:y_n,x_1:x_n$,1,1),1,2)
相关系数 r	INDEX(LINEST($y_1:y_n,x_1:x_n$,1,1),3,1)^0.5
因变量标准差 S_y	INDEX(LINEST($y_1:y_n,x_1:x_n$,1,1),3,2)
斜率标准差 S_b	INDEX(LINEST($y_1:y_n,x_1:x_n$,1,1),2,1)
截距标准差 S_a	INDEX(LINEST($y_1:y_n,x_1:x_n$,1,1),2,2)
残差平方和 S	INDEX(LINEST($y_1:y_n,x_1:x_n$,1,1),5,2)

2）直接求出拟合参量

Excel 软件还提供了直接求出截距、斜率、相关系数和因变量标准差等拟合参量的函数，函数的句型列于表 2－5－2 中。

表 2－5－2　Excel 软件中直接求拟合参量的函数

参　量	Excel 函数
斜率 b	SLOPE($y_1:y_n,x_1:x_n$)
截距 a	INTERCEPT($y_1:y_n,x_1:x_n$)
相关系数 r	CORREL($y_1:y_n,x_1:x_n$)
因变量标准差 S_y	STEYX($y_1:y_n,x_1:x_n$)

3）给出拟合参数

利用“图表”功能中的“添加趋势线”功能给出拟合参数。这种方法可以给出拟合直线的截距、斜率和相关系数等参数。具体做法是：选定数据 (x_i,y_i) 后，使用 Excel 软件工具栏或“插入”下拉菜单中“图表”功能中的“XY 散点图”中的“平滑散点图”作图，然后将鼠标移到图中的直线上，按鼠标右键选择“添加趋势线”，进而选择“添加趋势线”标签中“类型”栏中的“线性”与“选项”栏中的“显示公式”和“显示 R 平方值”两个选项，则在曲线图中就自动添加出方程 $y=kx+b$ 及相关系数 R^2 的值。

2.6 数据处理实例——测量圆柱体的密度

2.6.1 准备工作

测量工具：游标卡尺（精度为 0.02mm），千分尺（精度为 0.004mm），物理天平（感量为 20mg）。

2.6.2 数据处理过程

测量数据见表 2-5-3。

表 2-5-3 测量数据

被测量 测量值 序号	h/mm（游标卡尺）	d/mm（千分尺）	m/g（物理天平）
1	29.22	12.257	26.82
2	29.30	12.261	
3	29.24	12.256	
4	29.28	12.252	
5	29.26	12.257	
6	29.24	12.254	
7	29.28	12.258	
平均值	29.26	12.256	26.82

1. 密度 ρ 的平均值

依据公式 $\rho=\dfrac{4m}{\pi d^2 h}$，可求得

$$\rho=\frac{4\times 26.82\times 10^{-3}}{3.1416\times(12.256\times 10^{-3})^2\times 29.26\times 10^{-3}}=7.7696\times 10^3(\mathrm{kg/m^3})$$

2. 各直接测量量的标准不确定度

1）多次直接测量量 h 的合成标准不确定度

A 类 $u_A(h)$：

$$u_A(h)=S_{\bar{h}}=\sqrt{\frac{\sum_{i=1}^{n}(h_i-\bar{h})^2}{n(n-1)}}=0.0106(\mathrm{mm})$$

B 类 $u_B(h)$：由 GB/T 21389—2008《游标、带表和数显卡尺》查得游标卡尺示值误差 $\Delta_{ins}=$ 0.02mm，则

$$u_B(h)=\frac{0.02}{\sqrt{3}}\approx 0.0115(\mathrm{mm})$$

h 的合成标准不确定度为

$$u_h = \sqrt{u_A^{\ 2}(h) + u_B^{\ 2}(h)} = 0.0149(\text{mm})$$

h 的相对标准不确定度为

$$\frac{u_h}{\overline{h}} = \frac{0.0149}{29.26} = 0.000509 = 0.051\%$$

2)多次直接测量量 d 的合成标准不确定度

A 类 $u_A(d)$:

$$u_A(d) = S_{\overline{d}} = \sqrt{\frac{\sum_{i=1}^{n}(d_i - \overline{d})^2}{n(n-1)}} = 0.000109(\text{mm})$$

B 类 $u_B(d)$:由 GB/T 21389—2008 查得千分尺示值误差 $\Delta_{ins} = 0.004\text{mm}$,则

$$u_B(d) = \frac{0.004}{\sqrt{3}} \approx 0.00231(\text{mm})$$

d 的合成标准不确定度为

$$u_d = \sqrt{u_A^{\ 2}(d) + u_B^{\ 2}(d)} = 0.00255(\text{mm})$$

d 的相对标准不确定度为

$$\frac{u_d}{\overline{d}} = \frac{0.0026}{12.256} = 0.000212 = 0.021\%$$

3)单次直接测量量 m 的合成标准不确定度

对于单次直接测量量,其合成不确定度只计及 B 类不确定度,即

$$u_m = u_B(\text{m})$$

天平的感量近似作为仪器误差,即 $\Delta_{ins} = 20\text{mg}$,则

$$u_m = u_B(m) = \frac{20}{\sqrt{3}} = 11.55(\text{mg})$$

m 的相对不确定度为

$$\frac{u_m}{m} = \frac{11.55 \times 10^{-3}}{26.82} = 0.000431 = 0.043\%$$

3. 间接测量量 ρ 的标准不确定度 u_ρ

先求 ρ 相对不确定度,再由相对不确定度的定义求 ρ 的标准不确定度 u_ρ 较为方便。

$$\frac{u_\rho}{\overline{\rho}} = \sqrt{\left(\frac{u_m}{m}\right)^2 + 2^2\left(\frac{u_d}{\overline{d}}\right)^2 + \left(\frac{u_h}{\overline{h}}\right)^2}$$

$$= \sqrt{0.000431^2 + 4 \times 0.00255^2 + 0.000509^2} = 0.00514 = 0.51\%$$

$$u_\rho = \frac{u_\rho}{\overline{\rho}} \times \overline{\rho} = 0.00514 \times 7.7696 \times 10^3 = 0.03994 \times 10^3(\text{kg/m}^3)$$

4. 测量密度 ρ 的扩展不确定度

$$u_{0.95} = 2u_\rho = 0.07988 \times 10^3(\text{kg/m}^3)$$

5. ρ 的扩展相对不确定度

$$\frac{u_{0.95}}{\bar{\rho}} \approx 0.010 = 1.0\%$$

2.6.3 最终测量结果的表示

$$\rho = \bar{\rho} \pm u_P = (7.77 \pm 0.08) \times 10^3 (\mathrm{kg/m^3}) \quad (P = 0.95)$$

$$\frac{u_{0.95}}{\bar{\rho}} \approx 0.010 = 1.0\% \quad (P = 0.95)$$

第3章　常用仪器和器件

物理实验涉及很多种类的实验仪器，要做好每一个实验项目，学习好该门课程，就应该学会正确使用、维护仪器设备及相关器件，本章主要介绍实验课中一些常用仪器及器件。

3.1　长度测量仪器

日常生活中经常要用器具测量物体的长度，而米尺是最常见的普通型的测长器具。用米尺测量物体的长度时，可精确到毫米位，但在实际测量中有时要将被测的长度测量到百分之一甚至千分之一毫米位，这是用米尺测量无法做到的。为了提高长度测量的精度，在米尺的基础上，人们设计制造了多种测量长度的仪器，常用的为游标卡尺和螺旋测微器及读数显微镜。

3.1.1　游标卡尺

游标卡尺又称为卡尺或游标尺，是利用游标原理对两测量爪相对移动所分隔的距离进行读数的一种通用长度测量工具。可用来测量零部件的长度、宽度、深度和内径、外径尺寸。常用的游标卡尺有 0 ~ 125mm、0 ~ 150mm、0 ~ 200mm、0 ~ 300mm 等多种规格，分为 10 分度、20 分度及 50 分度三种，对应精度分别为 0.1mm、0.05mm 和 0.02mm 三种。

游标卡尺的基本构造如图 3 – 1 – 1 所示。它主要由主尺 D 和游标（副尺）E 组成。主尺是毫米分度尺，其上有固定的测量爪 A 与 A′。游标可沿主尺滑动，其上也装有测量爪 B 与 B′。测量物体的外部尺寸时用外测量爪 AB；测量物体的内径时用内测量爪 A′B′；测量物体内部深度时，用深度尺杆 C。当量爪测量面合并时，游标的“零”刻线刚好应与主尺的“零”刻线对齐。图 3 – 1 – 1 所示是 50 分度游标卡尺，其主尺上刻有毫米分格（每格长度 1mm）。而游标上刻有 50 分格，但它的总长等于主尺 49 分格（49mm），所以每格是 0.98mm，即游标与主尺每分格长度值差为 0.02mm，此刻度值称为游标尺的“分度值”。

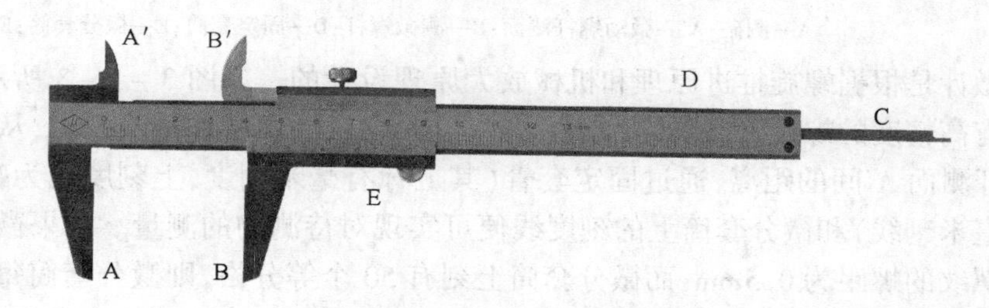

图 3 – 1 – 1　游标卡尺结构示意图

在测量读数时，要求测量爪和被测件接触松紧恰当，读数时首先按游标尺“零”刻线所在的位置读出主尺刻度的整数部分，然后找出游标尺哪条刻线与主尺刻线对齐，用卡尺的分度值乘以游标尺的该刻线序数即可读出它的小数部分，再将主尺读数与小数相加即为被测件尺寸（视频 3 – 1 – 1）。如图 3 – 1 – 2为测物体长度时的读数，通过游标尺“零”刻线在主尺上读出

42mm(图中“4”表示该位置为 40mm 刻度),游标尺的第 16 刻度线和主尺某一刻度线重合(图中“3”表示该位置为第 15 刻线),由于该尺分度值为0.02mm,所以该测量值为:42.0mm + 0.02 × 16mm = 42.32mm。

使用游标卡尺时应注意:

(1)物体被测量的部位必须与游标尺本身平行,切忌挪动已被夹紧的物体。

(2)使用时,轻轻推动游标,把物体卡住并固定螺钉即可读数。

(3)注意保护量爪,不得磨损刀口和钳口,不允许用游标尺测量粗糙的物体。

(4)用完后应立即放回盒内,禁止与潮湿物相接触。

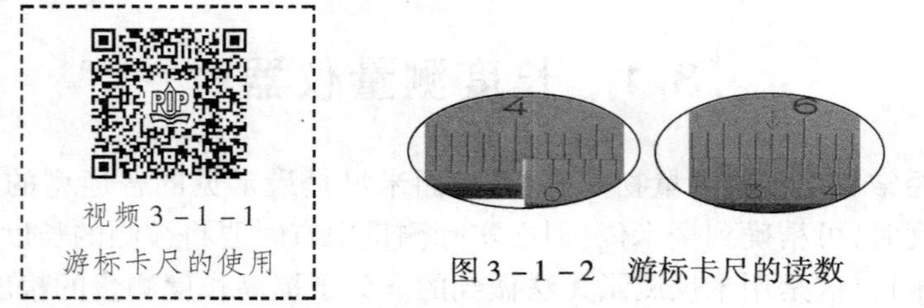

图 3-1-2　游标卡尺的读数

3.1.2　螺旋测微计

螺旋测微计是比游标卡尺更精密的测长量具,常用来测量细丝直径、薄板厚度等精密零件的尺寸。其结构如图 3-1-3 所示,主要由主尺(一根精密的测微螺杆 B 和固定套筒 D 组成)和副尺(一个具有 50 分度的微分套筒 E)组成,尾部带有棘轮旋柄 F(视频 3-1-2)。

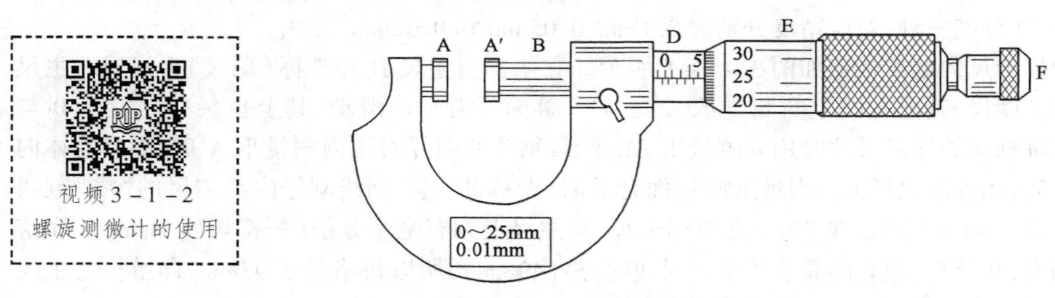

图 3-1-3　螺旋测微计示意图

A—测砧;A′—微动螺杆测面;B—测微螺杆;D—固定套筒;E—微分套筒;F—棘轮旋柄

螺旋测微计是根据螺旋推进原理和机械放大原理设计的。如图 3-1-3 所示,由于微动螺杆上加工有高精度的螺纹,当旋转微分套筒时,可实现微动螺杆前进或后退,从而控制测砧 A 和微动螺杆测面 A′间的距离,通过固定套管(其上标有毫米刻度,上刻度线为毫米刻线,下刻度线为半毫米刻线)和微分套筒上的刻度线便可实现对待测物的测量。如果微动螺杆上加工的高精度螺纹的螺距为 0.5mm,而微分套筒上刻有 50 个等分格,则微分套筒每旋转一个分格,微动螺杆沿轴线前进(或后退)$\frac{0.5}{50}$mm = 0.01mm,从而使沿轴线方向的微小长度用圆周上较大的弧度精确地表示出来。这种螺旋测微计的精度为 0.01mm,测量时估读到 0.001mm。

在使用时,将待测物置于测砧和微动螺杆测面之间,旋转棘轮柄,使微动螺杆前进,当微动螺杆测量面与待测物相接触时即可从固定刻度套管和微分筒上读出待测量的数值。在读数时,先由微分套筒的前沿在固定套筒上的位置读出整数格,再从固定套筒上的横线所对活动套筒上的分格线读出小数部分，注意要估读，同时读数时一定要注意微分套筒的前沿有没有漏出标尺上表示半毫米的刻度。例如，图 3-1-4(a)没有露出 3.5mm 的刻线，所以读数为 3.243mm；而图 3-1-4(b)中，因露出了 3.5mm 的刻线，所以读数为 3.743mm。

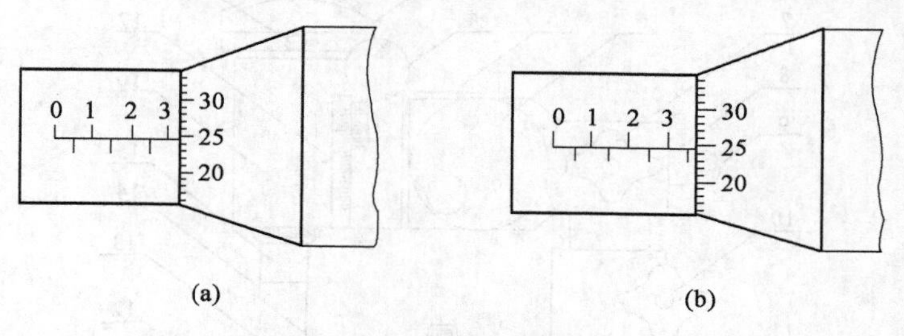

图 3-1-4　螺旋测微计读数

使用螺旋测微计时应注意：

(1)测量前应检查并记录零点读数。零点读数是测砧与微动螺杆测面刚好接触时的读数。测量时，测出的读数应减去零点读数，才应为待测物长度。

(2)测量及记录零点读数时，应轻轻转动棘轮旋柄推进螺杆，而不要直接转微动螺杆或微分套筒。棘轮旋柄靠棘轮的摩擦力带动微动螺杆前进，当微动螺杆接触并压紧待测物体时，其压力大于棘轮摩擦力，这时将发生"嘎嘎"声，此时应停止旋动棘轮柄，以保证每次测量时待测物压力均相等，从而保证测量的准确性，并避免因压力过大而损坏测微螺旋或待测物体。

(3)测量完毕应使两个测量面保留一定的间隙，以免仪器自身热膨胀而损坏。若长期保存应进行防锈处理。

3.1.3　读数显微镜

读数显微镜是用来测量微小距离或微小距离变化的光学仪器，例如进行孔距、直径、刻线距离即刻线宽度等量测量。它由机械部分和光学部分两部分组成，结构如图 3-1-5 所示。光学部分是一个长焦距显微镜，装在一个由丝杆带动的滑动台上(滑动台安装在底座上)，滑动台和显微镜可以按不同方向安装，可对准前方，上下、左右移动或对准下方，左右移动。机械部分是根据螺旋测微原理制造的，一个与螺距为 1mm 的丝杠联动的刻度圆盘上有 100 个等分格，读数显微镜的测量精度为$\frac{1}{100}$mm = 0.01mm，估读误差为$\frac{1}{1000}$mm = 0.001mm，因此，它的分度值是 0.01mm，可读到 0.001mm。读数显微镜的量程一般为几厘米。

用读数显微镜进行测量时，首先将被测物体放置在载物台上并用压片固定，调节读数显微镜，对准被测物体。再调节读数显微镜的目镜，使目镜内分化平面上的叉丝(或标尺)清晰。其次调节显微镜的聚焦情况或移动整个仪器，使被测物成像清楚，并消除视差(眼睛上下移动时，看到叉丝与待测物的像之间无相对移动)。然后先让叉丝对准被测物上一点(或一条线)，记下读数；转动丝杆，对准另一点，再记下读数，最后将两次读数求差即为被测物的长度。

使用读数显微镜的注意事项如下：

(1)调节聚焦过程应使目镜筒自下而上缓慢聚焦，严禁将镜筒反向调节聚焦。

(2)使用时，显微镜的方向和被测两点间连线要平行。

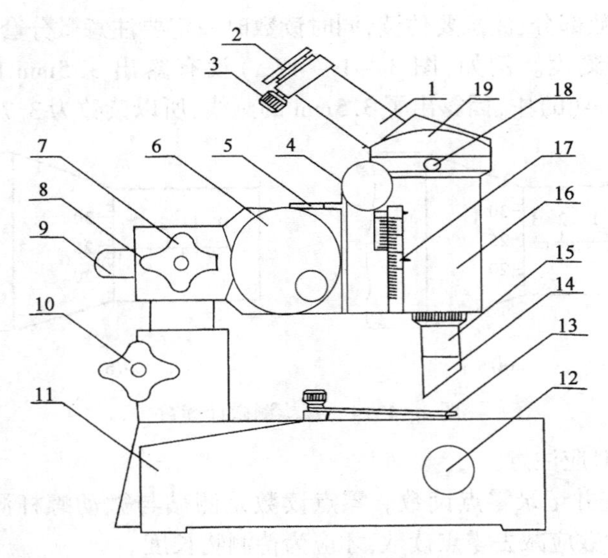

图 3-1-5　读数显微镜结构图

1—目镜接筒；2—目镜；3—锁紧螺钉；4—调焦手轮；5—标尺；6—测微手轮；7—锁紧手轮Ⅰ；8—接头轴；9—方轴；10—锁紧手轮Ⅱ；11—底座；12—反光镜旋轮；13—压片；14—半反镜组；15—物镜组；16—镜筒；17—刻尺；18—锁紧螺钉；19—棱镜室

(3) 在测量时应向同一方向转动鼓轮，使叉丝和各目标对准，当移动叉丝超过了目标时，就要多退回一些，重新再向同一方向转动鼓轮去对准目标以避免螺距产生的空程误差。

(4) 物镜和目镜勿用手触摸，只能用镜头纸轻拭。

3.2　时间测量仪器

时间是物理学中的一个基本概念之一，在现代科学中是不可缺少的基本量，特别在当今的无线电广播、计量技术、雷达测距、卫星发射与回收、计算机应用、自动控制等方面都需要精确的时间和时钟标准。在国际单位制中，时间的基本单位为秒(s)，定义为铯(^{133}Cs)原子在基态的两个超精细能级间跃迁辐射周期的9192631770倍的时间。在时间测量中，按测量内容可分为时段测量和时刻测量。物理实验中常涉及的是时段测量，多使用机械秒表、电子秒表及数字毫秒计等。

3.2.1　机械秒表

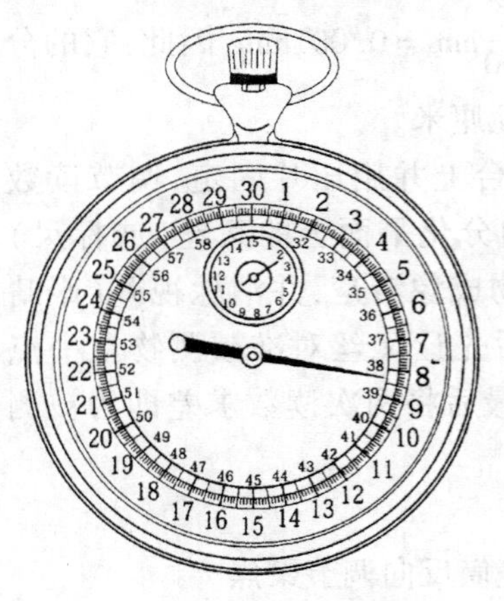

图 3-2-1　机械秒表实物图

机械秒表可分为单针和双针两种。单针秒表只能测量一个过程所经历的时段；双针秒表可分别测量两个同时开始但不同时结束的过程所经历的时段。机械秒表由频率较低的机械振荡系统、锚式擒纵调速器、操纵秒表启动、制动和指针回零控制机构(包括按钮、发条及齿轮)等机械零件组成，有的秒表还有暂停按钮，用来进行累积计时。一般秒表的表盘最小分度值为0.1s或0.2s，测量范围是15min或30min。

机械秒表(图3-2-1)使用时，应先上紧发条(转动带滚花的按钮，不宜过紧)；然后按一下按钮开始计时，再按一下停止计时,这时秒表指示的时间为终止时刻到起止时刻的差值。对于无暂停机构的秒表,按一下,指针又复位到零。

秒表工作时的准确与否对计时影响很大,短时间测量(几十秒内),误差主要来源于启动、制动停表时的操作误差,其值约为 0.2s,有时还会更大一些。长时间测量,测量误差除了掐表操作误差外,还有秒表的仪器误差。所以在实验前,须将秒表与标准电子计时仪进行校对。

3.2.2 电子秒表

电子秒表是以石英振荡器的振荡频率作为时间基准实现计时的,并采用 6 位数的液晶显示器,具有精度高,显示清楚,使用方便,功能较多等优点。图 3-2-2 是 SE-1 型电子秒表外形,使用过程中通过控制 S_1 和 S_3 按钮可以实现基本秒表显示、累加计时和取样等常用功能,控制 S_2 可以实现计时、计历和星期显示等功能键的切换。电子秒表计时误差主要来源于启动、制动停表时的操作误差。

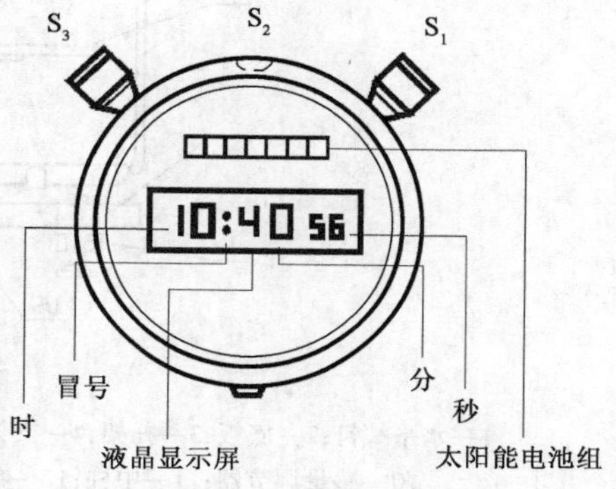

图 3-2-2 电子秒表示意图

在实现基本秒表显示时,当 S_3 在秒表功能时,应先使其复零,按 S_1 秒表计数开始,再按 S_1 秒表计数停止,再按 S_3 秒表复零;在累加计时,按一下 S_1 秒表计数开始,再按一下 S_1 即累加计时,如此可以重复继续累加;在取样时,按一下 S_1 秒表计时开始,再按一下 S_3 液晶显示器上的数字立刻停止,并在右上角出现"□"的记录信号,冒号仍在闪动,这时读数数字即为取样计时,要取消"□"再按一下 S_3 即可。

3.2.3 数字毫秒计

数字毫秒计又称电子计时仪,它利用高精度的石英振荡器输出的方波作为计时信号,因而计时准确度较高、测量范围较广。数字毫秒计一般由整形电路计数门、计数器、译码器、振荡器、分频器、复原系统、触发器等组成。数字毫秒计工作时,石英振荡器输出的信号频率可以为 1MHz、100kHz 和 10kHz,则数字毫秒计的标准时间单位可以是 0.001ms、0.01ms 或 0.1ms。

在工作原理上,首先由光电元件产生控制自动计时器开始计时和停止计时的信号,然后由脉冲信号在开始计时到停止计时的时间间隔内推动计数器计数,计数器所显示的脉冲个数就是以标准时间为单位的被测时间。在上述过程中,"光控"有两种计时方法:一种是记录光敏二极管的光照被遮挡时间;另一种是记录两次遮光信号的时间间隔,即遮挡一下光敏二极管开始计时,再遮挡一下计数器停止计时,两次遮光信号的时间间隔由数码管显示出来。

3.3 质量称衡仪器

物理天平是物理实验中常用的测量质量的仪器。天平的规格主要由最大称量和感量(或灵敏度)来标识。最大称量是天平允许称量的最大质量。天平感量是指天平指针偏转一最小分格所需增(或减)的砝码量,即天平的精度。一般来说,感量的大小应该与天平砝码(游码)

读数的最小分度值相适应(例如相差不超过一个数量级)。感量的倒数称为天平的灵敏度。天平感量与最大称量之比定义为天平的精度级别。天平的仪器误差一般取感量的$\frac{1}{2}$。

图3-3-1所示是一种双盘悬挂等臂式天平,主要由底座、支柱、横梁和秤盘四大部分组成。横梁上部刻有游码标尺数。

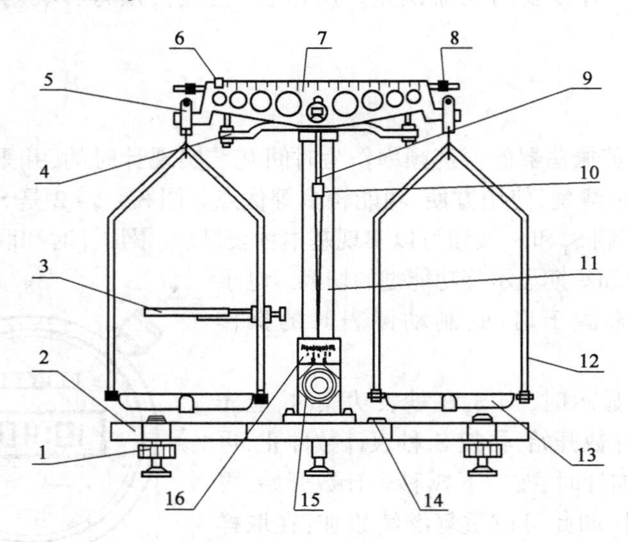

图3-3-1　物理天平

1—水平螺钉;2—底板;3—托架;4—支架;5—吊耳;6—游码;7—横梁;8—平行调节螺母;9—读数指针;10—感量调节器;11—中柱;12—盘梁;13—秤盘;14—水准器;15—制动旋钮;16—读数标尺

天平的调节按以下顺序调节:第一步,水平调节。调节水平螺钉1,观察底座上的水准器14,若气泡在中央位置,说明天平底座水平。第二步,零点调节。将游码6拨到刻度“0”处,将吊耳5置于横梁7两端刀口上,并将两秤盘悬挂端挂到吊耳挂钩上,再顺时针旋转制动旋钮15,支起中间支点,观察指针平衡与否。当指针在标尺的中线位置,即可认为零点调好,否则逆时针旋回制动旋钮使之处于制动位置,调整平行调节螺母,再次顺时针旋转制动旋钮,重复上述步骤,直至调好零点。

在称衡物体质量时,先把待测物体放在左盘中央粗略估计它的质量,用镊子夹取砝码放在右盘中央;然后轻轻起动横梁,观察横梁是否平衡,若不平衡,判断是否应加(或减)砝码,操作时先止动天平,再加(减)砝码,重复测试,直到横梁平衡。然后慢慢制动天平,盘中砝码质量与游码指示值(游码移动一大格为$\frac{1}{10}$g)之和即为被测物体的质量,再把待测物体放在右盘中央,砝码放在左盘中央,重复以上步骤使天平重新平衡,此时砝码质量与游码指示值之差即为被测物体的质量,取两次测量值乘积的开方值为其测量值,这样可以消除天平的不等臂误差。称衡完毕,将制动旋钮放到制动处,托盘挂口摘离刀口,记录物体质量,将砝码用镊子放回砝码盒中。

使用物理天平的注意事项:

(1)酸、碱、油脂或其他化学药物不能直接放在托盘上称量。

(2)砝码切忌用手拿,应用镊子夹持,以免沾汗锈蚀,用完后应随即放回砝码盒中。

(3)天平的载荷量不能超过天平的最大称量。

(4)取放物体、砝码和移动游码或调节天平时,必须切记应在天平制动后进行,以免损坏刀口和其他部件。

(5)刀口处、游码杆和螺丝等处应常滴钟表油,以防生锈。

(6)天平应放在干燥、清洁而稳定的环境中,不要经常搬动。

3.4 温度测量仪器

温度是物体冷热程度的表示,是基本物理量之一,许多物质的特征数都与温度有着密切的联系。所以在一些科学研究和工农业生产中,温度的控制和测量显得特别重要。物理实验中常用液体温度计、热电偶温度计、热敏元件温度计和干湿球温度计等。

3.4.1 液体温度计

液体温度计是以液体为测温物质,利用液体的热胀冷缩性质来测量温度的。常见的测温物质有水银和酒精等,以水银应用最为广泛,主要是因为水银在标准大气压和 -38.87 ~ +356.58℃的温度下,其热膨胀系数变化很小,体积的改变量与温度改变量基本成正比,热传导性能良好,且与玻璃管壁不相黏附,是一种较精密的测温范围广的液体温度计。

常用水银温度计可分为标准温度计和普通汞温度计等规格。标准汞温度计可分为一等标准汞温度计和二等标准汞温度计,主要是用来校正其他各类温度计的。一等标准汞温度计测温范围在 -30 ~300℃之间,其分度值为0.05℃,每套由9支或13支测温范围不同的温度计组成,用于检定或校准二等标准汞温度计。二等标准汞温度计测温范围也是 -30 ~300℃,分度值为0.1℃或0.2℃,是校准各种常用玻璃液体温度计的标准仪表。标准温度计出厂时,每支温度计均有检定证书。普通汞温度计测温范围为0 ~50℃、0 ~100℃、0 ~150℃等多种,分度值一般为1℃,在实验室和日常生活中多使用此种温度计。

测量温度时,应使温度计头部(充有液体部分)与被测物有充分的接触,避免由于热传递产生的滞后性给测温带来一定的影响。读数时,注意采用正确的读数方法,即要正视玻璃管上的温度刻线。由于玻璃管温度计易碎,使用完毕一定要保管好。

3.4.2 热电偶温度计

热电偶温度计是利用温差电动势与温度的比例关系进行温度测量的。在结构上,它是把两种不同的金属或不同组分的合金两端彼此焊接(或熔接)成一个闭合回路,如图3-4-1所示。

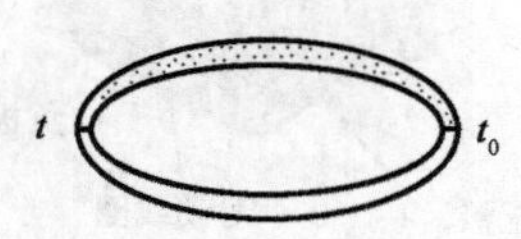

图3-4-1 热电偶结构图

当两接点保持在不同的温度 t 和 t_0 时,则回路中便产生温差电动势,温差电动势的大小依赖于两种材料的性质和温差($t-t_0$)。一般说来,电动势与温差的关系很复杂,但取第一级近似式,可写成 $\varepsilon=C(t-t_0)$,式中 t 为热端温度,t_0 为冷端温度,C 为温差系数(或称电偶常数),其大小取决于组成电偶的材料。若对热电偶进行定标以后,便可以根据温差电动势的大小确定未知温度。在实际应用中,一些标准热电偶(如铂铑—铂热电偶、铜—康铜热电偶等)都有现成的温差系数表可查阅,无须进行定标,只要测出温差电动势,便可根据热电偶一端的已知温度得出另一端的未知温度。

热电偶的优点是温差电动势与热电偶端部的体积无关,探头可以很小,消除了探头的热容和温度测量的时间滞后性。与普通温度计的测量范围相比,热电偶测量范围更大,可以测

1000℃以上的高温,因此常用于工业生产中。

3.4.3 热敏元件温度计

热敏元件温度计是以半导体热敏元件为温度探头,通过测量电阻变化进行温度测量的。根据相关理论,热敏元件的阻值与温度有关系:$R_T = ae^{\frac{b}{T}}$(a、b 为待定常数),只要事先确定未知常数,便可根据关系进行温度测量。热敏元件温度计由于使用方便,被广泛应用于自动控制仪器、仪表及测温集成电路等。

3.4.4 干湿球温度计

干湿球温度计由两只相同的温度计 A 和 B 组成。温度计 B 的储液球上包裹着纱布,纱布的下端浸在水槽内。由于水蒸发的吸热作用,使温度计 B 所指示的温度低于温度计 A 所指示的温度。当环境空气的湿度小时,由于水蒸气蒸发得快,吸收的热就多,所以两温度计所指示的温度差就越大,反之则两温度计所指示的温度差就越小。根据所指示的温度差,通过转动干湿球温度计中间的转盘便可查找出该温度差对应的环境的相对湿度。

3.5 电磁测量仪器

电磁学实验中,经常用到的如电流表、电压表等各种电测器具和电源、电阻以及开关等电磁器件统称为电磁测量仪器。

3.5.1 电表

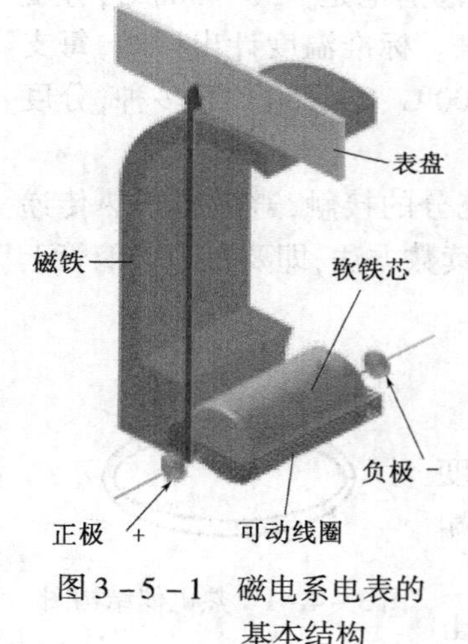

图 3-5-1 磁电系电表的基本结构

物理实验中经常用的电表主要是电流表、电压表和检流计。由于电磁系仪表具有准确度高、稳定性好以及受外磁场和温度影响较小等优点,所以在物理实验中所涉及的电表大都是直流磁电系仪表。因此,主要介绍实验室用的测量直流电的磁电系电表。

磁电系电表是根据载流线圈在磁场中受安培力产生磁力矩原理制成的。磁电系电表的基本结构如图 3-5-1 所示,其表头的主要部分是固定的永久磁铁和活动的通电线圈。当线圈中有稳定电流通过时,就会受磁力矩作用发生转动,可以证明线圈偏转的角度与流过线圈的电流成正比。只要将一个指针固定在安装线圈的轴上,当线圈转动时指针便随之旋转。如果在轴与支座之间再装一个螺旋状的弹簧(俗称游丝),当线圈旋转时,这个弹簧因旋转而产生一个反力矩以阻止它旋转,当两个力矩平衡,线圈与指针就停止在某一转角状态,如果把指针扫过的弧长以相应的电流标度,那么它的示值就可以定量地表示线圈中流过电流的大小。

视频 3-5-1 电表的使用

磁电系仪表具有灵敏度高、刻度均匀、便于读数等优点,但仅用于直流测量,如果要进行交流测量,则需要进行整流,变交流为直流。实验室常用的磁电系电表有检流计、电压表和电流表(视频 3-5-1)。

检流计,专门用来检验电路有无微小电流。它的特征是指针零点在刻度的中央，便于检出不同方向的直流电流。检流计的主要规格有电流计常数，即偏转一小格代表的电流值，一般约 10^{-5}A/小格。检流计的标读盘上通常标有字母"G"。

电压表，是测量电路中两点间电压的大小，它的主要规格参量有量程（即指针偏转满度时的电压值）和内阻（即电表两端的电阻）。同一电压表不同量程内阻不同，可以用下式计算量程的内阻：

$$内阻 = 量程 \times 每伏欧姆数$$

不同型号的电压表每伏欧姆数不同，表盘上都标有具体数值。

电流表，是测量电路中电流大小。它的主要规格参量有量程（即指针偏转满度时的电流值）和内阻（一般电流表内阻都在 0.1Ω 以下，毫安表、微安表内阻可达一二百欧到一二千欧）。

使用电表时应注意以下几点：

(1) 电表的联法。电流表必须串联在电路中，电压表应当与被测电压两端并联。

(2) 电流方向。直流电表的偏转方向与所通过的电流方向有关，所以接线时必须注意电表上接线柱的"+"标记表示电流流入端，"-"标记表示电流流出端。

(3) 量程的选择：测量时应事先估计待测量的大小，选择稍大的量程，试测一下，如不合适，选用更合适的量程。

(4) 读数时注意消除视差。

[补充说明]：

电器仪表的准确度（精度级别）：根据国家标准，仪表的准确度（精度）分别为 0.1、0.2、0.5、1.0、1.5、2.5、5.0 七个级别。

在规定条件下，对于单向标度尺的指示仪表使用时测量的最大绝对误差为

$$\Delta_{\max} = \pm x_n S\% \tag{3-5-1}$$

式中，x_n 为仪表的量程；S 为精度级别。测量时，某一示值 x 的最大相对误差为

$$k = \frac{\Delta_{\max}}{x} = \frac{x_n S\%}{x} \tag{3-5-2}$$

由此可见，这类仪表只有在满量程附近时，其示值的精度才接近于仪表的精度。因此使用时，一般要使指针在表盘的 2/3 到满刻度之间工作。

3.5.2 电路元器件

1. 滑线变阻器

滑线变阻器的结构如图 3-5-2 所示。电阻丝是采用经过氧化绝缘处理的康铜导线，将其密绕于瓷管上，外装金属保护支架，电阻丝的表面上有可以做调节使用的滑动导电接触电刷。在图 3-5-2 中 A、B 为两个固定端，C 为滑动端，在电路中可以做限流、分压和代替未定阻值的可变电阻器应用。滑线变阻器的主要技术指标为全电阻和额定电流（功率），应根据外接负载的大小和调节要求选用，尤其要注意，通过变阻器任一部分的电流均不允许超过其额定电流（视频 3-5-2）。

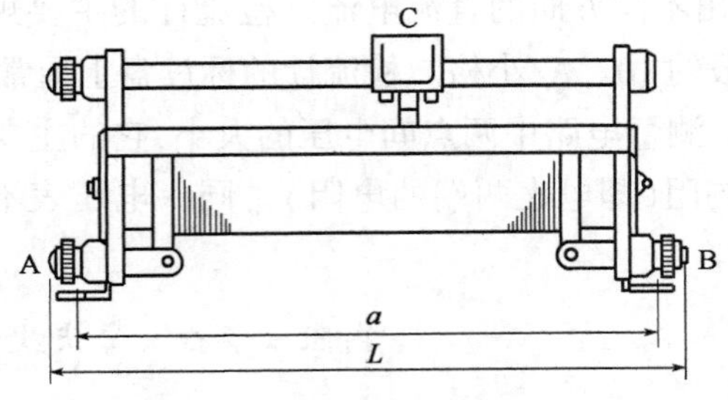

图 3－5－2　滑线变阻器结构图

实验室常用滑线变阻器来改变电路中的电流或电压，分别连接成限流电路和分压电路。图 3－5－3 所示为限流电路，当滑动 C 时，改变了整个回路电阻，电流随之改变。在接通电源前，一般应使 C 滑动到 B 端使电流最小，在接通电源后，可以逐步减小电阻，使电流增至所需值。图 3－5－4 所示为分压电路，当改变滑动端 C 的位置时，负载两端电压随之改变。同限流电路一样，为保证安全，在接通电源前，一般应使 C 滑动到 B 端使输出电压最小，接通电源后，可逐步减小电阻使电压增至所需值。

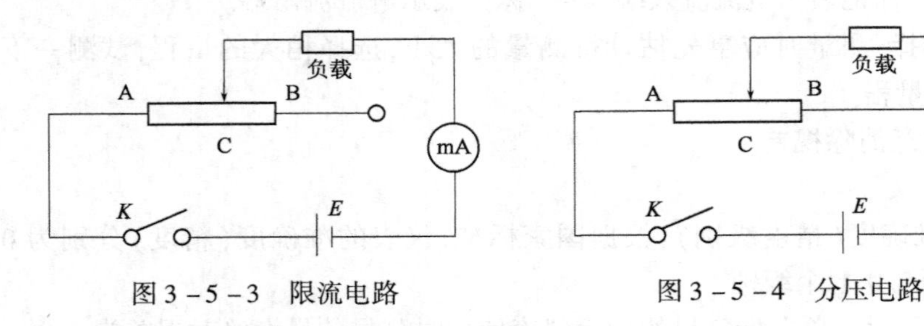

图 3－5－3　限流电路　　　　图 3－5－4　分压电路

2. 电阻箱

图 3－5－5 和图 3－5－6 是旋转式电阻箱内部电路及其外形示意图（视频 3－5－3）。电阻采用高稳定锰铜合金线制成，具有较低的零值电阻。转动转换开关，可以改变两接线柱间的电阻值，可用在直流电路中调节阻值。

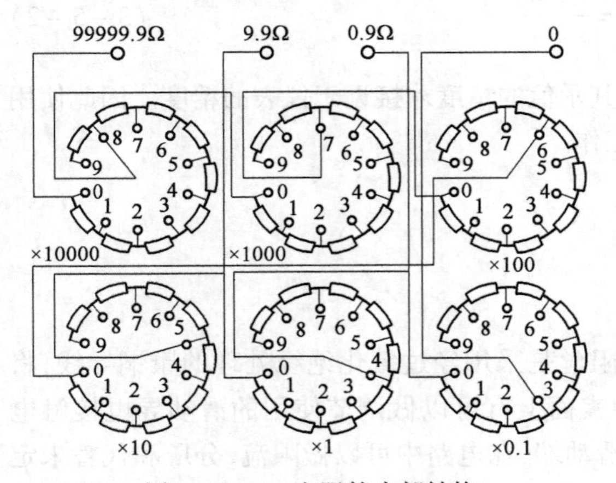

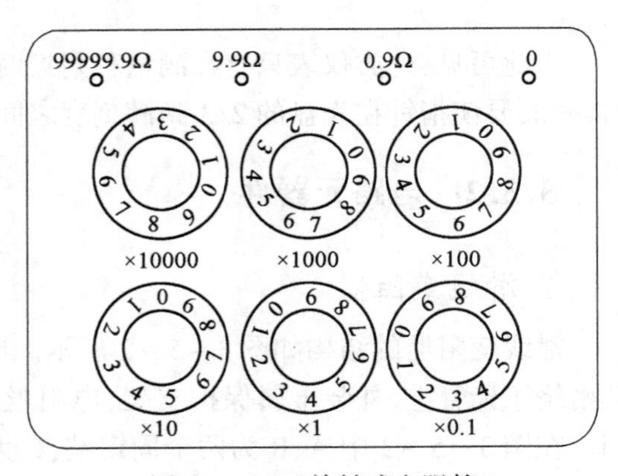

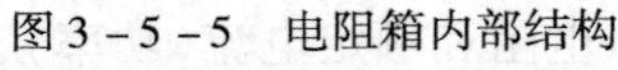

图 3－5－5　电阻箱内部结构　　　　图 3－5－6　旋转式电阻箱

电阻箱的规格由总电阻（即最大电阻）和额定功率（有的电阻箱给出额定电流）表示。根据 JB/T 8225—1999《实验室直流电阻器》标准，电阻器的准确度等级分为 9 级，用 a 以基准值

的百分数表示为 $a = 0.0005$、0.001、0.002、0.005、0.01、0.02、0.05、0.1、0.2，对于多十进电阻器（电阻箱）还有 4 级，即 $C = 0.5$、1、2、5。电阻箱的仪器误差限为

视频 3-5-3 电阻箱的使用

$$\Delta_{\text{ins}} = \sum_{i=1}^{n} R_i \times a_i\% + R_0$$

其中，R_i、a_i分别是电阻箱各旋钮的示值和准确度等级。例如，ZX21 型直流多值电阻器，6 个旋钮的倍率分别为 ×0.1、×1、×10、×100、×1000、×10000，对应的准确度等级分别为 $a = 5$、0.5、0.2、0.1、0.1、0.1，其零位误差为 $R_0 = 0.03\Omega$。当环境温度、相对湿度、消耗功率、位置都满足参考条件时，若电阻器的示值为 1684.3Ω，则仪器误差限：

$$\Delta_{\text{ins}} = 1600 \times 0.1\% + 80 \times 0.2\% + 4 \times 0.5\% + 0.3 \times 5\% + 0.03 = 1.8(\Omega)$$

电阻箱使用时，应使电路电流小于各挡步进电阻所规定的额定电流值。要根据测量需要选择合适的接线柱即测量量程，以避免电阻箱其余连接部分的导线电阻。

3.5.3　电源

实验中用的电源有直流和交流两种。交流电源用符号“～”或“AC”表示，实验中若对电压稳定性要求较高，通常需接稳压器。直流电源用符号“－”或“DC”表示，通常用的直流电源有直流稳压电源、干电池和蓄电池等。一般用“＋”或红色表示正极，用“－”或黑色表示负极，使用中不能将正负极接反，严禁将正负极短路。

3.5.4　开关

在电学实验中，常用开关来实现电路的导通、断路或改变电流的方向。实验室常用的开关有单刀单掷、单刀双掷、双刀双掷开关、双刀换向开关和按钮开关等。其符号如图 3-5-7 所示。

(a)单刀单向　(b)单刀双向　(c)双刀双向　(d)双刀换向　(e)按钮开关

图 3-5-7　实验室常用开关符号

3.6　常用电子仪器

3.6.1　直流稳压电源

实验室里使用的直流电源除了蓄电池和干电池以外，还广泛应用晶体管稳压电源。实验室用的直流稳压电源是将 220V 的交流电经过降压、整流和稳压之后输出的从 0 到某一电压值连续可调的直流电压，也有只输出不可调单一电压值的。它是一种通用的电子仪器。直流稳压电源的最大特点是输出性能好，内阻小，纹波系数小，稳定度高。有的直流稳压电源有双路输出，适当地连接，可以分别获得正负电源电压，在电子线路实验中是非常方便的；有的直流稳压电源具有短路（或过载）自动保护装置，当输出电流超过额定值时，它会自动保护，将输出电压降到 0，只有将故障排除并“复位”之后，才能正常工作。

晶体管稳压电源在使用时，应先接入 220V 交流电压，再接通电源开关，然后拨动粗调开关至所需电压挡，调节细调旋钮，将电压表读数调到所需电压。使用时要注意正、负端，正确接线。

3.6.2 万用表

万用表(万用电表的简称)是电子制作中必备的测试工具。它具有测量电流、电压和电阻等多种功能,有的万用表还可以测量晶体管的主要参数及电容器的电容量等。常用的万用表有指针式万用表和数字万用表两种。指针式万用表是以表头为核心部件的多功能测量仪表,测量值由表头指针指示读取;数字万用表的测量值由液晶屏幕直接以数字的形式显示,有些还带有语音提示功能。

量程和灵敏度是万用表的主要性能参数。万用表表头的灵敏度是指表头指针由零刻度偏转到满刻度时,动圈中通过的电流值。一般万用表中均不直接标明表头的灵敏度为多少微安,而用其满刻度指示时测试电流的倒数Ω/V来表示。Ω/V越大表明表头满刻度电流值越小,表头灵敏度越高。满刻度时电流的大小对电子电路测量准确度至关重要。万用表的量程是指仪表最小到最大的测量范围,半导体器件的发展使万用表电压、电流量程范围更大,最低电压量程有2.5V挡位,最小电流量程有25~50μA挡位,一般万用表交、直流电压最高量程为500V,特殊型号可达2.5 kV。

1. 指针式万用表

指针式万用表种类很多,外形各异,但基本结构和使用方法是相同的。一般来讲,万用表面板上都有表头、选择开关、欧姆挡调零旋钮及表笔插孔。

万用表的表头是一个灵敏电流计,表盘上印有多种符号、刻度线及数值。符号A-V-Ω表示这只电表是可以测量电流、电压和电阻的多用表,符号"-"或"DC"表示直流,"~"或"AC"表示交流,"≃"表示交流和直流共用。刻度线下的几行数字是与选择开关的不同挡位相对应的刻度值,其中右端标有"Ω"的是电阻刻度线(右端为零、左端为无穷大),刻度值分布是不均匀的。万用表的选择开关是一个多挡位的旋转开关,用来选择测量项目和量程。

1)测量方法

一般的万用表测量项目包括直流电流、直流电压、交流电压及电阻等多种物理量,每个测量项目又划分为几个不同的量程以供选择。不同量的测量方法如下:

(1)直流电流测量:将红色表笔插入有"+"号的插孔,黑色表笔插入有"-"号的插孔。转动转换旋钮至电流功能挡和所需量程,将表笔按正确方向串联于待测电路中。

(2)交、直流电压测量:同上述方法插好表笔,转动转换旋钮到所需位置,将表笔两端并接于待测电压的两端。

(3)电阻测量:同上述方法插好表笔,转换开关旋钮调至电阻挡,先使两表笔短路,指针向满刻度偏转,轻轻调动"Ω调零电位器"调节旋钮,使指针指零欧姆,再将表笔两端接于待测电阻的两端。

2)注意事项

(1)每次测量前应把万用表水平放置,观察指针是否指零,指针不指零时用旋具稍微调整表头的机械零点螺钉,使指针指零。红、黑色表笔应正确插入万用表插孔。转换旋钮应放置在所要测量电参量的量程挡上,决不可误放。

(2)如果不清楚所测电压、电流值的大概范围,应首先用表上的最大电压挡、最大电流挡估测,然后再改用适当的量程测量。

(3)如果不清楚被测电路的正、负极性,可将转换旋钮调至最大量程,测量时用表笔轻轻碰一下被测电路，同时观察指针的偏转方向，从而确定出电路的正、负极。

(4)如果不清楚所要测量的电压是交流还是直流电压，可先用交流电压的最高挡来估测，得到电压的大概范围，再用适当量程的直流电压挡进行测量，如果此时表头表针不发生偏转，断定此电压为交流电压，若有读数则为直流电压。

(5)测量电流、电压时，正确的量程应该使表头指针指示在大于量程一半以上的位置，此时所得结果误差较小。

(6)测量高阻值电阻时，不要用双手接触电阻的两端，以免将人体电阻并联到待测电阻上。测量装在仪器上的电阻时应关掉仪器电源，将电阻的一端与电路焊开再进行测量。如电路待测部分有容量较大的电容存在，应先将电容放电后再测量。

(7)测量电阻时，每改变一次量程，都要重新调整零欧姆旋钮。如发现调整零欧姆旋钮不能使指针指零欧姆，不应使劲扭旋钮，而应更换新电池。读数时两眼垂直观察指针，不应斜视。

(8)实验完毕，应把转换旋钮放到交流电流电压最高挡处。长时间不用时应将电池从表中取出，把万用表放置在干燥、通风、清洁的环境中。

2. 数字万用表

数字万用表是用途广泛的测量仪表。其工作原理是将测量的模拟信号(如电压、电流、频率等)，经过处理(如放大、整流、分压、分流等)后，再转换为数字信号(A/D 转换)，显示在屏幕上。数字万用表与一般指针式万用表相比，具有功能多、输入阻抗高、灵敏度高、读数方便、体积小、重量轻、携带方便等优点。缺点是随着测量精度及分辨率的提高，测量反应速度明显下降。

数字万用表的显示位表示其能读出的最多有效数字的多少。能读出的有效数字越多，说明其测量精度越高。如 4½(称为四位半数字)，其最大显示位数是 199.99 (共五位)，但最高位只能显示“0”或“1”，称为半位。其他四位均可显示 0 ~ 9 的数字，所以称其为四位半数字表。如果此时是在 200V 直流电压挡读数，由表 3 − 6 − 1 可查出 200V 直流电压挡的准确度为：±(0.1% 读数 ±2 个字)。其测量结果计算如下：

$$U = \text{读数} \pm (0.1\%\,\text{读数} \pm 2\,\text{个字}) = 199.99 \pm (0.1\% \times 199.99 \pm 0.02)$$
$$= 199.99 \pm (0.40 \pm 0.02) = 199.9 \pm 0.4\,(\mathrm{V})$$

结果表示为$(1.999 \pm 0.004) \times 10^2\,(\mathrm{V})$。

表 3 − 6 − 1　常见电气仪表面板上的标记

名　　称	符　　号	名　　称	符　　号
指示测量仪表的一般符号	○	磁电系仪表	⋂
检流计	⊕	静电系仪表	÷
安培表	A	直流	—
毫安表	mA	交流(单相)	~
微安表	μA	直流和交流	≃
伏特表	V	以标度尺量限百分数表示的准确度等级，例如 1.5 级	1.5
毫伏表	mV	以指示值的百分数表示的准确度等级，例如 1.5 级	(1.5)

续表

名　称	符　号	名　称	符　号
千伏表	kV	标度尺位置为垂直的	
欧姆表	Ω	标度尺位置为水平的	
兆欧表	MΩ	绝缘强度试验电压为2kV	2
负端钮	–	接地用的按钮	
正端钮	+	调零器	
公共端钮	*	二级防外磁场及电场	

1)使用方法

(1)测交直流电压、电流、电阻、频率、二极管及通断时,黑表笔插入 COM 插孔,红表笔插入 VΩHz 插孔。

(2)测交直流电流时,黑表笔插入 COM 插孔,当被测电流在 200mA 以下时,红表笔插入 A 插孔;如被测电流在 200mA ~20 A 之间,则将红表笔移至 20 A 插孔。

(3)测电容、三极管有其相应的插座,必要时应注意极性。并选好测量功能和挡位。如不知被测量的范围时应选高量程挡逐步调低。在测量不断变化的信号时,可用 HOLD 键使数据保持,以方便读数。

2)注意事项

(1)被测值高于所选挡位时,只在最高位显示“1”,表示超量限。这时应选高一量程的挡位。不要测高于测量功能最高挡位的电压、电流值。

(2)在最前面位显示“ – ”时,表示此时红表笔接的是信号负端,黑表笔接的是信号正端。

(3)用电阻挡、二极管挡测量时,红表笔为表内电池正端,黑表笔端为表内电池负端,这正好与指针式万用表相反。

(4)在测量电容时,要对电容放电后才能进行测量(特别是大容量电容)。

(5)数字万用表用完后应将量程置最大电压挡并关闭电源。

3.7　常用光学仪器

光学实验是物理实验不可分割的一部分。在实验中,常常使用读数显微镜、标尺望远镜、测量显微镜、分光计和迈克尔逊干涉仪(见实验 20)等光学仪器。

3.7.1　标尺望远镜

标尺望远镜是观测远处标尺读数的一种光学仪器。标尺望远镜由底座、内调焦望远镜、可调毫米尺组件和反光镜 M 等组成。内调焦望远镜结构如图 3 – 7 – 1 所示。为了调节和测量,在物镜和目镜之间装有叉丝分划板和内调焦透镜,叉丝和分划板固定在筒 B 上,内调焦透镜由微调旋钮带动齿条,使其在 A 中沿轴线前后移动,目镜 C 则装在筒 B 里,并可沿筒前后移动以改变目镜与叉丝分划板的距离。

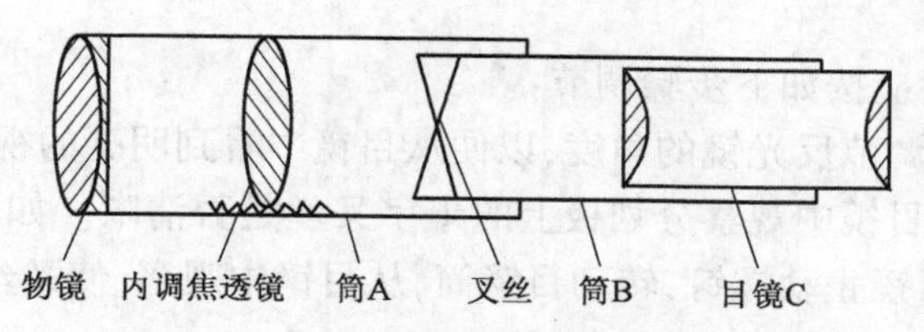

图 3－7－1　内调焦望远镜结构

标尺望远镜常与光杠杆配套使用,使用方法详见实验 21。使用过程中应注意:勿用手抚摸目镜与物镜;调试仪器时,一定要用手先托住望远镜的移动部分,然后再松开锁紧手轮,以免望远镜沿立柱下滑撞击底座受损;各手轮及螺旋和可动部分如发生阻滞不灵现象,应查明原因。在原因未查清之前,切勿过分扭扳,以防损坏仪器或构件。

3.7.2　测量显微镜

测量显微镜也是光学仪器的一种,它结构简单,操作方便,可以用它在直角坐标系中测量长度,如测定孔距、刻线距离、刻线宽度、狭缝宽度等,还可以用转动度盘测量角度,也可以用作观察显微镜。

测量显微镜(JLC 型)的结构及其光路原理图分别如图 3－7－2、图 3－7－3 所示,其中显微镜原理与一般其他显微镜原理相同。操作时,当旋转 x 轴测微鼓轮时,载物台将沿 x 轴方向移动,由于测微鼓轮边上有刻线为 100 等分,螺杆螺距是 1mm,所以测微鼓轮每转动一格相当于移动 0.01mm 距离。y 轴方向测微器边上刻线为 50 等分,而螺距为 0.5mm,每转一格也相当于移动 0.01mm。测量工作台圆周上刻有角度值,绕垂直轴旋转后有游标读数,游标每格示值为 6′。反光镜装在底座上,根据光源方向可以四面转动,以得到明亮的视场。

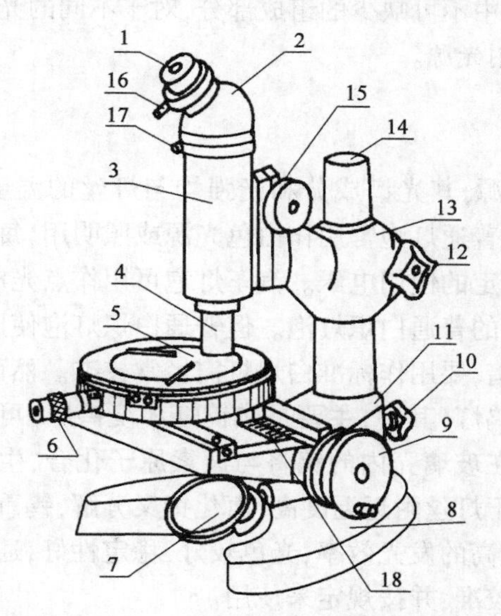

图 3－7－2　JLC 型测量显微镜外形图

1—目镜;2—棱镜座;3—镜筒;4—物镜;5—载物台;
6—Y 轴测微器;7—反光镜;8—底座;9—X 轴测微器;
10—平台固定螺钉;11—平台;12—支架固定螺钉;13—支架;
14—立柱;15—调焦手轮;16—目镜止动螺钉;
17—棱镜座止动螺钉;18—X 轴螺杆

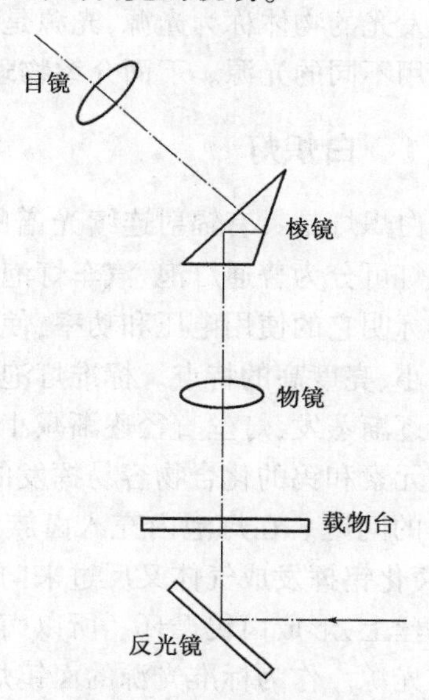

图 3－7－3　JLC 型测量显微镜光路图

测量显微镜使用时应按如下步骤调节：

(1)调节反光镜。调节反光镜的角度，以便从目镜中看到明亮的视场。

(2)调节叉丝。从目镜中观察分划板上的十字叉丝是否清晰。如不清晰，微微转动目镜，使叉丝最清晰。松开目镜止动螺钉，转动目镜筒，从目镜中观察，使叉丝分别与 x、y 轴平行，然后止动螺钉。

(3)调焦。把被测物放在载物台上，先从侧面观察，旋转调焦手轮使显微镜下移接近待测物(但不能触及)。然后从目镜中观察，同时旋转调焦手轮使显微镜缓缓上移(不得向下移动，以防止物镜镜头与被测物接触而损坏)，直到看到被测物的清晰图像为止。

(4)测量与读数。例如测定小钢珠的直径，将待测钢珠放在载物台上，旋转 x 与 y 轴的测微鼓轮，使横竖叉丝与珠的像同时相切，即始终保证叉丝与钢珠的像沿着待测直径方向移动。从标尺和测微鼓轮上读出初始相切时的读数 x_0，然后移动 x 轴测微鼓轮，使叉丝与钢珠像的另一侧相切，读出读数 x，则钢珠直径为两次读数之差，即 $d = x - x_0$。

测量显微镜使用过程中应注意以下事项：

(1)测量时要单方向移动鼓轮，避免因存在螺纹间隙而产生的空程误差。

(2)物镜和目镜不得用手抚摸，不得用其他任何纸片或布块擦拭。

(3)在调节过程中要注意 x 轴、y 轴方向的量程，当载物台移到尽头时，绝对不能再继续旋转测微器的旋转柄和鼓轮，否则将要损坏螺杆的精密螺纹。

3.8 常用光源

发光的物体称为光源，光源是光学实验系统中不可缺少的组成部分，对于不同的光学实验常使用不同的光源。下面介绍物理实验中的常用光源。

3.8.1 白炽灯

白炽灯是具有辐射连续光谱的一种复色光源，其光谱成分和光强均与灯丝的温度有关。白炽灯可分为普通灯泡、汽车灯泡和标准灯泡。普通灯泡常用作白色光源或照明用，每个灯泡上都标明它的使用电压和功率，使用时应注意规定的使用电压。汽车灯泡可用作点光源，具有线度小、亮度高的特点。标准灯泡是经过校准后的普通白炽灯泡。但普通白炽灯泡使用后，钨丝会逐渐蒸发，灯丝直径逐渐减小，灯泡越用越暗，要用作标准白炽灯需经常校准。然而，利用卤素元素和钨的化合物容易挥发的特点制成卤钨灯(其中，主要是碘钨灯和溴钨灯)可以解决这样的问题。在灯泡内充入卤族元素后，沉积在玻璃壳内的钨将与卤素原子化合，生成卤化钨，卤化钨挥发成气体又反过来向灯丝扩散，由于灯丝附近温度高，卤化钨又分解，钨重新沉积在钨丝上，形成卤钨循环。所以卤钨灯能获得较高的发光效率，光色较好，稳定性好，通常作为标准光源。作为标准光源的卤钨灯仍然要经过校准，并按规定来使用。

3.8.2 水银灯

水银灯又称汞灯，是一种气体放电光源，发光物是水银蒸气。稳定后发出绿白色光，在可见光范围内的光谱成分是几条分离的谱线。

由于水银灯在常温下要有很高的电压才能点燃，因此在灯管内还充有辅助气体，如氖、氩等。通电时辅助气体首先被电离而开始放电,使灯管温度得以升高,汞逐渐气化而产生水银蒸气的弧光放电。弧光放电的伏安特性曲线有负阻现象,需要在电路中接入一定的阻抗以限制电流,否则电流急剧增加会把灯管烧坏。一般在交流 220V 电源与灯管的电路中串入一个扼流圈用来限流。不同的水银灯泡电流的额定值不同,所需扼流圈的规格也不同,不能互用,切忌弄混。水银灯点燃后一般要经 5 ~ 15min 后才能稳定发光,水银灯辐射的紫外线较强,不要用眼直接注视,以防眼睛受伤。

3.8.3 钠光灯

钠光灯也是一种气体放电光源。在可见光范围内有两条强谱线(5.890×10^{-7}m、5.896×10^{-7}m),因此是一种比较好的单色光源。

这种灯是将钠封闭在特种玻璃泡内,且充以辅助气体氩,发光过程类似水银灯,钠为难熔金属,冷时蒸气压很低,工作时钠蒸气压约为 0.133Pa,通电 15min 后可发出较强的黄色光。灯泡两端电压约为 20V(AC),电流为 1.0 ~1.3 A。电源用 220V(AC)并与水银灯一样需串入扼流圈或电子镇流器。

3.8.4 激光器

激光器是 20 世纪 60 年代发展起来的一种新型单色光源,具有发光强度大、单色性好、相干长度大和方向性好等优点。激光器由工作物质、激励装置和光学谐振腔三部分组成,按其工作物质可分为固体、液体、气体、半导体等类激光器。

氦氖激光器是实验室常用的一种气体激光器,输出波长为 6.328×10^{-7}m 的橙红色偏振光,输出功率在几毫瓦到几十毫瓦之间。激光管两端是多层介质膜片,管体中间有一毛细管,它们组成光学谐振腔,这是激光器的主要部分,使用时必须保持清洁。点燃时应严格按说明书的要求控制辉光电流的大小,不得超过额定值。若低于阈值则使激光闪烁或熄灭。在光学实验中,可以利用各种光学元件将激光管射出的激光束进行分束、扩束或改变激光束的方向满足不同的实验要求。

由于氦氖激光管两端加有高压(1200 ~8000V),操作时应严防触及,以免造成电击事故。由于激光束的能量高度集中,绝对不能迎着激光束的方向观察。照射到人眼中的未扩束的激光,将造成视网膜的永久损伤。

第4章 实验项目

实验1 用天平测量物体的质量和密度

[引言]

质量是国际计量大会规定的七个基本物理量之一。从质量的内涵来看,质量包括三方面的内容:一是表征物体惯性的大小;二是表征物体间引力的大小;三是表征物体能量的多少。国际单位制中质量的单位是千克(kg),本实验室用天平测定物体质量即天平法。目前在实验和教学科研中常用的测量质量方法主要有:用(各种)天平测定物体质量、运用牛顿第二定律测质量、用动量守恒定律测质量、用单摆和弹簧秤测质量、用万有引力定律推测天体质量、用匀强电场测带电粒子的质量、用质谱仪测带电微粒的质量、用理想气体状态方程测气体的质量等。

[实验目的]

(1)掌握物理天平和电子天平的调整和使用方法。

(2)用流体静力称衡法测固体的密度。

[预习问题]

(1)物理天平有哪几部分构成?各部分作用是什么?

(2)电子天平使用中要注意哪些问题?

[实验仪器]

物理天平、电子天平、待测物体、细线、烧杯、水、温度计。

[实验原理]

1. 测定规则形状固体的密度

设有天平测出固体质量为 m,用长度测量仪器测出长度,求出固体体积为 V,则该固体的密度为 $\rho=\dfrac{m}{V}$。

2. 测定不规则形状固体的密度

对于密度大于水且不溶于水的不规则形状固体,其质量可用天平来测定,但体积要用流体静力称衡法来确定。具体方法如下:

用天平称出物体在空气中的质量 m,然后用一根细线把物体拴好浸入水中(图 S1-1),用天平称出物体悬浮在水中的质量为 m_1,则物体在水中受到的浮力为

$$F=(m-m_1)g$$

根据阿基米德原理:浮力的大小等于物体排开同体积水的重量,即

$$F=\rho_{水}Vg$$

联立上两式，得物体体积为

$$V=\frac{m-m_1}{\rho_{水}}$$

由此可得物体的密度

$$\rho=\frac{m}{V}=\frac{m}{m-m_1}\rho_{水}$$

式中，$\rho_{水}$为水在 t℃时的密度，见书后附表 18。

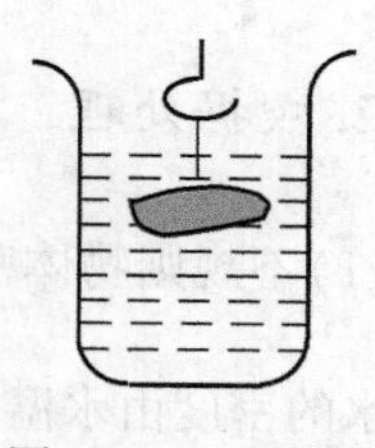

图 S1－1　不规则物体密度测量

［实验内容与步骤］

(1) 调节物理天平和电子天平。

(2) 测定规则形状固体的密度：将待测长方体放在左盘上，在右盘上加砝码，称出待测物质量 m；实验 1 已测出体积 V，代入公式即可求出该固体密度。

(3) 用流体静力称衡法测不规则形状固体的密度。

①将待测物体放在左盘上，在右盘上加砝码，称出待测物质量 m。

②将烧杯装上水，放在托盘 G 上，待测物用细线悬吊在水中，称出待测物在水中的质量 m_1。

③用温度计测出水的温度。

④将数据填入表 S1－1 中。

［注意事项］

(1) 砝码切忌用手拿，以免沾汗锈蚀，应用镊子夹持，用完后应随即放回砝码盒中。

(2) 调节底板水平，仔细调节底座上的三个水平螺钉，使水准器中的气泡位于中心。

(3) 天平的载荷量不能超过天平的最大称量值。取放物体、砝码和移动游码或调节天平时，必须切记应在天平制动后进行，以免损坏刀口和其他部件。旋转制动旋钮，应小心均匀地旋转，并应在天平指针接近标尺中间分度时进行。

(4) 刀口处、游码杆和螺钉等处应常滴钟表油，以防生锈。

(5) 酸、碱、油脂或其他化学药物不能直接放在秤盘上称量。

(6) 天平宜放在干燥、清洁而稳定的环境中，不要经常搬动。

(7) 电子天平使用时，不能频繁开启电源，调节按键时用力适当。

［数据及处理］

1. 数据记录（表 S1－1）

表 S1－1　用流体静力称衡法测不规则形状固体的密度　　单位：g

项目	空气中质量 m		浸入水中时质量 m_1	
次数	m_i	$\Delta m_i=m_i-\overline{m}$	m_{1i}	$\Delta m_i=m_{1i}-\overline{m_1}$
1				
2				
3				
4				
5				
平均值	$\overline{m}=$	$S_m=$	$\overline{m_1}=$	$S_{m_1}=$

2. 数据处理

(1) 不规则物体密度的平均值为： $\bar{\rho}=\dfrac{\bar{m}}{\bar{m}-\bar{m}_1}\rho_{水}$

水的密度由水温根据书后附表 18 查出。

(2) 不规则物体密度的标准不确定度 u_ρ。

先求相对不确定度，再求 u_ρ 较方便。

相对不确定度为： $\dfrac{u_\rho}{\bar{\rho}}=\sqrt{\dfrac{m_1^2}{m^2(m-m_1)^2}u_m^2+\dfrac{1}{(m-m_1)^2}u_{m1}^2+\dfrac{1}{\rho_{水}^2}u_{\rho水}^2}$

不规则物体密度的标准不确定度为： $u_\rho=\dfrac{u_\rho}{\bar{\rho}}\times\bar{\rho}$

(3) 测量结果的表示

$$\rho\pm 2u_\rho=\_\_\_\_\_\_\text{单位}\qquad (P=0.95)$$

$$\frac{2u_\rho}{\bar{\rho}}\times 100\%=\_\_\_\_\_\_\qquad (P=0.95)$$

3. 结果分析

通过用物理天平测量物体质量和密度的实验，你所得出的有关结论有哪些？

[问题讨论]

(1) 使用物理天平和电子天平应注意哪几点？怎样消除天平两臂不等而造成的系统误差？

(2) 分析造成本实验误差的主要原因有哪些？

实验 2　刚体转动惯量的测定

[引言]

转动惯量是描述刚体转动惯性大小的物理量，大小不仅取决于刚体的总质量，而且与刚体的质量分布、形状大小和转轴位置密切相关。对于几何形状规则，质量均匀分布的刚体，可以通过数学方法计算出它绕特定转轴的转动惯量，但对于形状复杂，或质量分布不均匀的刚体，用数学方法计算其转动惯量是非常困难，通常采用实验方法来测定其转动惯量。

本实验是从刚体转动定律出发，选定直接施以外力矩使刚体发生定轴转动的物理过程，建立起测量刚体转动惯量的实验方案，并对刚体绕定轴转动规律进行实验验证，同时测定刚体转动惯量。

[实验目的]

(1) 学习用恒力矩转动法测定刚体转动惯量的原理和方法。

(2) 观测刚体的转动惯量随其质量、质量分布及转轴不同而改变的情况，验证刚体转动定律。

(3) 学会用作图法进行线性拟合的数据处理方法。

[预习问题]

(1) 如何利用恒力矩转动法测定刚体转动惯量？

(2)什么是平行轴定理？本实验如何验证平行轴定理？

[**实验仪器**]

刚体转动惯量实验仪、砝码、电子秒表、米尺、螺丝刀等。

[**实验原理**]

1. 实验装置

刚体转动惯量实验仪如图 S2－1 所示，为减小转动时的摩擦力矩，绕线塔轮通过特制的轴承安装在主轴上。塔轮半径为 15mm、20mm、25mm、30mm 和 35mm 共 5 挡，可与砝码托及 5 个砝码组合，产生大小不同的力矩。载物台用螺钉与塔轮连接在一起，随塔轮转动。测试样品有 1 个圆盘、1 个圆环、两个圆柱，试样上标有几何尺寸及质量，便于将转动惯量的测试值与理论计算值比较。圆柱试样可插入载物台上的不同孔，这些孔到中心的距离分别为 45mm、60mm、75mm、90mm 和 105mm，便于验证平行轴定理。铝制小滑轮的转动惯量与实验台相比可忽略不计。一只光电门做测量，一只备用，可通过智能计时计数器上的按钮方便地切换。

2. 恒力矩转动法测定转动惯量

当刚体绕定轴转动时，根据转动定律，刚体的角加速度 β 与刚体所受的合外力矩 M 成正比，即

$$M = I\beta \qquad (S2-1)$$

式中，I 为刚体对该定轴的转动惯量。

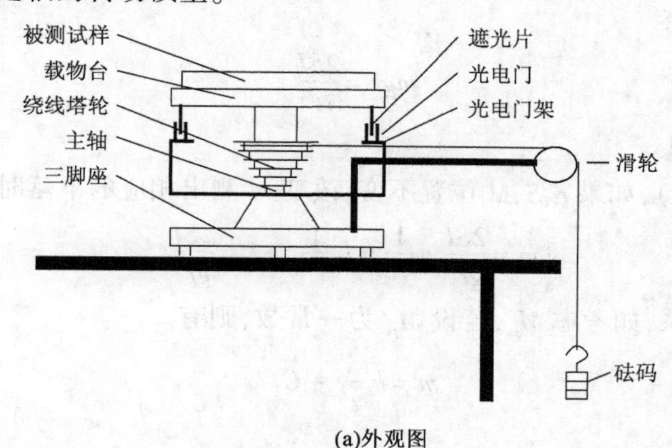

(a)外观图

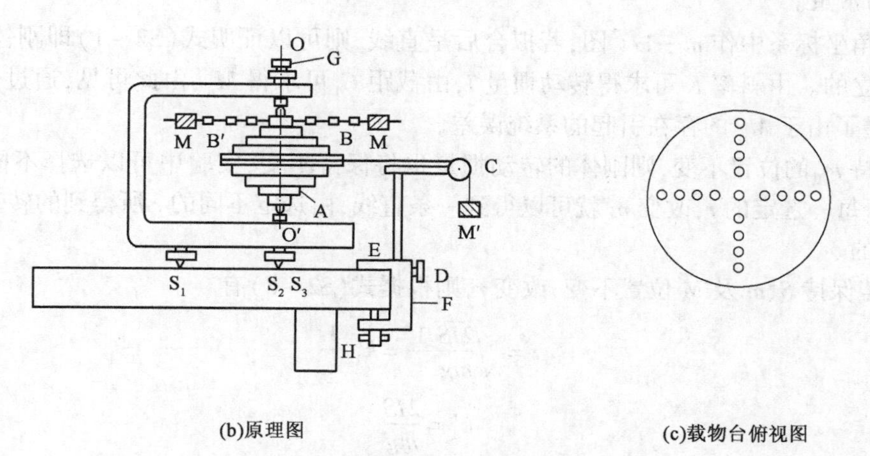

(b)原理图

(c)载物台俯视图

图 S2－1　刚体转动惯量实验仪

本实验装置中，刚体所受的合外力矩 M 为细线给予的力矩 T_r 和摩擦力矩 M_μ 之和，T_r 为细线的张力，与 OO′（塔轮中心轴）垂直，r 为塔轮的绕线半径，合外力矩为

$$M = T_r - M_\mu \tag{S2-2}$$

当略去细线、滑轮的质量及滑轮轴上的摩擦力，并认为细线的长度不变时，重物 M′（质量为 m）以匀加速度 a 下落，根据牛顿第二定律，则砝码下落的运动方程为

$$T = m(g - a) \tag{S2-3}$$

实验时，让重物 M′由静止开始自由下降，由于重物 M′作初速为 0 的匀加速运动，设重物 M′下降 S 距离的时间为 t，由运动学方程

$$S = \frac{1}{2}at^2 \tag{S2-4}$$

$$\beta = \frac{a}{r} \tag{S2-5}$$

因此根据转动定律：

$$m(g - a)r - M_\mu = \frac{2SI}{rt^2} \tag{S2-6}$$

在实验过程中，保持 $a \ll g$，则有

$$mgr - M_\mu \approx \frac{2SI}{rt^2} \tag{S2-7}$$

若 $M_\mu \ll mgr$，略去 M_μ 有

$$mgr = \frac{2SI}{rt^2} \tag{S2-8}$$

下面讨论两种情况：

(1)根据式(S2-8)，如果 r、S、M 位置不变，改变 m 测出相应的下落时间 t，有

$$m = \frac{2SI}{gr^2} \cdot \frac{1}{t^2} = k\frac{1}{t^2} \quad k = \frac{2SI}{gr^2} \tag{S2-9}$$

即有 m 与 t^2 成反比关系，如考虑 M_μ，若设 M_μ 为一常数，则有

$$m = k\frac{1}{t^2} + C_1 \tag{S2-10}$$

式中，C_1 为常量。

在直角坐标系中作 m—$1/t^2$ 图，若拟合后呈直线，则可以证明式(S2-1)即刚体定轴转动定律是成立的。由斜率 K 可求得转动惯量 I，由截距 C_1 可求得 M_μ，由此可见，通过作图（线性拟合）回避了由于 M_μ 的存在引起的系统误差。

若保持 m_0 的位置不变，则刚体的转动惯量也将保持不变，实验中可以选择不同的 r 进行实验，对于每一选定的 r，改变 m 就可以得到一条直线，由选定不同的 r 所得到的转动惯量 I 应该是相同的。

(2)若保持 S、m 及 M 位置不变，改变 r，则根据式(S2-8)有

$$r = \sqrt{\frac{2IS}{mg}}\frac{1}{t} = k_1'\frac{1}{t} \tag{S2-11}$$

其中
$$k_1' = \frac{2IS}{mg}$$

由式(S2-11)可知：r 与 t 成反比，若考虑 M_μ，并设 M_μ 不变，则有

$$r = \frac{2IS}{mg} \frac{1}{t^2 r} + \frac{M_\mu}{mg} = k_2' \frac{1}{t^2 r} + C_2 \qquad (S2-12)$$

其中

$$C_2 = \frac{M_\mu}{mg}$$

在直角坐标系中作 $r—1/t$ 图或 $r—1/(t^2 r)$ 图，若拟合为一直线，说明式（S2－1）即刚体转动定律是成立，由 k_1'、k_2' 可求 I，由 C_2 可得 M_μ。

［实验内容与步骤］

（1）安装调试实验装置：取下塔轮，换上铅直准钉，调节塔轮架的底角螺钉使转轴 OO′铅直，装上塔轮尽量减小摩擦，使其转动自如后用固定螺钉 G 固定。仪器调试好后，在实验中不得再移动，以保证在实验过程中摩擦力不变。

（2）将细线绕在 $r = 25$mm 的轮子上，绕线尽量密排，将圆柱试样 m_0 放置 105mm 处的小孔中。让物体 M 从 F 处由静止开始下落，下落高度预先确定在 75～100cm 间，物体质量从 10.00g开始以 5.00g 为增量逐次增至 40.00g（砝码和托盘质量均为 5.00g），用秒表记下物体下落时间，每加一个砝码，测 t 值 3 次，求其平均，要求三个 t 值中任何两个相差不大于 0.1s。

（3）改变 M 的位置，对称地将 M 向内移动到 75mm 处的小孔中，重复步骤（2）。

（4）保持物体下降距离 S 不变，同时物体质量保持在 $m = 20.00$g，改变 r，r 分别取值 10mm、15mm、20mm、25mm、30mm，对每一半径 r，分别测量物体下降时间 t，并且测量三次，求其平均值。

［数据及处理］

1. 数据记录（表 S2－1、表 S2－2）

表 S2－1　实验数据记录表格

m/g	t_1/s	t_2/s	t_3/s	$\bar{t}$/s	$\bar{t}^2$/s^2	$\frac{1}{\bar{t}^2}$/s^{-2}
10.00						
15.00						
20.00						
25.00						
30.00						
35.00						
40.00						

表 S2－2　实验数据记录表格

r/mm	t_1/s	t_2/s	t_3/s	$\bar{t}$/s	$\bar{t}^2$/s^2	$\frac{1}{r\bar{t}^2}$/$cm^{-1} \cdot s^{-2}$
15						
20						
25						
30						
35						

2. 数据处理

(1) 在同一坐标系中做出 M 在不同位置的两组数据的 $m—1/t^2$ 曲线(m 为纵坐标)。分别求出两条直线的斜率 K_1、K_2(在直线上任选两点计算,而不是用实验点计算),根据得出的斜率计算出刚体的转动惯量 I_1、I_2。

(2) 在直角坐标系中作出 $r—\frac{1}{rt^2}$ 直线(r 为纵轴),用直线上两点求斜率 K_2',并根据 K_2' 计算出刚体转动惯量 I。

3. 结果分析

对实验结果进行分析,并得出实验结论。

[注意事项]

(1) 用塔轮上的不同半径 r 做实验时,一定要上下调节滑轮的位置,以保证细线从塔轮绕出后总是与转轴 $\overline{OO'}$ 垂直,同时要使滑轮与细线在同一平面内。

(2) 塔轮转动时,存在摩擦力矩的,在实验中应尽量保持摩擦力不变,已调好的装置在实验中不能再任意变动。

(3) 调节仪器转轴与支承面垂直,调整滑轮支架的位置、高低,使塔轮绕线水平跨过滑轮,塔轮转轴不能固定太紧,也不能固定太松,以尽量减少摩擦。

视频 S2　刚体转动惯量的测定

(4) 砝码开始下落时,一定要做到初速度为零。且要保证 $g \gg a$ 这一实验条件。

[问题讨论]

(1) 为什么说 $m—1/t^2$ 曲线是直线就验证了刚体定轴转动定律?

(2) 试分析实验中存在哪些系统误差?

刚体的惯量的测定如视频 S2 所示。

实验 3　用拉伸法测量金属丝的杨氏模量

[引言]

固体材料受外力作用时会发生形变,而形变有弹性形变和塑性形变之分。弹性形变是指在外力作用撤销后能恢复原状,而塑性形变不能恢复原状。本实验测量金属丝伸长的纵向弹性模量,也称杨氏模量,它是表征固体材料的弹性性质的物理量。理论上是金属丝伸长应变为 1 时,单位面积所受的力。但是与伸长应变有关的伸长量是微小的变化量,仅为 10^{-2} mm 量级,因此,拉伸法测定金属丝杨氏模量的实验关键是借助光学放大法,即利用光杠杆将微小伸长量放大后进行测量的方法。

[实验目的]

(1) 学习测量杨氏模量的方法。

(2) 学习调整铅直水平、望远镜聚焦以及光学共轴等高调节等方法。

(3) 学习用逐差法、作图法及图解法处理数据的方法。

[预习问题]

(1) 如何调节望远镜和光杠杆反射镜的组合光路系统使其达到测量要求?

(2) 熟悉光杠杆测量微小长度变化量的原理和方法。

(3)本实验测量各种长度时,该如何选配测量器具?

[实验仪器]

YMC—Ⅱ型杨氏模量测定仪一套、光杠杆反射镜、望远镜及标尺、激光瞄准装置、照明标尺装置、砝码、钢丝、千分尺、游标卡尺、钢卷尺。

[实验原理]

实验装置如图 S3-1、图 S3-2 所示。一根原长为 L,截面积为 S,直径分布均匀的金属丝(或棒),受到长度方向的外拉力 F 的作用时而发生形变,伸长为 ΔL。根据胡克定律,在弹性限度内,其应力与应变成正比,即

$$\frac{F}{S}=E\frac{\Delta L}{L} \qquad (S3-1)$$

式中,比例系数 E 称为该金属材料的杨氏模量,它只取决于材料的弹性性质,而与其长度 L、截面积 S 无关。它的单位为 N/m^2(Pa)。设金属丝直径为 d,则截面面积 $S=\frac{1}{4}\pi d^2$,代入式(S3-1)得出杨氏模量为

$$E=\frac{4FL}{\pi d^2\Delta L} \qquad (S3-2)$$

式中,F、L、d 较容易测量,而 ΔL 是一个微小的长度变化量。对于这样一个微小伸长量,很难用普通的测长仪器测量,因而,本实验采用光杠杆的光学放大原理来测量 ΔL。

光杠杆法测量装置包括光杠杆反射镜、望远镜及标尺。光杠杆反射镜如图 S3-3 所示,镜架上装有平面全反射镜 M,镜架前下方有一刀口,后方装有主杠 T 型支脚。镜面倾角及主杠尖脚到刀口间的距离均可调整。测量时,刀口放在平台槽沟内,主杠尖脚放在固定金属丝的小圆柱面上。在平面反射镜 M 前方的测高仪上竖直安放一标尺,尺旁安置一望远镜,适当调节其左右位置并对望远镜聚焦后,便可从望远镜中看清由平面全反射镜反射回来的标尺像,并可读出与望远镜叉丝相重合的标尺刻度的指示值。

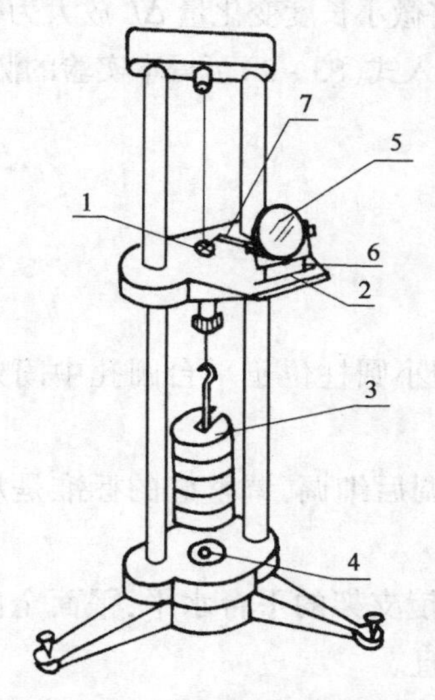

图 S3-1　杨氏模量测量装置

1—平台小孔;2—平台槽沟;3—砝码;4—调平气泡;
5—光杠杆小镜;6—刀口;7—主杠尖脚

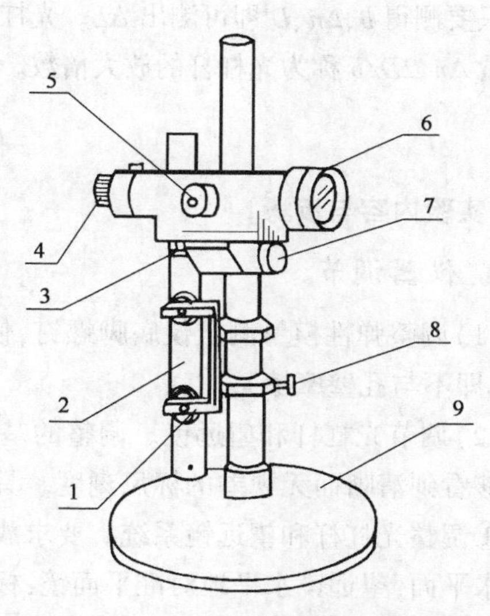

图 S3-2　JCW-1 型标尺望远镜

1—毫米尺组;2—标尺;3—微动手轮;4—视度圈;
5—调焦手轮;6—调焦望远镜;7,8—锁紧螺纹;9—底座

利用光杠杆放大法测量微小长度变化量原理如图 S3-4 所示，初始时，假定平面全反射镜 M 的法线和望远镜光轴在同一直线上，且望远镜光轴和标尺垂直，从望远镜中读得标尺读数为 n_0，当增加砝码时，金属丝受向下的拉力有微小伸长 ΔL，主杠尖脚随之下降 ΔL，使平面镜转过一个角度 θ，根据反射定律，反射线将转过角度 2θ。

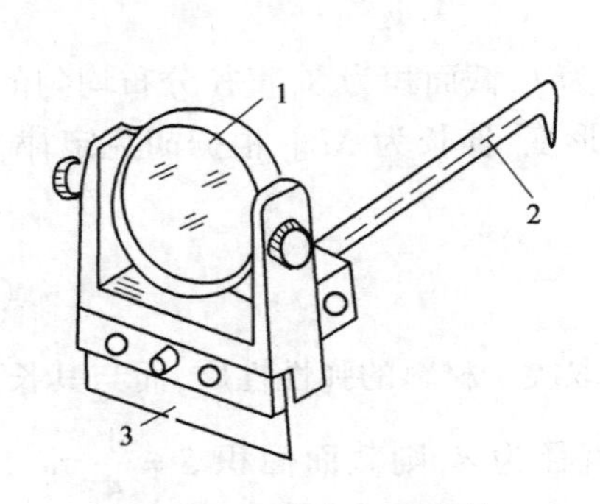

图 S3-3　光杠杆反射镜

1—平面全反射镜 M；2—主杠支脚；3—刀口

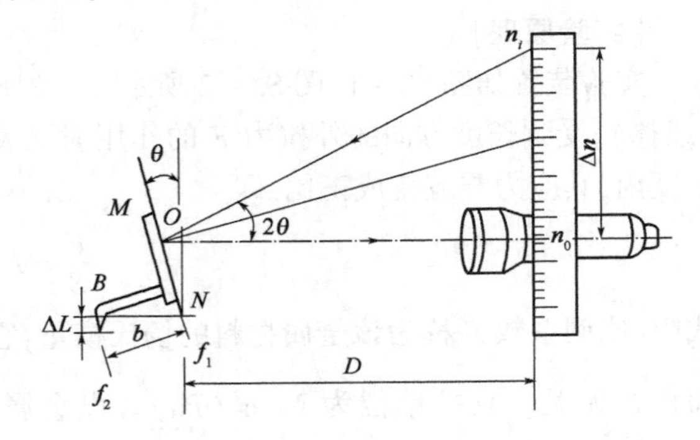

图 S3-4　光杠杆放大原理图

此时，在望远镜中可读到标尺的另一读数 n_i，设平面全反射镜到标尺的距离为 D，主杠尖脚到刀口间距离为 b，n_i 与 n_0 的距离为 Δn。

由于 $\Delta L \ll b$，所以 θ 也很小，故有

$$\frac{\Delta L}{b} = \tan\theta \approx \theta \tag{S3-3}$$

又因 $n_i - n_0 \ll D$，故有

$$\frac{n_i - n_0}{D} = \tan 2\theta \approx 2\theta \tag{S3-4}$$

两式相除得

$$\Delta L = \frac{b}{2D}\Delta n \tag{S3-5}$$

只要测得 b、Δn、D 即可测出 ΔL。光杠杆的作用在于将微小长度变化量 ΔL 放大为标尺上的位移量 Δn，$2D/b$ 称为光杠杆的放大倍数。将式(S3-5)代入式(S3-2)可得本实验的测量公式

$$E = \frac{8mgDL}{\pi d^2 b \Delta n} \tag{S3-6}$$

[实验内容与步骤]

1. 仪器调节

(1) 调整弹性模量测定仪底脚螺钉，使固定金属丝的小圆柱位于平台圆孔中间处于自由状态(即不与孔壁摩擦)。

(2) 调节光杠杆和望远镜。调整的基本原则是先粗调后细调，调整好的标准是从望远镜中能够看到清晰而无视差的标尺刻度。具体步骤如下：

①调整光杠杆和望远镜系统。要求放置平面全反射镜支架的平台水平，平面全反射镜垂直于水平面，望远镜水平地对准平面镜，标尺与望远镜垂直。

②调节望远镜与平面全反射镜中心共轴等高，并将望远镜水平叉丝对准标尺的零刻度线或者零刻度线附近。

③调节望远镜聚焦，使通过目镜能看到标尺成像清晰且无视差。

2. 测量

(1)测量前预加一个砝码,将钢丝拉直(不计标尺读数)。

(2)依次在砝码钩上加挂砝码(每次1kg,加到8kg),待砝码静止后,记下相应的标尺读数$n_1,n_2,n_3,\cdots,n_8$。依次减少砝码(每次1kg,减到1kg)。记下相应的标尺读数$n_8,n_7,n_6,\cdots,n_1$。

(3)取同一负荷下标尺读数的平均值$\bar{n}_i$,再用逐差法求出$\Delta\bar{n}_i$。

(4)用千分尺测量金属丝直径d,在上、中、下各部位分别测量3次,求其平均值$\bar{d}$。

(5)用钢卷尺单次测量标尺平面到光杠杆小镜镜面的距离D。

(6)用钢卷尺单次测量钢丝的有效长度L。

(7)取下光杠杆,将刀口及主杆尖脚印在纸上,用游标卡尺测量主杠尖脚至刀口间的距离b,测6次取平均值$\bar{b}$。

[注意事项]

(1)平面全反射镜上有灰尘、污迹时,用擦镜纸擦去,切勿用手触摸平面全反射镜镜面和望远镜镜头。

(2)切勿打碎反射镜。

(3)避免用力旋转望远镜调焦旋钮,尤其是在两端感觉已被限位时,不能用力,否则易损坏;不用时应用镜罩遮挡物镜、目镜。

(4)调试仪器时,切记要用手托住移动部分,然后旋松锁紧手轮,以免互相撞击。

(5)加减砝码应平稳,防止产生冲击力。

[数据及处理]

1. 数据记录(表S3-1~表S3-3)

表S3-1　光杠杆测钢丝微小长度变化数据记录表

单位:mm

序号	前半组				序号	后半组				$\Delta n_i=$
n	m/kg	增	减	n_i	n	m/kg	增	减	n_{i+4}	$n_{i+4}-n_i$
1					5					
2					6					
3					7					
4					8					

表S3-2　钢丝直径测量数据记录表

$d_0=$________,$\Delta_{仪}=$________

单位:mm

次数	上			中			下		
	1	2	3	4	5	6	7	8	9
$d=d_i-d_0$									

表S3-3　主杠尖脚至刀口间距离b测量数据记录表

$b_0=$________,$\Delta_{仪}=$________

单位:mm

次数	1	2	3	4	5	6
$b=b_i-b_0$						

2. 数据处理

计算出金属丝杨氏模量的测量不确定度，并给出测量结果的完整评价。

3. 结果分析

通过拉伸法测定金属丝的弹性模量的实验，你所得出的有关结论有哪些？

视频 S3　用拉伸法测量金属丝的杨氏模量

[问题讨论]

(1)分析讨论实验条件的满足程度对测量结果的影响。

(2)在实验条件完全满足的情况下，对长度为 1m 的金属丝进行拉伸实验，加载荷和卸载荷时，通过线性拟合得到 $F—\Delta n$ 曲线不相重合，试分析其产生的原因，归类于哪一种误差？如何消除其对测量结果的影响？

用拉伸法测量金属丝的杨氏模量如视频 S3 所示。

实验 4　用单摆测量重力加速度

[引言]

绕一个悬点来回摆动的物体，都称为摆，但其周期一般和物体的形状、大小及密度的分布有关。但若把尺寸很小的质块悬于一端固定的长度为 1 且不能伸长的细绳上，把质块拉离平衡位置，使细绳和过悬点铅垂线所成角度小于 10°，放手后质块往复振动，可视为质点的振动，其周期只和长度和当地的重力加速度有关，即周期与质块的质量、形状和振幅的大小都无关系，其运动状态可用简谐振动公式表示，称为单摆。如果振动的角度大于 10°，则振动的周期将随振幅的增加而变大，就不成为单摆了。如摆球的尺寸相当大，绳的质量不能忽略，就成为复摆，周期就和摆球的尺寸有关了。

[实验目的]

(1)学习用单摆测量重力加速度的原理。

(2)设计测定实验室所在地的重力加速度 g 的实验方案，并按实验精度要求选配仪器。

(3)巩固和加深对单摆周期公式的理解。

[预习问题]

(1)改变单摆的摆长和摆球的质量，对测量重力加速度有何影响？

(2)在不改变单摆摆长 l，且 $\theta<5°$的情况下，如何减小测量单摆周期的系统误差？

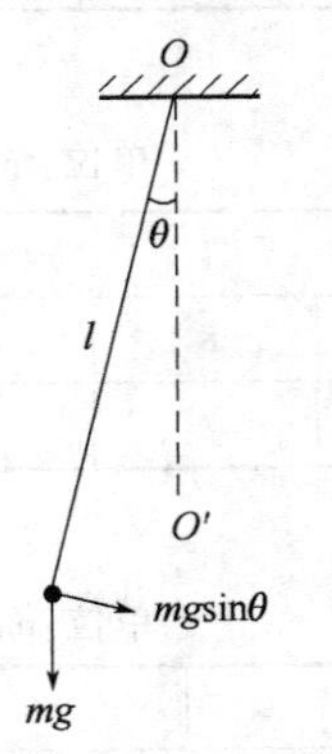

图 S4-1　单摆示意图

[实验原理]

用一不可伸长的轻线悬挂一小球，作幅角 θ 很小的摆动就是一单摆，如图 S4-1 所示。

设小球的质量为 m，其质到摆的支点 O 的距离为 l（摆长）。作用在小球上的切向力的大小为 $mg\sin\theta$，它总指向平衡点 O'。当 θ 角很小，则 $\sin\theta\approx\theta$，切向力的大小为 $mg\theta$，按牛顿第二定律，质点的运动方程为

$$ma_{切} = -mg\sin\theta$$

即

$$ml\frac{\mathrm{d}^2\theta}{\mathrm{d}t^2} = -mg\sin\theta$$

当 θ 角很小时，$\sin\theta\approx\theta$，所以

$$\frac{d^2\theta}{dt^2} = -\frac{g}{l}\theta \quad (S4-1)$$

这是一简谐运动方程(参阅普通物理学中的简谐振动),式(S4－1)的解为

$$\theta(t) = P\cos(\omega_0 t + \phi) \quad (S4-2)$$

$$\omega_0 = \frac{2\pi}{T} = \sqrt{\frac{g}{l}} \quad (S4-3)$$

式中,P 为振幅;ϕ 为幅角;ω_0 为角频率(固有频率);T 为周期。可见,单摆在摆角很小,不计阻力时的摆动为简谐振动,简谐振动是一切线性振动系统的共同特性,它们都以自己的固有频率作正弦振动,与此同类的系统有:线性弹簧上的振子,LC 振荡回路中的电流,微波与光学谐振腔中的电磁场,电子围绕原子核的运动等,因此单摆的线性振动,是具有代表性的。由式(S4－3)可知该简谐振动固有角频率 ω_0 的平方等于 g/l,由此得出

$$T = 2\pi\sqrt{\frac{l}{g}}, \quad g = 4\pi^2\frac{l}{T^2} \quad (S4-4)$$

由式(S4－4)可知,周期只与摆长有关。实验时,测量一个周期的相对误差较大,一般是测量连续摆动 n 个周期的时间 t,由式(S4－4)得

$$g = 4\pi^2\frac{n^2 l}{t^2} \quad (S4-5)$$

式中,π 和 n 不考虑误差,因此式(S4－5)的误差传递公式为

$$\frac{\Delta g}{g} = \frac{\Delta l}{l} + 2\frac{\Delta t}{t} \quad (S4-6)$$

从上式可以看出,在 Δl、Δt 大体一定的情况下,增大 l 和 t 对测量 g 有利。

[实验内容与步骤]

(1)测量摆长。

(2)测量震动周期。

[数据及处理]

将数据记录在表 S4－1 中。

小球的直径:d = ________ mm;

细线的长度:l = ________ cm;

30 个周期的时间:t = ________ ms。

表 S4－1　周期记录表　　单位:s

次数	1	2	3	4	5	6
$t=30T$						

[注意事项]

(1)选择材料时应选择细、轻又不易伸长的线,长度一般在 1m 左右,小球应选用密度较大的金属球,直径应较小,最好不超过 2cm。

(2)单摆悬线的上端不可随意卷在铁夹的杆上,应夹紧在铁夹中,以免摆动时发生摆线下滑、摆长改变的现象。

视频S4　用单摆测量重力加速度

(3)摆球摆动时,要使之保持在同一个竖直平面内,不要形成圆锥摆。

[问题讨论]

(1)在测量中为什么摆角需小于5°,若摆角偏大会对实验结果有什么影响?

(2)实验中摆长是如何确定的?

单摆运动规律的研究如视频S4所示。

实验5　液体动力黏度的测定

[引言]

液体缓慢流动时,平行于流动方向的各层流体速度不同,层间有力的作用,称为黏滞力。其方向平行于界面,与流动方向相反,其大小与速度及界面面积成正比,即$f=\eta\frac{\mathrm{d}u}{\mathrm{d}x}\cdot s$。其中,比例系数$\eta$称为动力黏度,它表示液体黏滞性的大小。各种液体具有不同程度的黏滞性。通常情况下,液体还表现出其动力黏度随温度升高而减小的重要特性。在化学、生物、石油等工业中都需要测量流体的黏度。测定液体动力黏度的方法有多种,本实验采用金属小球在被测液体中自由下落时,在运动过程中再现出小球由加速到匀速的运动规律,从中测得液体的动力黏度。这种方法简称为落球法。它适用于黏度大、透明度好的液体,如蓖麻油、甘油等。

[实验目的]

(1)通过对小球运动规律的观测,加深对液体内摩擦特性的理解。

(2)掌握利用斯托克斯定律测量液体的动力黏度方法。

(3)掌握用读数显微镜测微小长度及秒表的使用方法。

[实验仪器]

玻璃圆筒、读数显微镜、小钢球多于20粒、秒表、米尺、比重计、温度计、待测液体、镊子。

[实验原理]

由于液体具有黏滞性,固体在液体内运动时,附着在固体表面的一层液体和相邻层液体间有内摩擦阻力作用,即黏滞阻力的作用。对于半径为r的球形物体,在无限宽广的液体中以不太大的速度或小于声速v运动,并无涡流产生时,小球所受到的黏滞阻力F为

$$F=6\pi\eta rv \tag{S5-1}$$

式(S5-1)称为斯托克斯公式。其中η为液体的动力黏度,它与液体种类和温度有关。

如果让质量为m半径为r的小球在无限宽广的液体中(密度为ρ)自由下落,它将受到三个力的作用,即重力mg、液体浮力$\frac{4}{3}\pi r^3\rho g$、黏滞阻力$6\pi\eta rv$,这三个力作用在同一直线上,方向如图S5-1所示。起初,重力大于其余两个力之和,小球向下作加速运动;随着速度的增加,黏滞阻力也相应地增大,合力相应地减小。当小球所受合力为零时,即

图S5-1　受力图

$$mg-\frac{4}{3}\pi r^3\rho g-6\pi\eta rv_0=0 \tag{S5-2}$$

这时,小球以速度v_0向下作匀速直线运动,故v_0称为收尾速度。由式(S5-2)可得动力黏度为:

$$\eta = \frac{\left(m - \frac{4}{3}\pi r^3 \rho\right)g}{6\pi r v_0} \tag{S5-3}$$

当小球达到收尾速度后，通过路程 L 所用时间为 t，则 $v_0 = L/t$，将此公式代入式（S5－3）又得

$$\eta = \frac{\left(m - \frac{4}{3}\pi r^3 \rho\right)g}{6\pi r L} t \tag{S5-4}$$

上式成立的条件是小球在无限宽广的均匀液体中下落，但实验时，小球匀速下降深度是有限的，容器的内壁宽度也是有限的，故实验时作用在小球上的黏滞阻力将与斯托克斯公式给出的有偏差。为此在斯托克斯公式后面需加一项修正值，即可描述液体的动力黏度 。此时式（S5－4）变成：

$$\eta = \frac{\left(m - \frac{4}{3}\pi r^3 \rho\right)g}{6\pi r L\left(1 + 2.4\frac{r}{R}\right)} t \tag{S5-5}$$

式中，R 为玻璃圆筒的半径。式（S5－5）就是测量液体动力黏度的数学表式。

［实验内容与步骤］

（1）利用铅直锤调整玻璃圆筒到竖直方向，测出液体的初温度 T_1。

（2）用游标卡尺测圆筒内径 D。

（3）用米尺测出圆筒上下两条标记线之间的距离 L。

（4）用读数显微镜测小球不同位置处的直径 d，共测五次，小球的密度由实验给出。

（5）用比重计测量待测液体的比重，求出密度 ρ。

（6）用镊子夹起小球，从圆筒液面的中心处自由下落，用秒表测量小球通过两标记线之间的时间 t。

（7）测液体的末温度 T_2。

［注意事项］

（1）使小球沿玻璃管的中心线下落，切勿贴着管壁下落。

（2）管子内的液体应无气泡，小球表面应光滑无油污。

（3）油的黏度系数随温度变化，因此实验中不要用手摸玻璃管，确保温度不变。

（4）两横线间的距离 L 应尽量取在玻璃筒的下端，应使小球在上标线前已达到匀速运动。

（5）小球下落时，应保持液体处于静止状态。每下落一粒小球要间隔一定时间。

［数据及处理］

1. 数据记录（表 S5－1～表 S5－3）

表 S5－1　各直接测量量的测量值

圆筒内径 D/mm		液体初温 T_1/℃	
两标线之间的距离 L/mm		液体末温 T_2/℃	
重力加速度 g/m · s^{-2}		平均温度 T/℃	
液体密度（无水甘油）ρ/kg · m^{-3}	1260	小球密度（实验室给出）$\rho_{球}$/kg · m^{-3}	7874

表 S5-2 测量小钢球的直径 d 单位:mm

次数	1	2	3	4	5	$\bar{d}=\frac{1}{5}\sum_{i=1}^{5}d_i$
位置读数 x_{1i}						
位置读数 x_{2i}						
$d_i=\lvert x_{1i}-x_{2i}\rvert$						

表 S5-3 小球下落时间的测量 单位:s

次数	1	2	3	4	5	$\bar{t}=\frac{1}{5}\sum_{i=1}^{5}t_i$
t_i						

2. 数据处理

求出对液体动力黏度测量的不确定度,并给出测量结果的完整评价。

3. 结果分析

对实验结果进行分析,并得出实验结论。

[问题讨论]

(1)在判断小球通过圆筒上的两标线时,如何消除视差?

(2)实验中当小球的半径减小时,它的下落速度将如何变化?当小球的密度变化时,情况又如何?

(3)在温度不同的同种润滑油中,同一小球下降的速度是否相同?为什么?

(4)如何判断小球已经进入了匀速运动阶段?

实验6 冰的熔解热的测定

[引言]

单位质量的0℃冰吸收一定量的热量后,可以转变为同温度的水,所吸收的热量称为熔解热。在孤立系统中,使高温的水和0℃的冰混合,它们之间进行热交换,最终达到一个平衡温度。根据高温水放出的热量等于0℃冰熔解吸收的热量加0℃水升温到平衡温度吸收的热量,即可求出冰的熔解热。

[实验目的]

(1)了解量热和记录温度的基本方法。

(2)掌握用混合法测定冰的熔解热。

(3)学会一种散热修正的方法。

[预习问题]

(1)什么是冰的熔解热?什么是混合量热法?

(2)散热补偿法的基本思想是什么?

(3)混合法测冰的溶解热的实验中,水的初温选得太高或太低有什么不好?投入的冰块用大的好还是用小好?为什么?

[实验仪器]

量热器、物理天平、水银温度计、量筒、秒表、玻璃杯等。

[实验原理]

一定压强下晶体开始熔解的温度,称为该晶体在此压强下的熔点。单位质量的固体物质在熔点时变成同温度的液体所吸收的热量,称为该物质的熔解热。

1. 混合量热法测冰的熔解热

几个温度不同的物体相混合时,它们之间要进行热交换,其中高温物体要放出热量,温度降低,低温物体要吸收热量,温度升高。如果热交换仅在几个物体之间进行,即与外界没有热量交换,根据热平衡原理,各物体最终将具有相同的末温。由能量守恒定律知,高温物体放出的热量就应等于低温物体吸收的热量,即

$$Q_{放} = Q_{吸}$$

这就是混合量热法的基本原理。

若将质量为 m、温度为 T_0(在实验室条件下,冰的温度和冰的熔点均认为是0℃)的冰与质量为 m_1,温度为 T_1 的水在量热器内混合。冰全部熔解为水后,水的平衡温度为 T_2,有

$$Q_{吸} = m\lambda + mc_1(T_2 - T_0) \tag{S6-1}$$

$$Q_{放} = (m_1c_1 + m_2c_2 + m_3c_3 + 1.92V)(T_1 - T_2) \tag{S6-2}$$

则

$$\lambda = \frac{1}{m}(m_1c_1 + m_2c_2 + m_3c_3 + 1.92V)(T_1 - T_2) - c_1(T_2 - T_0) \tag{S6-3}$$

式中,λ 为冰的熔解热;c_1 为水的比热容;m_2、m_3 和 c_2、c_3 分别为量热器内筒及搅拌器的质量和比热容;$1.92V$ 为水银温度计浸入水中部分放出的热量,一般可以被忽略。测出式(S6-3)中其他各量,即可求出 λ。

2. 误差修正

1) 散热补偿法

只要实验系统与外界存在温度差,系统就不可能达到完全绝热的要求。本实验介绍一种粗略修正散热的方法—散热补偿法,其基本思想是设法使系统在实验过程中从外界吸收的热量等于系统散失的热量。

牛顿冷却定律指出,系统温度 T 与环境温度 θ 之差不大时,系统的散热速率(单位时间内散失的热量)与温度差成正比,即

$$\frac{\Delta Q}{\Delta t} = K(T - \theta) \tag{S6-4}$$

式中,K 称为散热系数,与系统表面积成反比,并随表面的热辐射本领而变。

由式(S6-4)可知,当 $T > \theta$ 时,$\frac{\Delta Q}{\Delta t} > 0$,系统向外界散热,当 $T < \theta$ 时,$\frac{\Delta Q}{\Delta t} < 0$,系统从外界吸热。

本实验量热器中水的温度随时间变化的曲线如图 S6-1 所示。在混合初,冰块大,水温高使冰块熔解快,系统温度降低快;随着冰的熔解,水温降低,冰块变小,熔解变慢,系统温度的降低也就变慢了。在 $t_1 \sim t_2$ 这段时间里,温度由 T_1 降为 θ,由式(S6-4)可得系统放出的热量

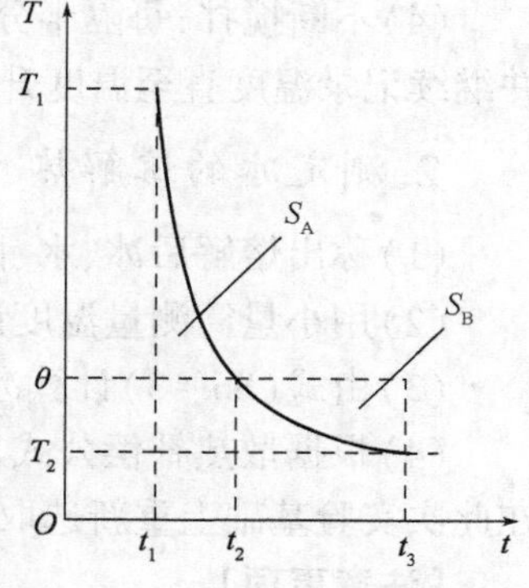

图 S6-1 温度—时间曲线

$$Q'_{放} = K\int_{t_1}^{t_2}(T-\theta)\,dt = KS_A \tag{S6-5}$$

水在 $t_2 \sim t_3$ 时间内系统温度低于环境温度 θ，系统从外界吸收热量

$$Q'_{吸} = K\int_{t_2}^{t_3}(\theta-T)\,dt = KS_B \tag{S6-6}$$

散热补偿法要求 $Q'_{吸} \approx Q'_{放}$，即只要使 $S_A = S_B$，系统对外界的吸热和散热就可以相互抵消，即系统吸收的热量可以补偿散失的热量，实现散热补偿的目的。

因此，用散热补偿法时，在实验中，应注意恰当地选择各实验参量（如温度、冰及水的质量等），以使 $T_1-\theta>\theta-T_2$，例如，初温高于室温 10～15℃，并且用足够多的冰，使末温低于室温 10℃左右，使冰水混合过程中，系统向环境散失的热量和吸收的热量基本上相互抵偿。

2）修正温度的外推法

在实际测定冰的熔解热时，有两个过程在同时进行，一是由于系统与环境的温差而导致的热交换（吸热或放热）使冰的温度发生的微小变化（升高或降低）；另一是由于冰投入后吸热而导致水温下降。因此，需用外推法对系统温度随时间变化的 T—t 图线进行处理，才可得到无热交换时系统温度的实际值，即单纯由于冰水混合而使水温改变的量。外推法的基本思想是，在处理数据时，把系统的热交换外推到进行得无限快（即使系统趋于与外界无热交换）的情况。

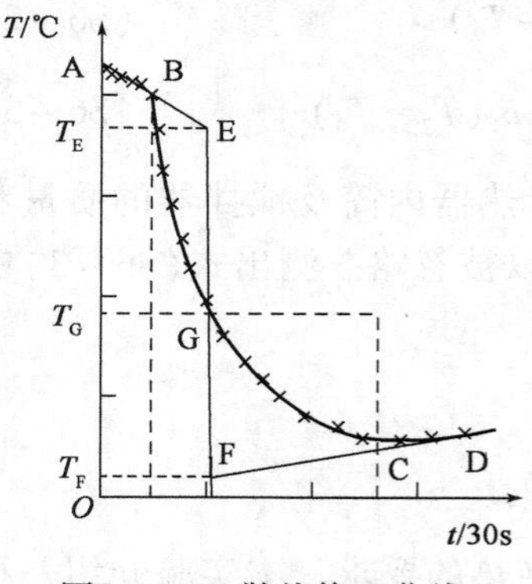

图 S6－2　散热修正曲线

在图 S6－2 中，ABCD 是系统温度随时间变化曲线，延长 AB 到 E，延长 DC 到 F，作平行于 T 轴的 EF 直线，交 ABCD 于 G，使面积 BEG 与面积 GFC 相等。因此，可将原来的 BGC 过程等效为 BE、EGF 和 FC 三个过程，其中 BE 和 FC 表示在整个过程中由于系统与环境的热交换而产生的温度变化。而 EGF 表示系统由于冰水混合而引起的温度变化，即等效为冰在瞬间全部熔化为水的过程。所以 T_E、T_F 为系统与环境无热交换时的初温和终温。

［实验内容及步骤］

1. 作温度—时间曲线，用外推法求初温和终温

（1）测量环境温度 θ。

（2）将量热器内筒和搅拌器用物理天平称出其质量 m_2、m_3。

（3）将高于室温 10℃左右的热水注入量热器内筒，约占内筒体积的 1/2，称出水的质量 m_1。

（4）不断搅拌，每隔半分钟记录一次温度，4min 后将擦干水珠的冰块投入量热器中，搅拌并继续记录温度直至温度升高。作温度—时间曲线，用外推法确定水的初温 T_1 和终温 T_2。

2. 测定冰的熔解热

（1）称出熔解后冰、水、内筒和搅拌器的总质量，求出冰的质量。

（2）用小量筒测量温度计浸入水中部分的体积 V。

（3）由式（S6－3）计算冰在 0℃时的熔解热。

（4）根据散热补偿公式，分析各参量的选择是否满足散热补偿要求，如补偿效果不佳，可在此次实验基础上重新选取初温和末温及水和冰块的质量，并重做实验。

［注意事项］

（1）投冰及搅拌时都要避免将水溅出，取出温度计时，也应避免带出水滴。

(2)测量时,要不停地搅拌,不要碰温度计和量热器。

(3)温度计不要接触量热器和冰块,应悬于水中。

(4)精细温度计较大易打碎,使用时应特别小心。

(5)实验过程中,不要直接用手接触量热器,不要在阳光照射下或空气流动太快的地方进行实验,量热器应远离热源。

[数据及处理]

1. 数据记录

1)投冰前(表 S6-1)

表 S6-1 投冰前量热器内温度的变化记录表

时间/min	0	0.5	1.0	1.5	2.0	2.5	3.0	3.5	4.0
温度/℃									

2)投冰后(表 S6-2)

表 S6-2 投冰后量热器内温度的变化记录表

时间/min	0	0.5	1.0	1.5	2.0	2.5	3.0	3.5	4.0	4.5	5.0
温度/℃											
时间/min	5.5	6.0	6.5	7.0	7.5	8.0	8.5	9.0	9.5	10.0	…
温度/℃											

2. 数据处理

(1)做出散热修正曲线,确定水的初温 T_1 和冰水混合后的终温 T_2。

(2)计算冰的熔解热 λ。

(3)将测量值与冰的熔解热的标准值(3.329×10^5 J/kg)进行比较,计算相对误差,并分析产生误差的原因。

视频 S6 冰的熔解热的测定

3. 结果分析

对实验结果进行分析,并得出实验结论。

[问题讨论]

(1)混合量热法的基本条件是什么?本实验是如何保证的?

(2)水的初温选得太高或太低有什么不好?

(3)如果冰块没有擦干就投入量热器内筒,将对实验结果有何影响?

(4)分析本实验产生误差的原因,应如何改进?

冰的熔解热的测定如视频 S6 所示。

[补充说明]

保持系统为一孤立系统,是混合量热法所要求的基本实验条件。量热器就是为满足这一要求而设计的。

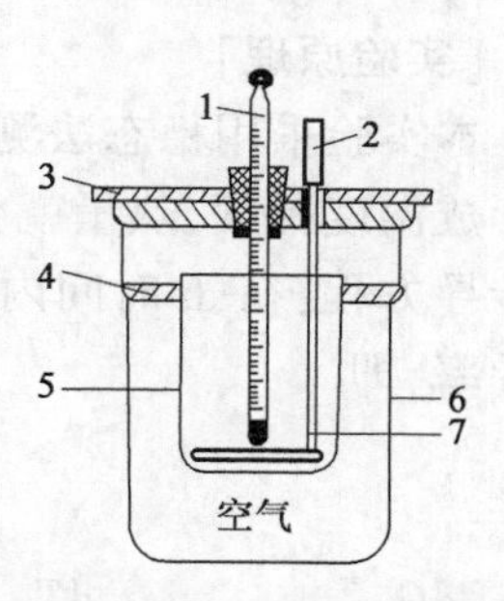

图 S6-3 量热器示意图

1—温度计;2—绝热柄;3—绝热盖;4—绝热架;5—内筒;6—外筒;7—搅拌器

量热器的结构如图 S6-3 所示,量热器的内筒和外筒由良导体制成,其表面光亮,以减少热辐射。内筒放在外筒内的绝热架上，外筒又用绝热盖盖住，使内筒与外界的热传导、热对流及热辐射都难以进行，即尽可能与外界绝热，绝热盖上有两个小孔，中间孔插温度计，旁边小孔插搅拌器，搅拌器与外界接触处有一绝缘柄。

内筒及放在内筒中的温度计、搅拌器、水及待测物等，就是所研究的实验系统。

实验7　不良导体热导率的测定

［引言］

当物体内部沿某方向有温度梯度存在时，就有热量从高温处向低温处传递，1882 年法国科学家傅里叶建立了热传导理论，并指出：在一定时间内通过物体热流动方向某截面的热量与导热面积、温度梯度及经历时间成正比，其中比例系数反映了该物体导热性能的好坏，称为物体的热导率。不同的物质，热导率值是不同的，热导率大的称为热的良导体，热导率小的称为热的不良导体。在实际应用时可由标准化的实用手册根据需要选择热导率值合适的材料，也可通过实验进行测定，其测量方法可归为两类：一类是稳态法，另一类是动态法，它们都是以傅里叶热传导定律作为基础的。稳态法是以样品两端达到稳定温度差时，样品的传热速率与散热盘从侧面和底面向周围散热的速率相等为依据，从而测出散热盘在稳定温度时的散热速率，求出不良导体的热导率。

［实验目的］

(1)了解热传导现象的物理过程。

(2)掌握用稳态法测量不良导体热导率的原理及方法。

(3)学会测定橡胶样品的热导率。

［预习问题］

(1)稳态法和动态法测量不良导体热导率的基本原理及方法有何不同?

(2)本实验中，影响测量不确定度的主要因素有哪些?

(3)如何正确使用 FD－TC－B 型导热系数测定仪?

［实验仪器］

FD－TC－B 导热系数测定仪(恒温控制温度：室温～80℃、温度计精度 ±0.1℃)、橡胶样品，游标卡尺(精度为 0.02mm、量程为 0～125mm)、ARA520 型电子天平(量程为 1500g、标准偏差为 0.01g、线性误差为 ±0.01g)等。

［实验原理］

本实验采用稳态法测量橡胶的热导率。测量时对系统加温以使测试样品沿其厚度方向形成温度梯度，热量就会从高温处传递到低温处，这种现象被称为热传导。根据傅里叶热传导方程，在 $\mathrm{d}t$ 时间内通过 $\mathrm{d}s$ 面积的热量 $\mathrm{d}Q$，正比于物体内的温度梯度，其比例系数为热导率，即

$$\frac{\mathrm{d}Q}{\mathrm{d}t}=-\lambda\frac{\mathrm{d}T}{\mathrm{d}x}\mathrm{d}s \tag{S7-1}$$

式中，$\frac{\mathrm{d}Q}{\mathrm{d}t}$为传热速率；$\frac{\mathrm{d}T}{\mathrm{d}x}$是与面积 $\mathrm{d}s$ 相垂直的方向上的温度梯度；负号表示热量从高温区域传向低温区域；λ 为物体的热导率，表示物体导热能力的大小，在国际单位制中 λ 的单位是 $\mathrm{W\cdot m^{-1}\cdot K^{-1}}$。对于各向异性材料，常用张量来表示各个方向的热导率。

图 S7－1 为热导率测定结构图，其中 A 为散热铜盘，C 为加热铜盘，样品 B 为一薄圆盘夹在 A、C 之间。当传热达到稳定状态时，样品上下表面的温度 $T_{上}$ 和 $T_{下}$ 不变。由于样品的厚度较薄，因此样品的侧面可以近似认为绝热，由式（S7－1）可知，在 $\mathrm{d}t$ 时间内通过样品的热量 $\mathrm{d}Q$ 满足下式：

$$\frac{\mathrm{d}Q}{\mathrm{d}t}=\lambda\frac{T_{上}-T_{下}}{h_{B}}S_{B}=\lambda\frac{T_{上}-T_{下}}{h_{B}}\times\frac{\pi D_{B}^{2}}{4} \qquad (S7-2)$$

式中，h_B 为样品厚度；S_B 为样品 B 的面积；D_B 为样品 B 的直径；$T_{上}-T_{下}$ 为样品上、下表面的温度差。

在实验中，要降低侧面散热的影响，就要尽量减小 h_B。而待测样品上下表面的温度 $T_{上}$ 和 $T_{下}$ 是用加热铜盘 C 的底面温度和散热铜盘 A 的温度来代替，所以还必须保证样品与加热铜盘 C 的底面和散热铜盘 A 的顶面热接触良好。

实验时，当传热达到稳定状态时（$T_{上}$ 和 $T_{下}$ 值恒定不变），可以认为加热铜盘 C 通过样品传递的热流量与散热铜盘 A 向周围环境的散热量相等。因此可以通过散热铜盘 A 在稳定温度 $T_{下}$ 时的散热速率 $\frac{\mathrm{d}T}{\mathrm{d}t}$ 来求出样品的传热速率 $\frac{\mathrm{d}Q}{\mathrm{d}t}$。

在读取稳态时的 $T_{上}$ 和 $T_{下}$ 之后，取走样品 B，让散热铜盘 A 直接与加热圆盘 C 的底面接触，加热散热铜盘 A，使其温度上升到比稳态时温度 $T_{下}$ 高 10℃左右，移开加热圆盘 C，让散热铜盘 A 在电扇作用下冷却，同时记录散热铜盘 A 的温度 T_A 随时间 t 的下降情况，直到 T_A 降至比 $T_{下}$ 低 20℃时为止，然后以时间为横坐标，以 T_A 为纵坐标，作散热铜盘 A 的冷却曲线如图 S7－2所示，过曲线上的温度等于 $T_{下}$ 的点作切线，则此切线的斜率就是散热铜盘 A 在 $T_{下}$ 温度时的冷却速率 $\left.\frac{\mathrm{d}T}{\mathrm{d}t}\right|_{T_A=T_{下}}$。

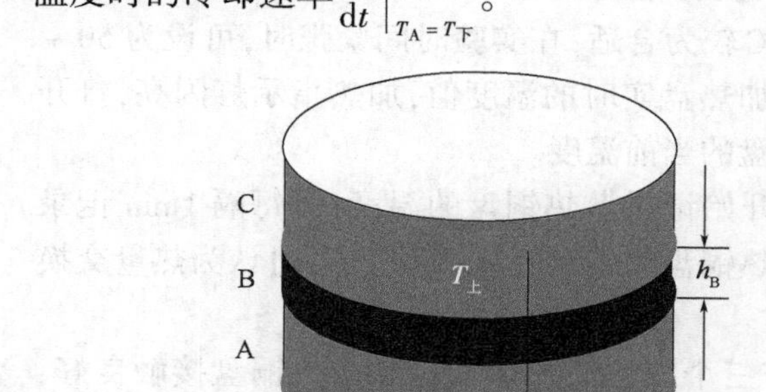

图 S7－1　热导率测定结构图

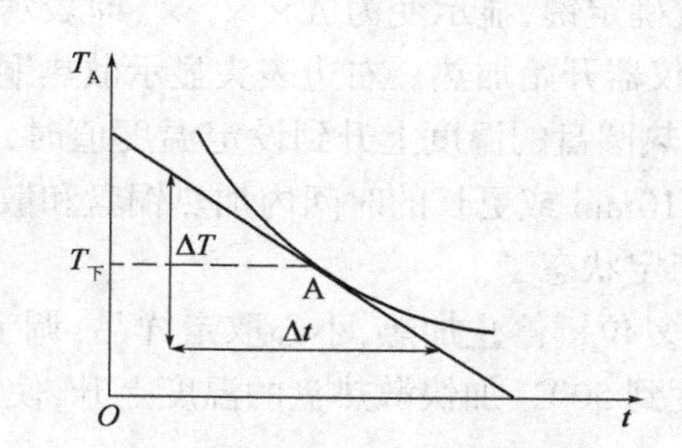

图 S7－2　铜盘 A 的冷却曲线

对散热铜盘 A，根据热容的定义，有

$$\frac{\mathrm{d}Q}{\mathrm{d}t}=mc\frac{\mathrm{d}T}{\mathrm{d}t}=m_{铜}c_{铜}\left.\frac{\mathrm{d}T}{\mathrm{d}t}\right|_{T_A=T_{下}} \qquad (S7-3)$$

式中，$m_{铜}$ 和 $c_{铜}$ 分别为散热铜盘 A 的质量和比热容。

对于散热铜盘 A，在达到稳态的过程中，其上表面并未暴露在空气中，散热的外表面积为 $\pi D_A^2/4+\pi D_A h_A$，其中 D_A、h_A 分别为散热铜盘 A 的直径和厚度。移去加热铜盘 C 后，散热铜盘 A 的散热外表面积为 $\pi D_A^2/2+\pi D_A h_A$，考虑到物体的散热速率与它的散热面积成比例，所以稳态时散热铜盘 A 的散热速率的表达式应作面积修正

$$\frac{dQ}{dt}=m_{铜}c_{铜}\frac{\pi D_A^2/4+\pi D_A h_A}{\pi D_A^2/2+\pi D_A h_A}\times\left.\frac{dT}{dt}\right|_{T_A=T_下}=m_{铜}c_{铜}\frac{D_A+4h_A}{2(D_A+2h_A)}\times\left.\frac{dT}{dt}\right|_{T_A=T_下} \quad (S7-4)$$

综合式(S7-2)和式(S7-4),可得出样品的热导率为

$$\lambda=\frac{2m_{铜}c_{铜}h_B(D_A+4h_A)}{\pi D_B^2(T_上-T_下)(D_A+2h_A)}\times\left.\frac{dT}{dt}\right|_{T_A=T_下} \quad (S7-5)$$

式中,$m_{铜}$、D_A、D_B、h_A、h_B、$T_上$ 和 $T_下$ 都可以由实验测出,$c_{铜}$ 可以通过查表得到,只要再求出 $\left.\frac{dT}{dt}\right|_{T_A=T_下}$,就可求出橡胶试样的热导率 λ。

[实验内容与步骤]

(1)用游标卡尺测量散热铜盘A和样品圆盘B的厚度与直径,多次测量取平均值。用电子天平测量散热铜盘A的质量 m。

(2)取下固定螺栓,将橡胶样品放在加热盘与散热盘之间,橡胶样品要求与加热盘、散热盘完全对准,上下绝热薄板要对准加热和散热盘。调节底部的三个微调螺栓,使样品与加热盘、散热盘接触良好,但注意不宜过紧或过松。

(3)如图S7-3所示,插好加热圆盘的电源插头,再将2根带有温度传感器的连接线的一端接入仪器后面板的相应位置,另一端的传感器分别插在加热铜盘和散热铜盘小孔中,要求传感器完全插入小孔中,可在传感器上抹一些硅油或者导热硅脂,以确保传感器与加热铜盘和散热铜盘热接触良好。在安放加热铜盘和散热铜盘时,还应注意使放置传感器的小孔上下对齐。

(4)接上导热系数测定仪的电源,开启电源后,左边表头首先显示为FDHC,然后显示当前温度。这时用户可以设定控制温度,按升温键左边表头显示B××.×,表示想要设定的温度值,可设范围为0.0~80.0℃。一般设定75~80℃较为合适,在实验时间紧张时,可设为50~55℃。再按确定键,显示变为A××.×,即表示加热盘实时的温度值,加热指示灯闪亮,打开电扇开关,仪器开始加热。右边表头显示散热铜盘的当前温度。

(5)加热圆盘的温度上升到设定温度值时,开始记录散热铜盘的温度,可每隔1min记录一次,待在10min或更长的时间内加热铜盘和散热铜盘的温度值基本不变,即可认为热量交换已经达到稳定状态了。

(6)按复位键停止加热,小心取走样品,调节三个螺栓使加热铜盘和散热铜盘接触良好,再设定温度到50℃,加快散热盘的温度上升,使散热盘温度上升到高于稳态时的 $T_下$ 值10℃左右即可。

(7)移去加热盘,让散热铜盘在风扇作用下冷却,每隔10s(或者30s)记录一次散热铜盘的温度示值,即右边表头的显示值,直至温度示值比稳定时的 $T_下$ 值低5℃以下即可停止记录。

(8)将所得的数据进行处理,计算出橡胶样品的热导率。

[注意事项]

(1)注意各仪器间的连线正确,加热铜盘和散热铜盘的两个温度传感器要一一对应,不可互换。

(2)温度传感器插入小孔时,要抹些硅油,并使传感器与散热铜盘热接触良好。

(3)导热系数测定仪铜盘下方的风扇做强迫对流换热用,可以减小样品侧面与底面的放

热比，增加样品内部的温度梯度，从而减小实验误差，所以实验过程中，风扇一定要打开。

(4)测量完毕，取出样品时，应先切断电源并注意防止烫伤。

[数据及处理]

1. 数据记录(表 S7-1 ~ 表 S7-3)

表 S7-1　散热铜盘 A 和样品圆盘 B 的厚度与直径数据记录表

游标卡尺(精度)：________　　量程：________　　零点读数：________

测量项目	h_A/mm		D_A/mm		h_B/mm		D_B/mm	
次数	直读值	消除零点值	直读值	消除零点值	直读值	消除零点值	直读值	消除零点值
1								
2								
3								
4								
5								
平均值	$\overline{h_A}$ =		$\overline{D_A}$ =		$\overline{h_B}$ =		$\overline{D_B}$ =	

散热铜盘 A 的质量：$m_{铜}$ = ________。

表 S7-2　稳态时样品 B 上下表面的温度

样品温度状态	上表面温度 $T_{上}$	下表面温度 $T_{下}$
温度值/℃		

表 S7-3　散热铜盘 A 在 $T_{下}$ 附近冷却时的温度示值

测量次序	1	2	3	4	5	6	7	8	9	…
冷却时间/s										
温度值/℃										

2. 数据处理

(1)根据表 S7-3 中的数据，作散热铜盘 A 的冷却曲线，用作图法求出散热铜盘在温度 $T_{下}$ 时的散热速率。

(2)将实验中测得的 $m_{铜}$、D_A、D_B、h_A、h_B、$T_{上}$ 和 $T_{下}$ 等量代入式(S7-5)，求出橡胶样品的热导率，并与标准值进行比较，求出百分相对误差。

3. 结果分析

对实验结果进行分析，并得出实验结论。

[问题讨论]

(1)比较稳态法和瞬态法测量热导率的联系与区别。

(2)热导率与温度有关，如何测量同一样品在不同温度下的热导率？

(3)分析本实验中各物理量的测量结果，哪一个对实验误差影响较大？

不良导体热导率的测定如视频 S7 所示。

视频 S7　不良导体热导率的测定

实验8 电源控制电路特性的研究

[引言]

在电学实验中，为了得到所需的电压或电流，需要把滑线变阻器连接成分压或分流形式，对电源进行控制与调节。负载和阻值会对电压和电流产生影响，本实验对这两种输出特性进行研究，以合理设计与选用控制电路。

[实验目的]

(1)了解控制电路的特点。

(2)学会根据实验条件与要求来合理选配控制电路。

[预习问题]

(1)在实验中如何选择变阻器的参数(如阻值、额定电流)?

(2)限流和分压电路形式有什么不同?

[实验仪器]

直流电源、滑线变阻器、电压表、电流表、电阻箱、电键、导线若干。

[实验原理]

电学实验的电路通常由电源、控制电路和测量电路三部分组成。测量电路往往需要电源提供的电压或电流在某一范围内数值可调，而电源输出一般是一个固定值。为解决这一矛盾，达到测量电路的要求，通常不同的控制电路中电压和电流的变化是通过不同规格的滑线变阻器来控制的。

控制电路有限流和分压两种基本接法，这两种接法的性能指标可用调节范围、细调程度和线性程度来表征，这三者的综合就是电源控制电路的输出特性，可以用控制电路的输出特性曲线表示。

1. 限流控制电路

限流控制电路如图 S8－1 所示，E 为电源电动势，电源内阻一般很小可忽略不计。S 是电键，R_L 为负载电阻，R_0 为滑线变阻器全电阻，R_1 为滑线变阻器未接入电路中的电阻，也即 AC 之间的电阻，R_2 为滑线变阻器接入电路部分的电阻，也即 BC 之间的电阻。将负载 R_L 与滑线变阻器的 BC 两端串联，移动滑线变阻器的滑动端 C 的位置就可以连续改变 BC 之间的电阻值 R_2，从而可以改变整个电路的电流，实现对负载电流的控制，使之符合要求。

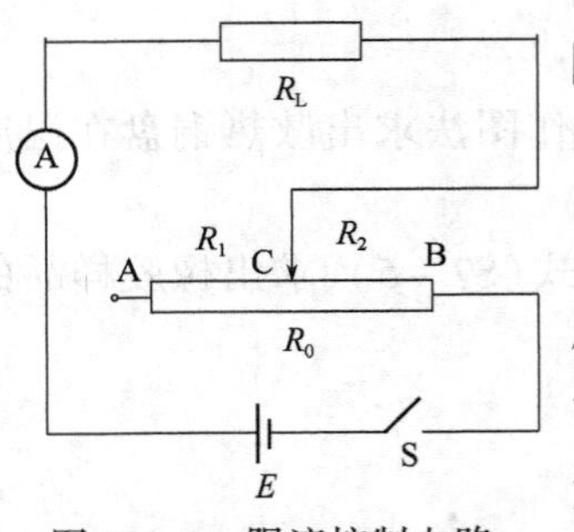

图 S8－1 限流控制电路

限流电路的输出特性曲线是控制电路输出电流 I，与其最大值 I_{max} 的比值 I/I_{max} 随变阻器滑动端位置改变而变化的曲线。它可以直观看出控制电路输出电流在调节范围内的变化规律。令负载电阻 R_L 与滑线变阻器全电阻 R_0 之比为电路特征系数 K，即 $K=R_L/R_0$，x 为滑线变阻器滑动端 C 在滑线变阻器上的相对位置，$x=\frac{R_1}{R_0}$。根据电路，控制电路的输出电流 $I=\frac{E}{R_L+R_2}$，当 $R_2=0$ 时，输出电流最大，即 $I_{max}=\frac{E}{R_L}$。则根据以上分析可得

$$\frac{I}{I_{max}}=\frac{E}{R_L+R_2}\cdot\frac{R_L}{E}=\frac{R_L}{R_L+R_2}=\frac{R_L/R_0}{R_L/R_0+R_2/R_0}=\frac{K}{K+1-x} \quad (S8-1)$$

通过实验可绘出不同 K 值下的限流特性曲线。

2. 分压控制电路

分压控制电路如图 S8－2 所示，E 为电源电动势，电源内阻一般很小可忽略不计。R_L 为负载电阻，R_0 为滑线变阻器全电阻。滑线变阻器的三个接线端的两个固定端 A、B 通过电源开关 S 分别与电源正负极相接，负载 R_L 的一端与滑动端 C 相接，另一端与固定端 A（或 B）相接。滑线变阻器与负载 R_L 并联部分电阻为 R_1，其余部分为 R_2。移动滑线变阻器的滑动端 C 的位置就可以连续改变 AC 之间的电压值，从而实现对负载电压的控制，使之符合要求。

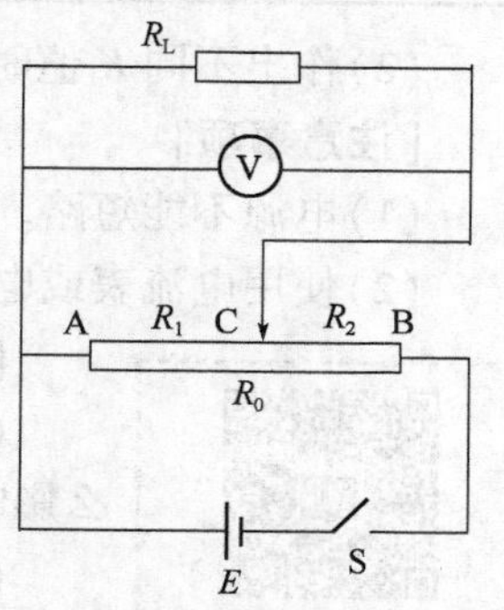

图 S8－2　分压控制电路

分压电路的输出特性曲线，是控制电路输出电压 U 与其最大值 U_{max} 的比值（U/U_{max}）随变阻器滑动端位置改变而变化的曲线，它可以直观看出分压电路输出电压在调节范围内的变化规律。电路的特征系数为 $K(K=R_L/R_0)$，变阻器滑动端 C 在滑线变阻器上的相对位置为 $x(x=R_1/R_0)$，则有

$$\frac{U}{U_{max}}=\frac{E}{\frac{R_LR_1}{R_L+R_1}+R_2}\cdot\frac{R_LR_1}{R_L+R_1}\cdot\frac{1}{E}=\frac{R_LR_1}{R_LR_1+(R_0-R_1)(R_L+R_1)}\cdot\frac{R_0^2}{R_0^2} \quad (S8-2)$$

$$=\frac{Kx}{Kx+K+x-Kx-x^2}=\frac{Kx}{K+x-x^2}$$

通过实验可绘出不同 K 值下的分压特性曲线。

[实验内容与步骤]

（1）分别令 $K=0.1,0.5,1,5,10$，$x=0,0.1,0.2,\cdots,1.0$，做出限流特性曲线。

（2）分别令 $K=0.1,0.5,1,5,10$，$x=0,0.1,0.2,\cdots,1.0$，做出分压特性曲线。

[数据及处理]

1. 限流控制电路特性研究

（1）按照图 S8－1 接线，分别使 $K=0.1,\ 0.5,\ 1,\ 5,\ 10$，在每一 K 值下，调节滑线变阻器，分别使 $x=0,\ 0.1,\ 0.2,\cdots,1.0$，记录输出电流的大小（表 S8－1）。

表 S8－1　限流电路特性数据记录　　$K=$______

x	0.0	0.1	0.2	0.3	0.4	0.5	0.6	0.7	0.8	0.9	1.0
I/A											
I/I_{max}											

（2）作出不同 K 值时的 I/I_{max}—x 的关系曲线，注意画在同一张图里。

2. 分压控制电路特性研究

（1）按照图 S8－2 接线，分别使 $K=0.1,\ 0.5,\ 1,\ 5,\ 10$，在每一 K 值下，调节滑线变阻器，分别使 $x=0,\ 0.1,\ 0.2,\cdots,1.0$，记录输出电压的大小（表 S8－2）。

表 S8－2　分压电路特性数据记录　　$K=$______

x	0.0	0.1	0.2	0.3	0.4	0.5	0.6	0.7	0.8	0.9	1.0
U/V											
U/U_{max}											

(2)作出不同 K 值时的 $U/U_{max}—x$ 的关系曲线,注意画在同一张图里。

[注意事项]

(1)电源不能短路。

(2)使用电流表或电压表时,不能超过其量程。

视频 S8　电源控制电路特性的研究

[问题讨论]

(1)在分压电路特性测量中,如电压表内阻不能远大于 R_L,会产生什么影响?

(2)在限流电路特性测量中,如电流表内阻不能远小于 R_L,会产生什么影响?

电源控制电路特性的研究如视频 S8 所示。

实验 9　等厚干涉——牛顿环和劈尖

[引言]

光的干涉是重要的光学现象之一,是光的波动性的重要实验依据。两列频率相同、振动方向相同且相位差恒定的相干光在空间相交区域发生相互加强或减弱现象,即光的干涉现象。获得相干光的两种方式是分波面干涉和分振幅干涉,其中分振幅干涉包括等倾干涉和等厚干涉。

[实验目的]

(1)理解等厚干涉产生的基本原理。

(2)观察等厚干涉条纹的特点。

(3)学会用等厚干涉法测量薄膜厚度(或微小细丝直径)、平凸透镜曲率半径的方法。

(4)掌握读数显微镜的正确使用。

[预习问题]

(1)读数显微镜如何正确使用(调整步骤、读数、空程误差及视差的消除方法)?

(2)牛顿环和劈尖产生的等厚干涉条纹有何特点?

(3)简述用读数显微镜测量凸透镜的曲率半径时读数的先后次序以及实验过程中应注意的关键问题是什么?

[实验仪器]

读数显微镜(JCD_3型,水平量限 50mm,分度值 0.01mm)、牛顿环装置、钠光灯、劈尖。

[实验原理]

1. 牛顿环

将一块曲率半径很大的平凸透镜 A 的凸面置于一块光学平玻璃板 B 上,在透镜凸面和平玻璃板间形成一层空气薄膜,且厚度以接触点 O 为中心向四周逐渐增厚,离 O 点等距离处厚度处处相同,如图 S9－1(b)所示。当入射光以平行单色光垂直入射时,入射光在薄膜上下两表面反射,在上表面相遇产生干涉。由于空气膜厚度相等的各处光程差相等(即相位相同),所以干涉条纹是以接触点为中心的一系列明暗相间的同心圆环,如图 S9－1(a)所示。

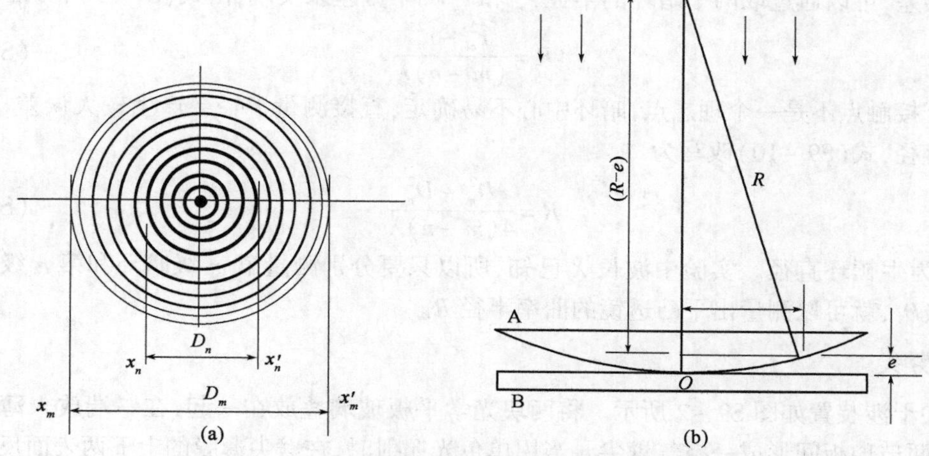

图 S9－1　牛顿环的干涉干涉条纹及干涉光路示意图

在图 S9－1(b)中，设 R 是平凸透镜的曲率半径，λ 是入射光的波长，r_k 为第 k 级干涉条纹的半径，e 为该环对应的空气薄膜的厚度，根据光的干涉原理，并考虑光从光疏媒质到光密媒质表面时产生半波损失，所以在空气膜层上、下表面反射光的光程差为

$$\delta_k = 2ne + \frac{\lambda}{2} \tag{S9-1}$$

n 为空气的折射率，空气的折射率近似为 1，则

$$\delta_k = 2e + \frac{\lambda}{2} \tag{S9-2}$$

由图 S9－1(b)的几何关系知

$$R^2 = r_k^2 + (R-e)^2 \tag{S9-3}$$

$$r_k^2 = 2eR - e^2 \tag{S9-4}$$

式中，R 为透镜的曲率半径。若空气薄膜厚度远小于透镜曲率半径，则有

$$e = \frac{r_k^2}{2R} \tag{S9-5}$$

根据干涉加强和减弱的条件，产生明、暗环的干涉条件为

明环公式　　$\delta_k = 2e + \frac{\lambda}{2} = k\lambda \quad (k = 1,2,3,\cdots)$　　(S9－6)

暗环公式　　$\delta_k = 2e + \frac{\lambda}{2} = (2k+1)\frac{\lambda}{2} \quad (k = 0,1,2,\cdots)$　　(S9－7)

将式(S9－5)代入明、暗环公式中，则

明环半径　　$r_k^2 = (2k-1)R\frac{\lambda}{2} \quad (k = 1,2,3,\cdots)$　　(S9－8)

暗环半径　　$r_k^2 = k\lambda R \quad (k = 0,1,2,\cdots)$　　(S9－9)

根据式(S9－8)和式(S9－9)，若入射光波长 λ 已知，钠光灯辐射波长为 5.893×10^{-7} m，测出各级暗环或明环的半径，则可求出曲率半径 R。

观察牛顿环时发现，牛顿环中心不是理想的一个接触点，而是一个不甚清楚的暗斑或亮斑。其原因是与透镜与平板玻璃接触时，接触压力引起玻璃弹性形变，而且接触处微小尘埃的存在，都会引起一个附加厚度产生附加光程差。因此很难准确判定级数 k 和测定出 r_k。为消

除这种误差，可以通过取两个暗环的半径 r_m 和 r_n 的平方差来实现，由式(S9－9)，可推得

$$R=\frac{r_m^2-r_n^2}{(m-n)\lambda} \tag{S9-10}$$

由于接触点不是一个理想点，暗环中心不易确定，直接测量半径会产生较大误差，故取测量暗环直径，式(S9－10)改写为

$$R=\frac{D_m^2-D_n^2}{4(m-n)\lambda} \tag{S9-11}$$

式中，D 为牛顿环直径。实验中波长 λ 已知，所以只要分别测量第 m 级暗环和第 n 级暗环的直径 D_m、D_n，就可以测量出平凸透镜的曲率半径 R。

2. 劈尖

劈尖干涉装置如图 S9－2 所示。将两块光学平板玻璃叠放在一起，在一端放入薄片或细丝，则在两玻璃板间形成一空气劈尖。当用单色光照射时，在劈尖薄膜的上下两表面反射的两束相干光相遇时发生干涉。干涉条纹是一组与玻璃板交线相平行的等间距明暗相间的平行直条纹，且形成在劈尖膜上表面附近，这也是一种等厚干涉条纹。

发生干涉时光程差为

$$\delta=2e+\frac{\lambda}{2} \tag{S9-12}$$

劈尖形成明纹的条件为

$$\delta=2e_k+\frac{\lambda}{2}=k\lambda \quad (k=1,2,3,\cdots) \tag{S9-13}$$

劈尖形成暗纹的条件为

$$\delta=2e_k+\frac{\lambda}{2}=(2k+1)\frac{\lambda}{2} \quad (k=0,1,2,\cdots) \tag{S9-14}$$

相邻两明条纹(或暗条纹)所对应的空气膜厚度差为

$$\Delta e=e_{k+1}-e_k=\frac{\lambda}{2} \tag{S9-15}$$

根据几何关系，有 $\frac{d}{L}=\frac{\Delta e}{l}$，故

$$d=\frac{L}{l}\Delta e=N\cdot\Delta e=nL\cdot\Delta e=nL\cdot\frac{\lambda}{2} \tag{S9-16}$$

式中，d 为劈尖厚度；L 为劈尖的长度；l 为相邻明条纹或暗条纹的间距；Δe 为相邻明条纹或暗条纹的高度差；N 为距离为 L 劈尖中明条纹或暗条纹数；n 为单位长度内明条纹或暗条纹数。

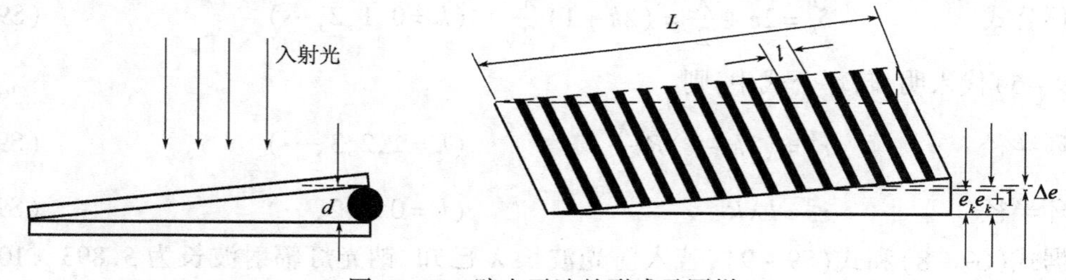

图 S9－2　劈尖干涉的形成及图样

［实验内容与步骤］

1. 利用牛顿环测量平凸透镜球面的曲率半径

(1)将牛顿环装置放在读数显微镜的载物平台上，并使直接目视能看到牛顿环部分置于

读数显微镜的正下方。

(2)点亮钠光灯,调节读数显微镜45°半反镜的角度,使钠光灯发出的水平光经45°半反镜反射后,向下垂直入射到牛顿环装置上,经空气薄膜反射后,向上到达读数显微镜中,调至显微镜视场中亮度最大。

(3)调节读数显微镜目镜,使十字叉丝清晰。调节显微镜的物镜,同时从目镜外观察,使物镜尽量贴近牛顿环,再从目镜中观察,转动显微镜调焦手轮,使目镜自下而上缓慢移动,直到目镜中能够看到清晰的干涉条纹,且环纹与叉丝无视差。微微移动牛顿环装置,使叉丝交点与牛顿环中心大致重合,并使一根叉丝与标尺平行。定性观察左右各35个暗环的干涉条纹是否清晰并且都在读数显微镜的读数范围之内,如果不清晰,此时可进一步调节45°半反镜的角度,并进一步调焦,使牛顿环干涉条纹清晰,无视差。

(4)从牛顿环的中心开始,单向转动测微鼓轮,使显微镜沿水平叉丝方向移动,则竖直叉丝将逐一扫过各条纹,数出移动过的暗环个数 m,当 $m=35$ 时,把测微鼓轮往相反方向转动到第30个暗环,并使叉丝对准暗环中间,记录读数(即环的位置读数),继续转动鼓轮,依次测出 $m=29$、28、27、26 和 $n=20$、19、18、17、16 各个暗环位置的读数。再继续旋转鼓轮,使镜筒经过圆心而到另一边,再依次测出 $n=16$、17、18、19、20 和 $m=26$、27、28、29、30 各个暗环位置的读数。将测量结果填入表 S9-1 中。

2. 用劈尖干涉法测细丝直径

(1)将待测细丝夹在两块平板玻璃的一端,两平玻璃板另一端直接接触,形成劈尖,然后置于读数显微镜载物台上,使读数显微镜对劈尖表面调焦,以看到清晰的干涉条纹。

(2)调节读数显微镜十字叉丝竖线与读数显微镜主尺方向垂直。为了测量准确,还应使细丝与干涉条纹平行。然后调节劈尖位置,使干涉条纹与十字叉丝竖线平行。

(3)用读数显微镜测出20条暗条纹间的垂直距离 l,再测出棱边到细丝所在处的总长度 L。重复测量5次,将数据填入自己设计的表格中,并求细丝直径的平均值。

[注意事项]

(1)实验时,为避免引入空程误差,测微鼓轮应单方向转动。

(2)严禁读数显微镜物镜自上而下移动而导致压碎牛顿环或劈尖装置。

(3)切勿用手触摸光学仪器表面。

(4)切勿随意调节牛顿环装置中的三个调节螺钉。

(5)切勿频繁开启钠光灯。

[数据及处理]

1. 数据记录(表 S9-1)

表 S9-1 牛顿环测量平凸透镜曲率半径 R 数据记录表格

环的级数	m	30	29	28	27	26
环的位置/mm	左					
	右					
环的直径 D_m/mm						
D_m^2/mm^2						
环的级数	n	20	19	18	17	16

续表

环的级数	n	20	19	18	17	16
环的位置/mm	左					
	右					
环的直径 D_n/mm						
D_n^2/mm²						
$(D_m^2-D_n^2)$/mm²						
$\Delta(D_m^2-D_n^2)$/mm²						

2. 数据处理

用逐差法处理数据,计算曲率半径 R 及不确定度,写出测量结果的表达式。

3. 结果分析

对实验结果进行分析,并得出实验结论。

[问题讨论]

(1)如何应用光的等厚干涉测量平凸透镜的曲率半径和金属细丝直径?

(2)牛顿环的中心斑在什么情况下是暗的?什么情况下是亮的?对实验结果是否有影响?为什么?

视频 S9　等厚干涉——牛顿环和劈尖

(3)如何调整读数显微镜?使用读数显微镜应注意哪些问题?

(4)从读数显微镜看到的是放大的牛顿环的像,测出的牛顿环的直径是否也是放大值?

(5)能否用牛顿环测气体或透明液体的折射率?如果能,如何测量?

等厚干涉——牛顿环和劈尖如视频 S9 所示。

实验 10　分光计的调整与玻璃三棱镜顶角的测量

[引言]

利用光的色散、干涉、衍射等现象可将复色光分成单色光。将复色光分成单色光的现象称为分光,实现分光的仪器称为分光计。通过用分光计测量入射光与出射光之间的偏转角,可间接测量折射率、光波波长、色散率以及进行光谱的定性分析。

棱镜分光计,就是将入射到三棱镜一侧表面上的复色光,在另一侧表面上分离出来,虽然入射角对各种波长的光来说是同一的,但出射角并不相同,表明对于不同波长的光,其折射率也不相同,这种折射率随波长变化的现象称为色散。

[实验目的]

(1)了解分光计的结构和各部件的功能,学习分光计的调节和使用方法。

(2)掌握测量三棱镜顶角的方法。

[预习问题]

(1)分光计由哪几部分组成?各部分的作用是什么?

(2)分光计有哪些锁紧螺钉?其作用各是什么?在什么情况下锁紧,又在什么情况下松开?

(3)分光计的调节过程有哪些?如何调节?调整好的分光计应满足哪些关系?

[实验仪器]

JJY1 型分光计(分度值:1′,精度值:1′,测量范围:0°~360°)、钠光灯、平面镜、三棱镜。

[实验原理]

1. 分光计的结构

分光计的型号很多,但结构基本相似,主要由平行光管、望远镜、载物台和读数装置四个部分组成,结构如图 S10-1 所示。

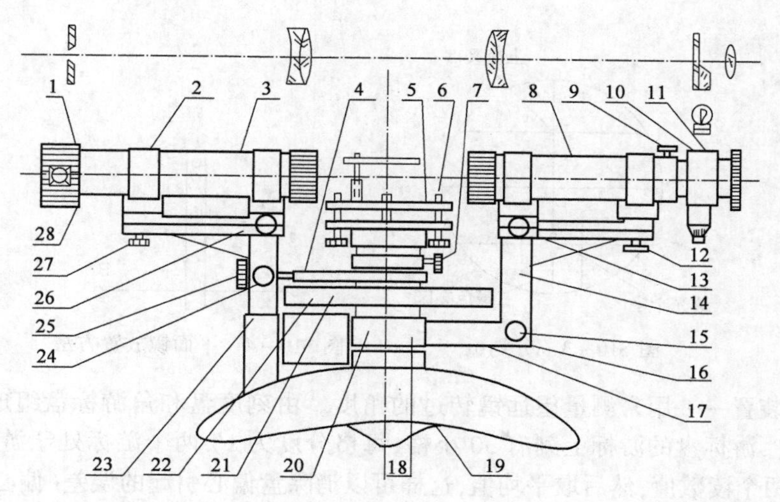

图 S10-1　分光计结构图

1—狭缝装置;2—狭缝锁紧螺钉;3—平行光管;4,18—制动架;5—载物台;6—载物台调平螺钉(3 只);7—载物台紧锁螺钉;8—望远镜;9—望远镜套筒锁紧螺钉;10—阿贝式自准直目镜;11—望远镜目镜焦距调解鼓轮;12—望远镜倾度调节螺钉;13—望远镜光轴水平调节螺钉;14—支臂;15—望远镜微调螺钉;16—主刻度盘止动螺钉;17—望远镜止动螺钉;19—底座;20—转座;21—游标盘;22—主刻度盘;23—立柱;24—游标盘微调螺钉;25—游标盘制动螺钉;26—平行光管光轴水平调节螺钉;27—平行光管倾度调节螺钉;28—狭缝宽窄调节螺钉

(1)平行光管——产生平行光的装置,管的一端装有会聚透镜,另一端装有一可伸缩的套筒,套筒末端有一狭缝装置。若将狭缝调到物镜的焦平面上,从平行光管出来的光就是平行光。狭缝的宽度可以用螺钉 28 来调节,平行光管的倾斜度可用平行光管下的螺钉 27 来调节。

(2)望远镜——用来观测的装置,由物镜、叉丝分化板和阿贝式目镜组成,它们分别置于三个彼此可以相互移动的套筒中,如图 S10-2 所示,目镜套筒 6 可沿套筒 3 前后移动,套筒 3 也可沿物镜套筒 2 前后移动。在套筒 3 里装有分化板,分化板的下方刻有透光十字,如图 S10-3所示。

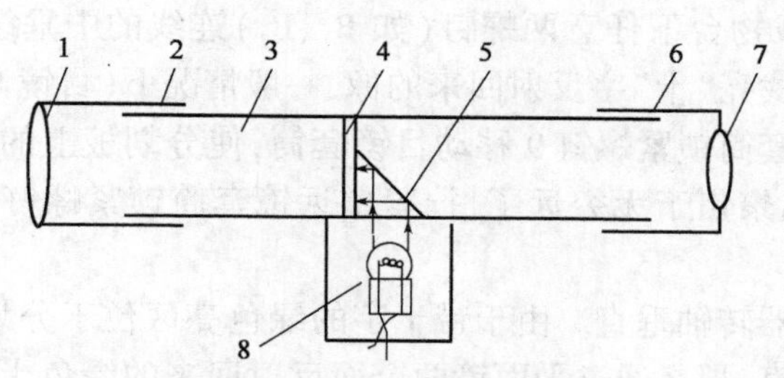

图 S10-2　望远镜结构图

1—望远镜物镜;2—物镜套筒;3—分划板固定套筒;4—分划板;5—45°直角三棱镜;6—目镜套筒;7—目镜;8—望远镜光源

分划板下方紧贴一块45°全反射小棱镜,棱镜与分划板的粘贴部分涂成黑色,仅留一个绿色的小十字形透光叉丝。光线从小棱镜的另一直角边入射,从45°反射面反射到分划板上,透光部分便形成一个在分划板上的明亮的十字窗。用这个十字作为物来调节望远镜,使其达到使用要求。

(3)载物台——用来放置平面镜、棱镜等光学元件。载物台下有三个螺钉 B_1、B_2、B_3(图S10-4),用来调节载物台的高度和倾斜度。图S10-1中螺钉7可将平台和游标盘固定在一起。

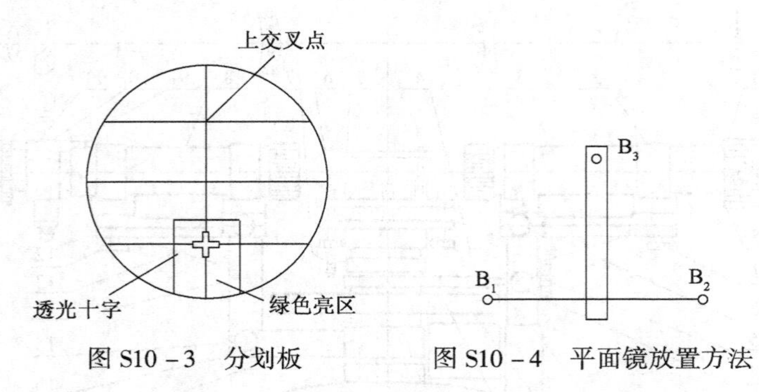

图S10-3　分划板　　图S10-4　平面镜放置方法

(4)读数装置——用来测量望远镜转过的角度。由刻度盘和角游标盘组成,刻度盘的最小刻度为0.5°,游标盘的游标上刻有30小格,每格分度为1′,两个游标处于游标盘的对称位置,测量读出两个读数值,然后取平均值,这样可以消除盘偏心引起的误差(偏心差)。

2. 分光计的调节

为保证准确测量,必须按要求对分光计进行调节。分光计调节好的标准是:平行光管能产生平行光;望远镜能接收平行光(即聚焦于无穷远);平行光管的光轴和望远镜的光轴均与仪器转轴垂直。

1)目测粗调。

如图S10-1所示,将望远镜与平行光管转至一条直线上,用眼睛粗略地估计,调节望远镜和平行光管的高低倾度调节螺钉12、27,使两者处于同一水平线上,调节载物台下的三个调平螺钉6,使载物台的高度适中并且大致处于水平状态,然后再对各部分进行精细调节。

2)分光计的精细调节

(1)望远镜的调节(自准法)。

使望远镜能接收平行光(即聚焦于无穷远)。接通电源,通过目镜观察分划板,同时转动目镜套筒直至看清分划板上的刻线和绿色十字亮区,如图S10-3所示。再将平面镜放于载物台上,并使其底座位于载物台下任意两螺钉(如 B_1、B_2)连线的中垂线上,如图S10-4所示。再通过目镜在分划板上找亮"十"字反射回来的像,一般情况下(目镜调节较好)可以看到一个亮斑,此时,松开望远镜套筒锁紧螺钉9移动目镜套筒,使分划板上的绿色十字清晰可见且没有视差,就说明望远镜已聚焦于无穷远了,拧紧望远镜套筒锁紧螺钉9将物镜与分划板位置固定。

使望远镜光轴与仪器转轴垂直。由于带十字的绿色亮区位于分划板刻线的下交叉点,如果望远镜与仪器转轴垂直,那么通过平面镜两个面反射回来的绿色十字像都应位于分划板刻线的上交叉点,如图S10-5所示,称为标准位置。但一般看到的清晰"十"字像并不处于标准

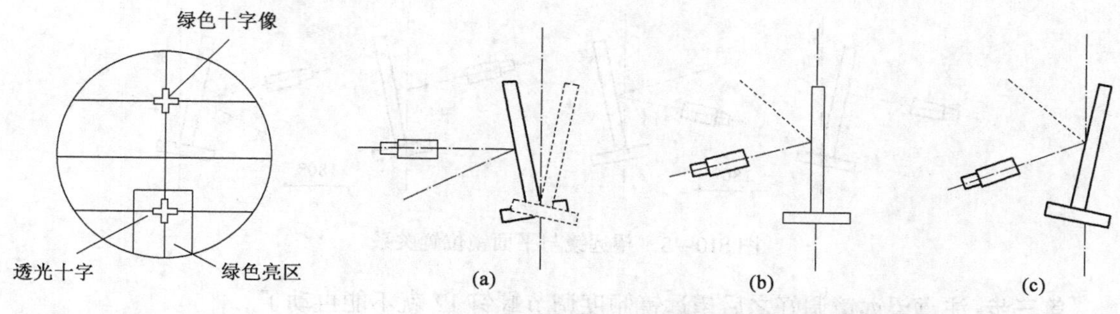

图 S10－5　标准位置

图 S10－6　望远镜与平面镜位置关系

位置，就是说望远镜光轴与平面镜并不垂直，此时，望远镜与平面镜存在以下三种可能位置关系：

图 S10－6 中：(a)表示望远镜光轴与仪器转轴垂直，而平面镜不与仪器转轴平行，那么此时看到的平面镜两面反射回来的绿色十字像位于标准位置的上下方，呈对称分布；(b)表示望远镜光轴与仪器转轴不垂直，而平面镜与仪器转轴平行，此时无论哪个面正对望远镜看到的绿色十字像始终位于标准位置的某一方（上方或下方）；(c)表示望远镜光轴与仪器转轴不垂直，而且平面镜与仪器转轴不平行，这是一种普遍情况，反射回来的像可位于标准位置的任一方，呈不对称分布。根据上面的分析，我们可以进行初步的调节，首先看到两个面反射回来的绿色十字像，假设如图 S10－7 所示，然后按下述步骤进行调节。

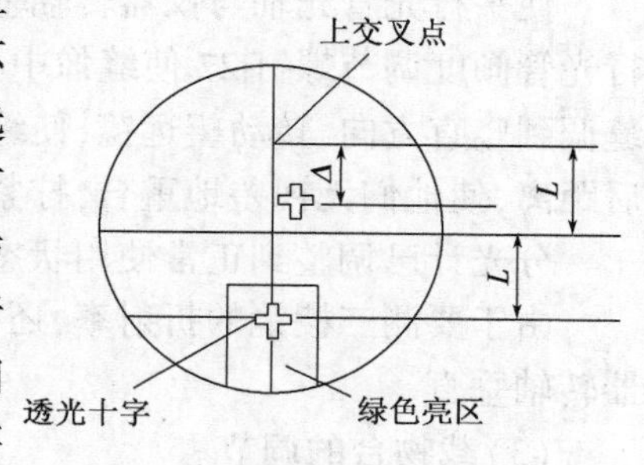

图 S10－7　绿色十字位置

第一步：调节望远镜倾度调节螺钉 12，使绿色十字像上升（或下降）$\Delta/2$ 距离，再调节螺钉 B_1，使像上升（或下降）$\Delta/4$ 距离，再调节螺钉 B_2，也使像上升（或下降）$\Delta/4$ 距离到达分划板的上横刻线。转动望远镜使绿色十字像处于标准位置，如图 S10－5 所示。

有时从望远镜视场中观察不到由平面镜反射回来的绿十字像，这种情况说明目测粗调未调到位，说明入射到平面镜的光线的入射角（或反射角）太大。解决的办法是，稍微将载物台转过一小角度，眼睛沿望远镜镜筒外反射光的方位对准反射镜找到绿十字像，并使视线与望远镜光轴保持在同一水平面，判断出绿十字像的高低，若高于此水平面，则使像降低；若低于此水平面，则使像上升（上升或下降均要求将调平螺钉和望远镜倾度调节螺钉配合调节，不能只调节 B_1，或只调望远镜倾度调节螺钉），经过筒外观察粗调后，转动载物台直至从望远镜视场中找到绿十字像后，再按上述方法将绿十字像调到与分划板上十字重合。

第二步：将载物台旋转 180°，在分划板上找到绿色十字像，无论像在分划板的什么位置，均采取与第一步同样的方法进行调节，使绿色十字像处于标准位置。再将载物台旋转 180°，如此反复调节，直至在不调节任何螺钉的情况下，平面镜两个面反射回来的绿色十字像都可以处于标准位置。这时望远镜光轴与仪器转轴垂直。

将载物台旋转 180°后，有时找不到反射回来的绿色十字像，其原因可能如图 S10－8 所示，这时需要判断出绿十字像的高低，若像高（低）于“水平面”，将载物台转过 180°（即平面镜转回到原来与绿色十字像重合的位置）后，则沿使绿十字像沿上升（下降）的方向调节 B_1，使原来重合的绿十字像向上（下）偏离分划板上交叉点（注意：勿移出视场外），然后调节望远镜倾度调节螺钉 12，使二者再次重合，再将载物台转过 180°，进行逼近调节。

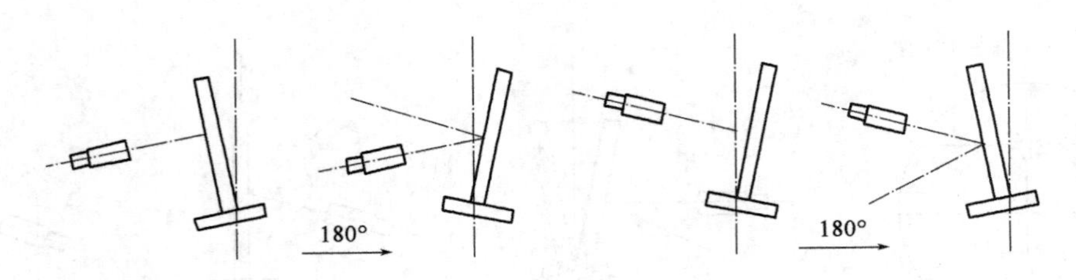

图 S10－8 望远镜与平面镜位置关系

第三步:注意望远镜调好之后望远镜倾度调节螺钉 12 就不能再动了。

(2)平行光管的调整。

平行光管产生平行光。以聚焦于无穷远的望远镜为标准,接通钠光灯照亮狭缝,调节狭缝宽窄调节螺钉 28 使狭缝宽窄适中。松开狭缝锁紧螺钉 2,前后移动狭缝体使狭缝清晰地成像于分划板上,此时狭缝处于物镜的焦平面上,此时平行光管产生平行光。

使平行光管光轴与仪器转轴垂直。用已调好的望远镜为标准,使狭缝成水平位置,调节平行光管倾度调节螺钉 27 使缝像中心与分划板中央的水平刻线重合,如图 S10－9 所示;再将狭缝调到竖直方向,转动望远镜,使缝像中心与分划板中央的竖直刻线重合,再微微调整狭缝前后距离,使他们无视差地重合,拧紧狭缝锁紧螺钉 2。

分光计已调整到正常使用状态。

由于要测三棱镜的折射率,还须对放上三棱镜的载物台进行调节,使三棱镜的主截面与仪器转轴垂直。

(3)载物台的调节。

为了便于调节,将三棱镜按图 S10－10 所示放于载物台上,首先使三棱镜的一个光学面 AB 正对望远镜,找到 AB 反射回的绿十字像,若像与分划板的上交叉点距离 Δ,则调 B_3 使像移到上交叉点(或上横刻线)。然后转动平台使三棱镜另一个光学面 AC 对准望远镜,按前述方法调节 B_1,使绿色十字像处在上交叉点位置。反复调节数次,使两个光学面反射回来的绿色像在不调任何螺钉的情况下均处在分划板的上交叉点位置(即标准位置),此时三棱镜的主截面就与仪器转轴垂直了。同时,载物台也调好了,此种方法称为自准法。

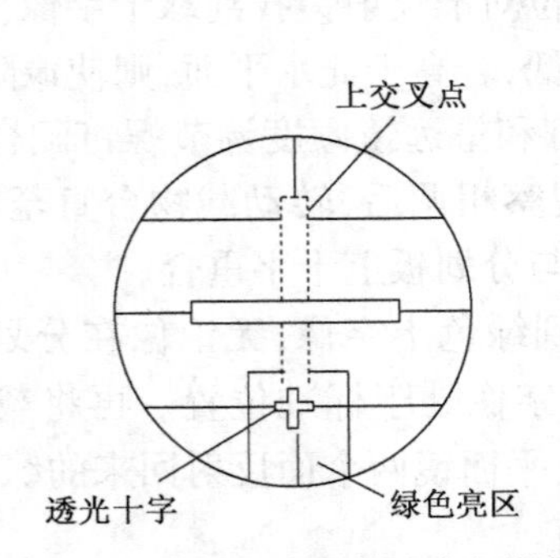

图 S10－9 平行光管光轴垂直仪器转轴

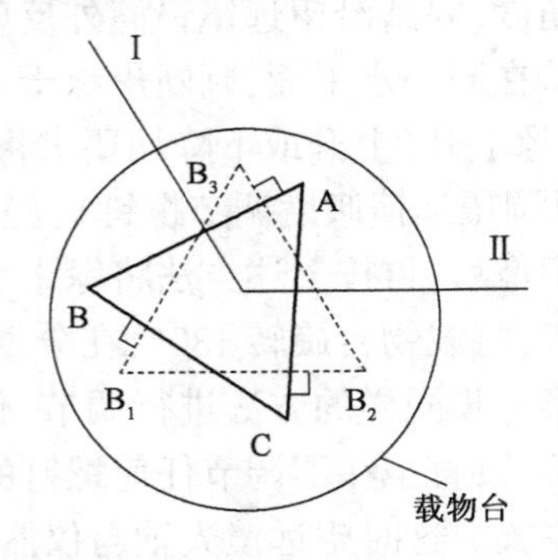

图 S10－10 三棱镜的放法

至此,分光计处于测量状态,可根据要求进行三棱镜折射率的测量。

[实验内容与步骤]

(1)按分光计调整步骤调节望远镜和平行光管(见分光计的调整部分),使分光计处于使用状态。

(2)调整载物台,使三棱镜主截面与仪器转轴垂直(见分光计的调整部分)。

(3)用自准法测量三棱镜顶角。

①把望远镜转到如图 S10-10 中的Ⅰ位置,使 AB 面反射的绿十字像处在标准位置(即分划板的上交叉点),记下此时两个角游标的读数 θ_1、θ_2。

②锁紧螺钉 16、17 和 7,松开游标盘制动螺钉 25,转动载物台,使棱镜的 AC 面对准望远镜,图 S10-10 中的Ⅱ位置,使 AC 面反射回来的绿十字像与叉丝的上交叉点重合,再记下此时两个角游标孔的读数 θ'_1 和 θ'_2(注意:θ_1 和 θ'_1 是同一个游标孔的两个读数,θ_2 和 θ'_2 也一样,不要记录错)。

③由几何关系得

$$\varphi = \frac{1}{2}\left[\,|\theta_1 - \theta'_1| + |\theta_2 - \theta'_2|\,\right] \qquad (S10-1)$$

则三棱镜的顶角为

$$A = 180° - \varphi \qquad (S10-2)$$

[注意事项]

(1)勿用手触摸光学元件光学面,若光学元件有尘埃,应用专用镜头纸轻轻擦拭。

(2)调节望远镜时,应遵从先粗调后细调的原则。

(3)分光计上的各个螺钉在未搞清其作用之前,不要随意扭动。

(4)在未松动某些固定螺钉的情况下,勿转动望远镜、平行光管、载物台、读数盘以及聚焦系统。

[数据及处理]

数据记录表格见表 S10-1。

表 S10-1　测三棱镜顶角数据记录表

次数＼数值	θ_1	θ_2	θ'_1	θ'_2	φ	A_i	$\bar{A}$
1							
2							
3							
4							
5							

[问题讨论]

(1)在调节分光计望远镜光轴与仪器转轴垂直时,为什么要使载物台上平面镜反射回来的绿色十字像与望远镜分划板上十字重合?

(2)为什么有时从望远镜视场中看不到由平面镜反射回来的绿色十字?如何解决?

(3)测三棱镜顶角时,可能得到类似以下的数据:

$\theta_1 = 271°31'$　$\theta_2 = 91°33'$　$\theta'_1 = 31°32'$　$\theta'_2 = 211°30'$

分析一下为什么会出现这种状况?应如何处理?

分光计的调整与玻璃三棱镜顶角的测量如视频 S10 所示。

视频 S10　分光计的调整与玻璃三棱镜顶角的测量

实验 11　用分光计测量玻璃三棱镜折射率

[引言]

光线入射到光学元件(如平面镜、三棱镜、光栅等)上时会发生反射、折射、色散或衍射等现象,光的反射定律、折射定律定量地描述了光线在传播过程中方向发生偏折时角度间的相互关系,而光在传播过程中的衍射、散射等物理现象也都与角度有关,一些光学量如折射率、光波波长、色散率等都可以通过直接测量相关的角度来确定。

[实验目的]

(1)进一步了解分光计的结构和各部件的功能,掌握分光计的调节和使用方法。

(2)掌握测量不同波长光的最小偏向角的方法。

(3)学会测量三棱镜的折射率。

[预习问题]

(1)分光计由哪几部分组成?各部分的作用是什么?

(2)分光计有哪些锁紧螺钉?其作用各是什么?在什么情况下锁紧,又在什么情况下松开?

(3)分光计的平行光管怎么调节?

(4)如何判断最小偏向角?

[实验仪器]

JJY1 型分光计(分度值:1′,精度值:1′,测量范围:0°~360°)、钠光灯、平面镜、三棱镜。

[实验原理]

光线由一种介质进入另一种介质时,在其介面上会发生折射现象。折射率是反映介质特性的一个重要物理量。光的折射满足折射定律:$n_1/n_2=\sin i/\sin\gamma$,$n_1$、$n_2$分别表示第一种介质和第二种介质的折射率。可见精确测量折射率就可以确定物质对光的折射程度。

测定折射率的方法很多,棱镜法是常用的方法之一,其测量原理如图 S11-1 所示,其中,ABC 表示三棱镜,AB 和 AC 是透光光学面,其夹角 A 称为三棱镜的顶角,BC 为毛玻璃面一束单色平行光入射在三棱镜 AB 面上,经三棱镜两次折射后,从 AC 面上出射,设入射角为 α、出射角为 β,出射光线与入射光线之间的夹角为 δ,称为偏向角,根据几何关系有:

$$\delta=(\alpha-\angle 1)+(\beta-\angle 2)$$

$$A=\angle 1+\angle 2$$

所以

$$\delta=\alpha+\beta-A \tag{S11-1}$$

由式(S11-1)可见,δ 是 α 和 β 的函数,而出射角 β 又只是入射角 α 的函数,故对于给定的三棱镜及入射波长的光线而言,δ 只是入射角 α 的函数。理论与实验都证明当入射角改变时,偏向角 δ 存在最小值,称为最小偏向角。令 $\dfrac{\partial\delta}{\partial\alpha}=0$,可求出 $\delta_{\min}$ 所对应的值 α,即当 $\delta=\delta_{\min}$ 时,$\alpha=\beta$,$\angle 1=\angle 2$ 代入式(S11-1),可得 $\alpha=\dfrac{1}{2}(\delta_{\min}+A)$。

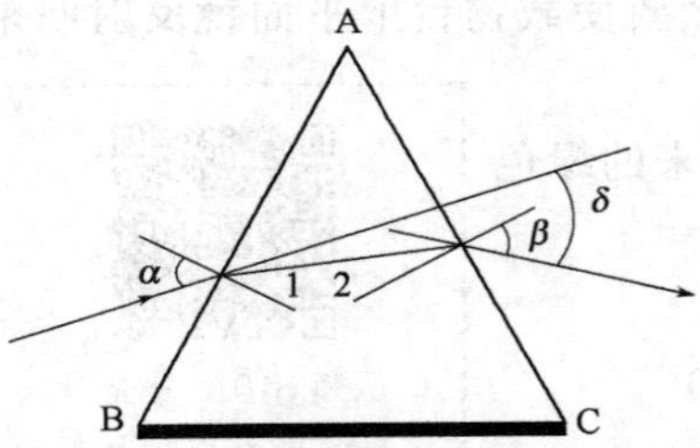

图 S11-1　偏向角光路图

根据折射定律,$\sin\alpha=n\sin\angle 1$,又因为 $A=\angle 1+\angle 2=$

$2\angle 1$，所以三棱镜对该单色光的折射率为

$$n=\frac{\sin\alpha}{\sin\angle 1}=\frac{\sin\frac{1}{2}(\delta_{\min}+A)}{\sin\frac{A}{2}} \tag{S11-2}$$

由此可见，要测量三棱镜对某种单色光的折射率，只要测得三棱镜的顶角 A 和该单色光穿过三棱镜的 $\delta_{\min}$，再按式（S11－2）可求出折射率。

［实验内容与步骤］

（1）按分光计调整步骤调节望远镜和平行光管（见分光计的调整部分），使分光计处于使用状态。

（2）调整载物台（见分光计的调整部分）。

（3）测最小偏向角。

①将棱镜置于载物台上，松开载物台紧锁螺钉 7 和望远镜止动螺钉 17，让 AB 面的法线与平行光管轴约成 60°角，如图 S11－2 所示。

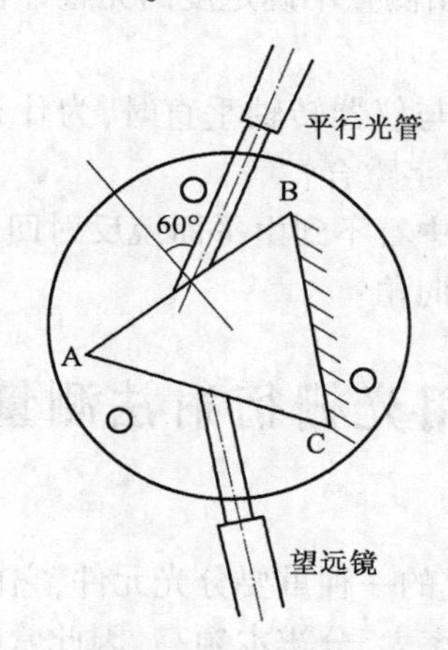

图 S11－2　最小偏向角的测定

②缓慢转动载物台，使 α 变小，用望远镜跟踪缝像，可看到缝像移至某一位置后将反向移动，这说明最小偏向角存在。

③用望远镜观察缝像将要改变而未改变移动方向为止，转动望远镜，使分划板中央的竖直刻线与缝像中心重合，拧紧螺钉 7 和 25，在重合位置附近可以通过望远镜微调螺钉 15 进行细微调节（在使用该螺钉时必须先固定螺钉 17），记下游标盘的读数 θ_A 和 θ_B。

④取下三棱镜，转动望远镜使缝像的中心与分划板的中央竖线再次重合。测出此时的角位置 θ'_A 和 θ'_B，则

$$\delta_{\min}=\frac{1}{2}\left[\,|\theta_A-\theta'_A|+|\theta_B-\theta'_B|\,\right] \tag{S11-3}$$

［注意事项］

（1）勿用手触摸光学元件光学面，若光学元件有尘埃，应用专用镜头纸轻轻擦拭。

（2）调节望远镜时，应遵从先粗调后细调的原则。

（3）分光计上的各个螺钉在未搞清其作用之前，不要随意扭动。

（4）在未松动某些固定螺钉的情况下，勿转动望远镜、平行光管、载物台、读数盘以及聚焦系统。

[数据及处理]

1. 数据记录(表 S11-1)

表 S11-1 测最小偏向角数据记录表

次数 \ 数值	θ_A	θ_B	θ'_A	θ'_B	δ_{min}	$\bar{\delta}_{min}$
1						
2						
3						
4						
5						

2. 数据处理

计算测量结果的不确定度,给出测量不确定度的完整评价。

[问题讨论]

(1)在调节分光计望远镜光轴与仪器转轴垂直时,为什么要使载物台上平面镜反射回来的绿色十字像与望远镜分划板上十字重合?

(2)为什么有时从望远镜视场中看不到由平面镜反射回来的绿色十字?如何解决?

(3)在实验中如何确定最小偏向角?

实验 12 用光栅衍射法测量光的波长

[引言]

光栅是根据多缝衍射原理制成的一种重要分光元件,它由大量等宽、等间距的平行狭缝组成。光栅衍射条纹狭窄细锐、色散率大、分辨本领高,因此常被用来精确地测定光波波长以及进行光谱分析。根据光栅衍射的基本原理,利用光栅将光源中含有不同波长的光按波长顺序分成许多光谱线,不同波长具有不同的衍射角,通过测量衍射角进而可测得光波波长。光栅衍射原理也是晶体 X 射线结构分析和近代频谱分析与光学信息处理的基础。

[实验目的]

(1)观察光栅衍射现象,加深对光的衍射及光栅分光原理的理解。

(2)进一步熟悉和巩固分光仪的调整和使用方法。

(3)利用光栅衍射测定汞灯谱线波长。

[预习问题]

(1)分光计调节好的标准是什么?

(2)测衍射角时,如何使望远镜中分划板上的黑色十字叉丝准确对准各谱线?

[实验仪器]

JJY 型分光计、汞灯、平行平面镜、光栅等。

[实验原理]

1. 光栅方程

光在传播过程中,能绕过障碍物的边缘改变它的传播方向,并使其强度发生重新分布的现象称为光的衍射。当一束平行光与光栅法线成角度 i 入射于光栅平面上时就会产生衍射，如图 S12－1 所示。

从 A 点作 AC 垂直于入射线 CB，作 AD 垂直于衍射线 BD，BD 与光栅平面法线所夹的角为衍射角 θ，如果在这个方向上的平行衍射光线用会聚透镜会聚起来，由于相互干涉使光振动加强，则在 F 点产生一亮线，其光程差 $CB+BD$ 必等于入射光波长 λ 的整数倍，即

$$d(\sin\theta \pm \sin i)=k\lambda, \quad (k=0,\pm1,\pm2,\cdots) \qquad (S12-1)$$

入射光线与衍射光线在光栅平面法线同侧时，式（S12－1）中 $\sin i$ 前取正号，两者分别在法线两侧时取负号。式（S12－1）称为光栅方程。

当一束平行光垂直于光栅平面入射的情况下，即 $i=0$，则光栅方程变为

$$d\sin\theta=k\lambda, \quad (k=0,\pm1,\pm2,\cdots) \qquad (S12-2)$$

其中

$$d=a+b$$

式中，d 为光栅常数；a 为透光部分宽度；b 为不透光部分宽度；k 为衍射级次。

2. 光栅方程的讨论

（1）如果入射光是单色光，则由式（S12－2）可知，波长 λ 相同，同级衍射条纹的衍射角 θ 也相同。具有相同衍射角 θ 的光束经过透镜会聚后发生干涉，此时我们在屏上可以看到光栅衍射图像，如图 S12－2 所示，各级正负级衍射条纹关于零级衍射条纹是对称分布的，且随级次 k 的增大，衍射角增大。不同光栅（d 不同），能看到的衍射条纹的更高级次不同，光栅常数 d 越大，能见到的条纹越多。

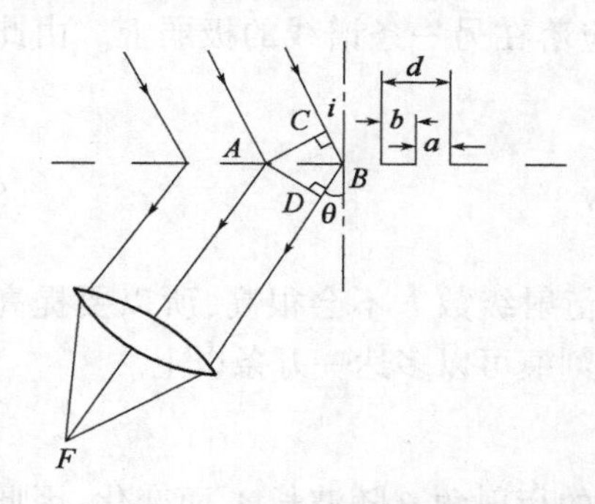

图 S12－1　光线斜射于光栅平面的衍射

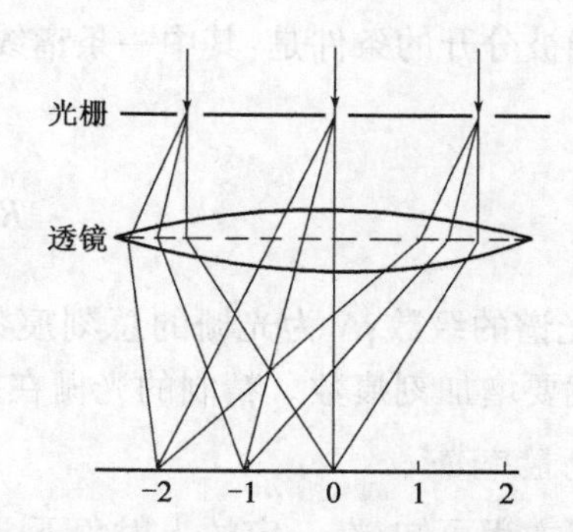

图 S12－2　单色光垂直入射于光栅平面的衍射

（2）如果入射光是复色光，则由式（S12－2）可知，波长 λ 不同，衍射角 θ 也随之不同，于是复色光被分解为单色光，在中央 $k=0$、$\theta_k=0$ 处，各色光仍重叠在一起，组成中央亮线。在中央亮线两侧对称分布着 $k=1,2,3,\cdots$ 各级谱线，所形成的彩色光谱称为衍射光谱，如图 S12－3 所示。

由光栅方程（S12－2）可以看出，不同波长的同级光谱（谱线）出现在不同方位，长波的衍射角大，短波的衍射角小。

由光栅方程（S12－2）可知：如果已知光栅常数 d，通过测出 k 级条纹的衍射角 θ_k，就可求出入射光波长 λ。

光栅方程（S12－2）的适用条件为入射光必须是平行光且必须垂直入射光栅平面。

3. 光栅的分辨本领和角色散率

表征光栅特征的除光栅常数 d 外，还有光栅的分辨本领 R 和角色散率 ψ。

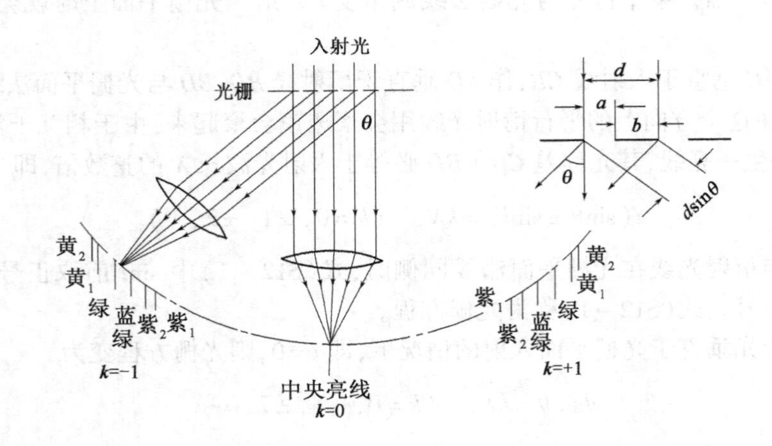

图 S12-3　汞灯的光栅衍射光谱

1）光栅的分辨本领 R

光栅的分辨本领 R 是指把波长靠得很近的两条谱线分辨清楚的本领。通常把恰能分辨的两条谱线的平均波长 λ 与这两条谱线的波长差 $\delta\lambda$ 之比定义为分辨本领 R，即

$$R=\frac{\lambda}{\delta\lambda} \tag{S12-3}$$

R 越大，表明刚刚能被分开的波长差 $\delta\lambda$ 越小，光栅分辨能力越强。按照瑞利准则，所谓两条谱线恰能被分开的条件是：其中一条谱线的极强应落在另一条谱线的极弱上。由此可推出分辨本领

$$R=\frac{\lambda}{\delta\lambda}=kN \tag{S12-4}$$

式中，k 为光谱的级数；N 为光栅的总刻痕数。一般衍射级数 k 不会很高，所以要提高光栅的分辨本领就要增加刻痕数。精制的光栅在一厘米内刻痕可以多达一万条以上。

2）角色散率 ψ

由光栅方程可知，在一定的入射角下，各光波线的衍射角 θ 随波长 λ 而变化，因此定义角色散率 ψ 为两条谱线衍射角之差与其波长差之比，即

$$\psi=\frac{\Delta\theta}{\Delta\lambda} \tag{S12-5}$$

将光栅方程微分并代入 ψ 的定义式可得

$$\psi=\frac{\Delta\theta}{\Delta\lambda}=\frac{k}{d\cos\theta} \tag{S12-6}$$

可见，角色散率 ψ 与光栅的总刻痕数 N 无关，仅决定于光栅常数 d 和光谱级数 k。为了得到大的角色散率，光栅常数必须很小，即光栅每单位长度内的刻痕数要尽量多。

从以上的讨论可以看出，提高光栅的分辨本领和角色散率的关键是减小光栅常数，提高单位长度内的栅线条数。

[实验内容与步骤]

(1)按分光计的调整要求(参阅实验 10 中分光计的结构和调整),认真调整分光计至正常使用状态,即平行光管能产生平行光;望远镜能接收平行光;望远镜光轴和平行光管光轴都与仪器转轴垂直。

(2)光栅的调节。要求光栅平面(刻痕所在的面)与平行光管光轴垂直并与仪器转轴平行;光栅刻痕与仪器转轴平行。

①调节光栅平面与平行光管光轴垂直,并平行于仪器转轴。将平行光管狭缝转到垂直方向,并对准汞灯,把狭缝宽度调到适中,用汞灯照亮平行光管的狭缝,按图 S12-4 将光栅置于载物台上,使光栅面正对望远镜。转动望远镜对准平行光管,止动望远镜,调节望远镜微调螺钉,使狭缝像的中心线与分划板十字的垂直线相重合。转动载物台,使光栅平面与平行光管光轴垂直,用望远镜观察光栅平面反射回来的绿十字像,通过轻微转动载物台及调节载物台下的 B_1 或 B_2 螺钉,使从光栅面反射回来的绿十字像的中心位于分划板叉丝的上交叉点上,即绿十字像与分划板上方黑色十字叉丝相重合,这时,光栅平面就与平行光管光轴严格垂直并平行于仪器转轴了。

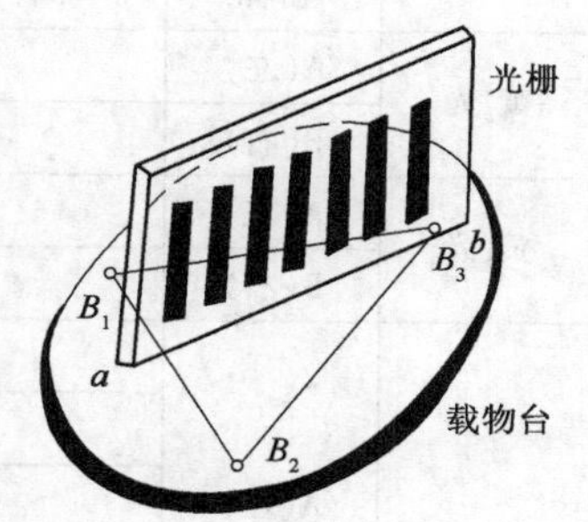

图 S12-4　光栅方法

②调节光栅刻痕与仪器转轴平行。在上述调节的基础上,进一步调节望远镜微调螺钉(以止动望远镜为前提),使狭缝像(即第 0 级光谱)与分划板上纵向中心刻度线严格重合。拧紧载物台、游标盘和刻度环的紧固螺钉,松开望远镜固定螺钉,转动望远镜,可观察到中央明条纹两侧的正负一级、正负二级等衍射谱线,观察这些谱线是否等高,若观察到的谱线有高低变化,说明狭缝与光栅刻痕不平行,此时调节载物台下的螺钉 B_3,直到中央明条纹两侧的衍射条纹基本上等高为止。这时,光栅刻痕与仪器转轴就平行了。

(3)测量谱线的衍射角。

先将望远镜正对平行光管,使零级谱线与分划板上纵向中心刻线严格重合,此时绿色十字的反射像的纵向也与谱线重合。然后再将望远镜转至右侧,测量 $k=+1$ 级各谱线的角位置。为了消除分光计的偏心差,使用左右两个对称的游标进行读数,分别记为 θ_A^{+1}、θ_B^{+1},然后将望远镜转至左侧,测出 $k=-1$ 级各谱线的角位置,读数记为 θ_A^{-1}、θ_B^{-1}。$\theta_A^{-1}-\theta_A^{+1}=2\theta_A$,$\theta_B^{-1}-\theta_B^{+1}=2\theta_B$,由于偏心差的存在,使衍射角 θ_A 和 θ_B 有差异,求其平均值 $\theta=\frac{\theta_A+\theta_B}{2}$,便消除了偏心差。所以,各谱线的衍射角为

$$\theta=\frac{\theta_A^{-1}-\theta_A^{+1}+\theta_B^{-1}-\theta_B^{+1}}{4}$$

测量时,从最右端的黄$_2$光开始,依次测到最左端的黄$_2$光。对绿光重复测量五次。望远镜要依次向一个方向运动,不要来回转动。为了使各级谱线都能与分划板的纵丝重合,读数时注意使用望远镜微调螺钉。

[注意事项]

(1)光栅是精密光学器件,严禁用手触摸光学表面。

(2)望远镜固定螺钉在紧固情况下,严禁硬性转动望远镜。

(3)严禁将汞灯刚启动,又立即关闭,或刚关闭又立即启动。

(4)汞灯的紫外光较强，切勿直视，以免伤害眼睛。

[数据及处理]

1. 数据记录(表 S12－1)

表 S12－1　汞灯谱线波长测量数据记录表格

谱线级数 $k=\pm 1$，光栅常数 $d=$________

谱线特征	分光计读数			$\theta=\frac{\theta_A^{-1}-\theta_A^{+1}+\theta_B^{-1}-\theta_B^{+1}}{4}$	测量值 λ/Å	标准波长 λ_0/Å	对 λ_0 的相对误差/%
	游标	θ^{+1}(右)	θ^{-1}(左)				
黄$_2$光	A(左)						
	B(右)						
黄$_1$光	A(左)						
	B(右)						
绿光	A(左)						
				$\theta_1=$			
				$\theta_2=$			
				$\theta_3=$　$\overline{\theta}=$			
				$\theta_4=$			
	B(右)			$\theta_5=$			
蓝绿光	A(左)						
	B(右)						
紫$_2$光	A(左)						
	B(右)						
紫$_1$光	A(左)						
	B(右)						

2. 数据处理

求出汞灯谱线中绿光的波长、对测量绿光波长的不确定度进行评定并表示出测量结果。

3. 结果分析

对实验结果进行分析，并得出实验结论。

[问题讨论]

(1)光栅分光和棱镜分光所产生的光谱有何区别？

(2)入射平行光不垂直光栅平面而引入的误差对测量结果有怎样的影响？

(3)实验中所用的通常是复制光栅或全息光栅，一般把涂有药膜的一面对着望远镜，如果把光栅面放反了是否会引入误差？

(4)用式(S12－2)测 d 或 λ 时，实验要保证什么条件？如何实现？

(5)当用 $\lambda = 5.893 \times 10^{-7}\text{m}$ 的钠光灯垂直入射到1mm内有500条刻痕的平面透射光栅上时，试问最多能看到几级谱线？并说明理由。

实验13 示波器的使用

[引言]

示波器是一种用途十分广泛的电子测量仪器。它能把肉眼看不见的电信号变换成看得见的图像，便于人们研究各种电现象的变化过程。示波器利用狭窄的、由高速电子组成的电子束，打在涂有荧光物质的屏面上，就可产生细小的光点。在被测信号的作用下，电子束就好像一支笔的笔尖，可以在屏面上描绘出被测信号的瞬时值的变化曲线。利用示波器能观察各种不同信号幅度随时间变化的波形曲线，还可以用于测试电压、电流、频率、相位差等。

[实验目的]

(1)了解示波器的工作原理和使用方法。

(2)学习用示波器测量交流电的电压、周期和频率的方法。

[预习问题]

(1)阅读本实验所用的示波器和低频信号发生器的使用说明，回答示波器的主要功能是什么？

(2)如何通过示波器和低频信号发生器观测李萨如图形？

(3)示波器的扫描原理是什么？如何控制扫描？示波器的整步作用是什么？

[实验仪器]

通用示波器(MOS－620型)、低频信号发生器(XD7型)、连接线等。

[实验原理]

1. 示波器的基本结构

示波器的主要部分有示波管、带衰减器的 Y 轴放大器、带衰减器的 X 轴放大器、扫描发生器(锯齿波发生器)、触发同步和电源等，其结构方框图如图S13－1所示。为了适应各种测量的要求，示波器的电路组成是多样而复杂的，这里仅就主要部分加以介绍。

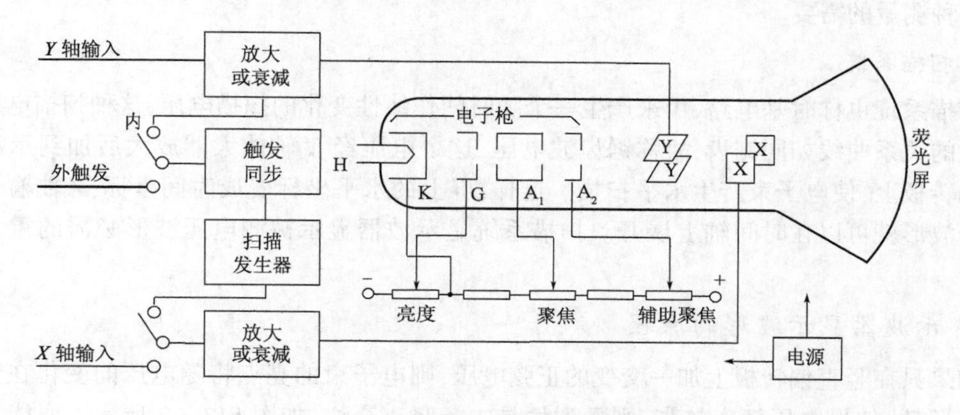

图S13－1 双通道通用示波器原理框图

1）示波管

如图 S13 - 1 所示，示波管主要包括电子枪、偏转系统和荧光屏三部分，全都密封在玻璃外壳内，里面抽成高真空。下面分别说明各部分的作用。

（1）荧光屏：它是示波器的显示部分，当加速聚焦后的电子打到荧光上时，屏上所涂的荧光物质就会发光，从而显示出电子束的位置。当电子停止作用后，荧光剂的发光需经一定时间才会停止，称为余辉效应。

（2）电子枪：由灯丝 H、阴极 K、控制栅极 G、第一阳极 A_1、第二阳极 A_2 五部分组成。灯丝通电后加热阴极。阴极是一个表面涂有氧化物的金属筒，被加热后发射电子。控制栅极是一个顶端有小孔的圆筒，套在阴极外面。它的电位比阴极低，对阴极发射出来的电子起控制作用，只有初速度较大的电子才能穿过栅极顶端的小孔然后在阳极加速下奔向荧光屏。示波器面板上的"亮度"调整就是通过调节电位以控制射向荧光屏的电子流密度，从而改变了屏上的光斑亮度。阳极电位比阴极电位高很多，电子被它们之间的电场加速形成射线。当控制栅极、第一阳极、第二阳极之间的电位调节合适时，电子枪内的电场对电子射线有聚焦作用，所以第一阳极也称聚焦阳极。第二阳极电位更高，又称加速阳极。面板上的"聚焦"调节，就是调第一阳极电位，使荧光屏上的光斑成为明亮、清晰的小圆点。有的示波器还有"辅助聚焦"，实际是调节第二阳极电位。

（3）偏转系统：它由两对相互垂直的偏转板组成，一对垂直偏转板 Y，一对水平偏转板 X。在偏转板上加以适当电压，电子束通过时，其运动方向发生偏转，从而使电子束在荧光屏上的光斑位置也发生改变。

容易证明，光点在荧光屏上偏移的距离与偏转板上所加的电压成正比，因而可将电压的测量转化为屏上光点偏移距离的测量，这就是示波器测量电压的原理。

2）信号放大器和衰减器

示波管本身相当于一个多量程电压表，这一作用是靠信号放大器和衰减器实现的。由于示波管本身的 X 轴及 Y 轴偏转板的灵敏度不高（约 0.1 ~ 1mm/V），当加在偏转板的信号过小时，要预先将小的信号电压加以放大后再加到偏转板上。为此设置 X 轴及 Y 轴电压放大器。衰减器的作用是使过大的输入信号电压变小以适应放大器的要求，否则放大器不能正常工作，使输入信号发生畸变，甚至使仪器受损。对一般示波器来说，X 轴和 Y 轴都设置有衰减器，以满足各种测量的需要。

3）扫描系统

扫描系统也称时基电路，用来产生一个随时间作线性变化的扫描电压，这种扫描电压随时间变化的关系曲线如同锯齿，故称锯齿波电压，这个电压经 X 轴放大器放大后加到示波管的水平偏转板上，使电子束产生水平扫描。这样，屏上的水平坐标变成时间坐标，Y 轴输入的被测信号波形就可以在时间轴上展开。扫描系统是示波器显示被测电压波形必需的重要组成部分。

2. 示波器显示波形的原理

如果只在竖直偏转板上加一交变的正弦电压，则电子束的亮点将随电压的变化在竖直方向来回运动，如果电压频率较高，则看到的是一条竖直亮线，如图 S13 - 2 所示。要能显示波形，必须同时在水平偏转板上加一扫描电压，使电子束的亮点沿水平方向拉开。这种扫描电压的特点是电压随时间呈线性关系增加到最大值,最后突然回到最小,此后再重复地变化。这种扫描电压即锯齿波电压,如图 S13－3 所示。当只有锯齿波电压加在水平偏转板上时,如果频率足够高,则荧光屏上只显示一条水平亮线。

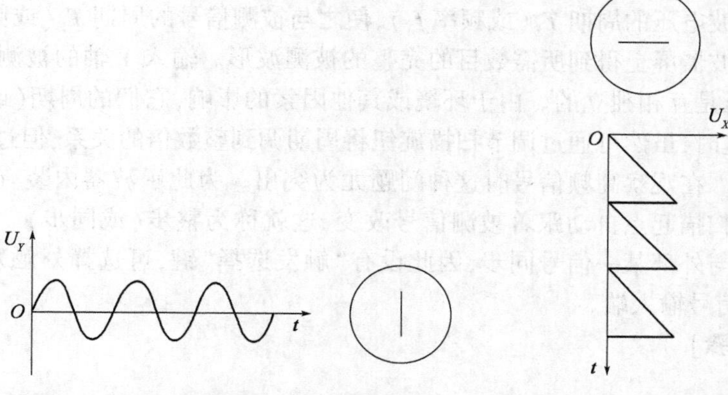

图 S13－2　交变的正弦电压变为竖直亮线　　　图 S13－3　锯齿波电压

如果在竖直偏转板上(简称 Y 轴)加正弦电压,同时在水平偏转板上(简称 X 轴)加锯齿波电压,电子受竖直、水平两个方向的力的作用,电子的运动就是两相互垂直的运动的合成。当锯齿波电压比正弦电压变化周期稍大时,在荧光屏上将能显示出完整周期的所加正弦电压的波形图,如图 S13－4 所示。

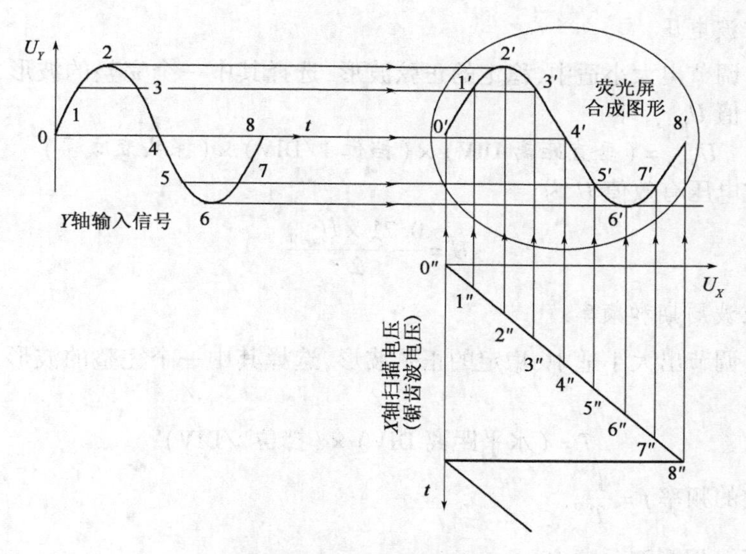

图 S13－4　扫描合成过程

3. 同步的概念

如果正弦波和锯齿波电压的周期稍微不同,屏上出现的是一移动着的不稳定图形。设锯齿波电压的周期 T_X 比正弦波电压周期 T_Y 稍小,如 $T_X/T_Y = 7/8$。在第一扫描周期内,屏上显示正弦信号 0～4 点之间的曲线段;在第二周期内,显示 4～8 点之间的曲线段,起点在 4 处;第三周期内,显示 8～11 点之间的曲线段,起点在 8 处。这样,屏上显示的波形每次都不重叠,好像波形在向右移动。同理,如果 T_X 比 T_Y 稍大,则好像在向左移动。以上描述的情况在示波器使用过程中经常会出现。其原因是扫描电压的周期与被测信号的周期不相等或不成整数倍，以致每次扫描开始时波形曲线上的起点均不一样所造成的。为了使屏上的图形稳定，必须使 $T_X/T_Y = n(n = 1,2,3,\cdots)$，$n$ 是屏上显示完整波形的个数。

为了获得一定数量的波形，示波器上设有“扫描时间”（或“扫描范围”）旋钮、“扫描微调”旋钮，用来调节锯齿波电压的周期 T_X（或频率 f_X），使之与被测信号的周期 T_Y（或频率 f_Y）成合适的关系，从而在示波器屏上得到所需数目的完整的被测波形。输入 Y 轴的被测信号与示波器内部的锯齿波电压是互相独立的。由于环境或其他因素的影响，它们的周期（或频率）可能发生微小的改变。这时，虽然可通过调节扫描旋钮将周期调到整数倍的关系，但过一会儿又变了，波形又移动起来。在观察高频信号时这种问题尤为突出。为此示波器内装有扫描同步装置，让锯齿波电压的扫描起点自动跟着被测信号改变，这就称为整步（或同步）。有的示波器中，需要让扫描电压与外部某一信号同步，因此设有“触发选择”键，可选择外触发工作状态，相应设有“外触发”信号输入端。

[实验内容与步骤]

1. 实验内容

1）观察信号发生器波形

（1）将信号发生器的输出端接到示波器 Y 轴输入端上。

（2）开启信号发生器，调节示波器（注意信号发生器频率与扫描频率），观察正弦波形，并使其稳定。

2）测量正弦波电压

在示波器上调节出大小适中、稳定的正弦波形，选择其中一个完整的波形，先测算出正弦波电压峰——峰值 U_{p-p}，即

$$U_{p-p} = (\text{垂直距离 DIV}) \times (\text{挡位 } V/\text{DIV}) \times (\text{探头衰减率})$$

然后求出正弦波电压有效值 U 为

$$U = \frac{0.71 \times U_{p-p}}{2}$$

3）测量正弦波周期和频率

在示波器上调节出大小适中、稳定的正弦波形，选择其中一个完整的波形，先测算出正弦波的周期 T，即

$$T = (\text{水平距离 DIV}) \times (\text{挡位 } t/\text{DIV})$$

然后求出正弦波的频率 $f = \dfrac{1}{T}$。

4）利用李萨如图形测量频率

设将未知频率 f_Y 的电压 U_Y 和已知频率 f_X 的电压 U_X（均为正弦电压），分别送到示波器的 Y 轴和 X 轴，则由于两个电压的频率、振幅和相位的不同，在荧光屏上将显示各种不同波形，一般得不到稳定的图形，但当两电压的频率成简单整数比时，将出现稳定的封闭曲线，称为李萨如图形。根据这个图形可以确定两电压的频率比，从而确定待测频率的大小。

图 S13－5 列出各种不同的频率比在不同相位差时的李萨如图形，不难得出：

$$\frac{\text{加在 } Y \text{ 轴电压的频率 } f_Y}{\text{加在 } X \text{ 轴电压的频率 } f_X} = \frac{\text{水平直线与图形相交的点数 } N_X}{\text{垂直直线与图形相交的点数 } N_Y}$$

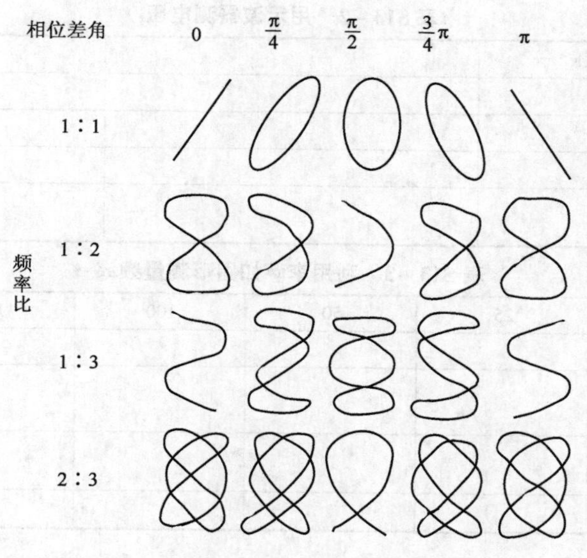

图 S13-5　李萨如图形

所以未知频率为

$$f_Y = \frac{N_X}{N_Y} f_X$$

应指出水平、垂直直线不应通过图形的交叉点。

2. 实验步骤

(1)将一台信号发生器的输出端接到示波器 Y 轴输入端上，并调节信号发生器输出电压的频率为50Hz，作为待测信号频率。把另一信号发生器的输出端接到示波器 X 轴输入端上作为标准信号频率。

(2)分别调节与 X 轴相连的信号发生器输出正弦波的频率 f_X 约为25Hz、50Hz、100Hz、130Hz、200Hz 等。观察各种李萨如图形，微调 f_X 使其图形稳定时，记录 f_X 的确切值，再分别读出水平线和垂直线与图形的交点数。由此求出各频率比及被测频率 f_Y，记录于下表中。

(3)观察时图形大小不适中，可调节"V/DIV"和与 X 轴相连的信号发生器输出电压。

[注意事项]

(1)仪器输出、输入的信号端与接地端千万不可搞混，否则会造成信号短路，损坏仪器。

(2)示波器使用时要注意辉度适中，光点不宜长时间停留一点，以免损坏荧光屏。

(3)旋转各旋钮时，应轻缓，用力适当，严禁用力过猛损坏旋钮或选出挡位。

[数据及处理]

1. 数据记录(表 S13-1 ~ 表 S13-3)

表 S13-1　用示波器测频率

信号发生器输出频率/Hz	200	1000	5000	25000
扫描时间				
一个周期的格子数				
周期				
测量频率				

表 S13-2 用示波器测电压

信号发生器输出电压		
电压灵敏度		
Y 轴偏转格数		
U_{p-p}		
有效值$\frac{U_{p-p}}{2\sqrt{2}}$		

表 S13-3 利用李萨如图形测量频率

标准信号频率f_X/Hz	25	50	100	150	200
李萨如图形(稳定时)					
频比$=\frac{水平线交点数 N_X}{垂直线交点数 N_Y}$					
待测电压频率$f_Y=f_X\frac{N_X}{N_Y}$					
f_Y的平均值/Hz					

2. 数据处理

利用相关公式计算示波器所测频率值和电压值。

3. 结果分析

视频 S13
示波器的使用

对实验结果进行分析,并得出实验结论。

[问题讨论]

(1)示波器为什么能显示被测信号的波形?

(2)荧光屏上无光点出现,有几种可能的原因?怎样调节才能使光点出现?

(3)荧光屏上波形移动,可能是什么原因引起的?

示波器的使用如视频 S13 所示。

[补充说明]

1. 示波器前面板介绍

一般通用示波器面板分为三部分:示波管部分;X 轴,Y_1(通道 1)和 Y_2(通道 2);触发扫描部分。下面以 MOS-620 型示波器前面板(图 S13-6)为例,介绍有关旋钮功能。

1)示波管部分

1—校准信号[CAL(U_{p-p})]:提供幅度为 $2U_{p-p}$、频率 1kHz 的方波信号,用于校正 10:1 探头的补偿电容器和检测示波器垂直与水平的偏转因数。

2—辉度(INTEN):控制光点和扫线的亮度。

3—聚焦(FOCUS):将扫描线聚焦最清晰。

4—轨迹旋转(TRACE ROTATION):半固定的电位器用来调整水平轨迹与刻度线的平行。

5—发光二极管。

6—电源(POWER):主电源开关,当此开关开启时发光二极管 5 亮。

2)X 轴、Y_1(通道 1)和 Y_2(通道 2)

7,22—垂直衰减开关:调节垂直灵敏度从 5mV/div ~ 5V/div 分 10 挡。

8—CH_1(X)输入;在 $X-Y$ 模式下,作为 X 轴输入端。

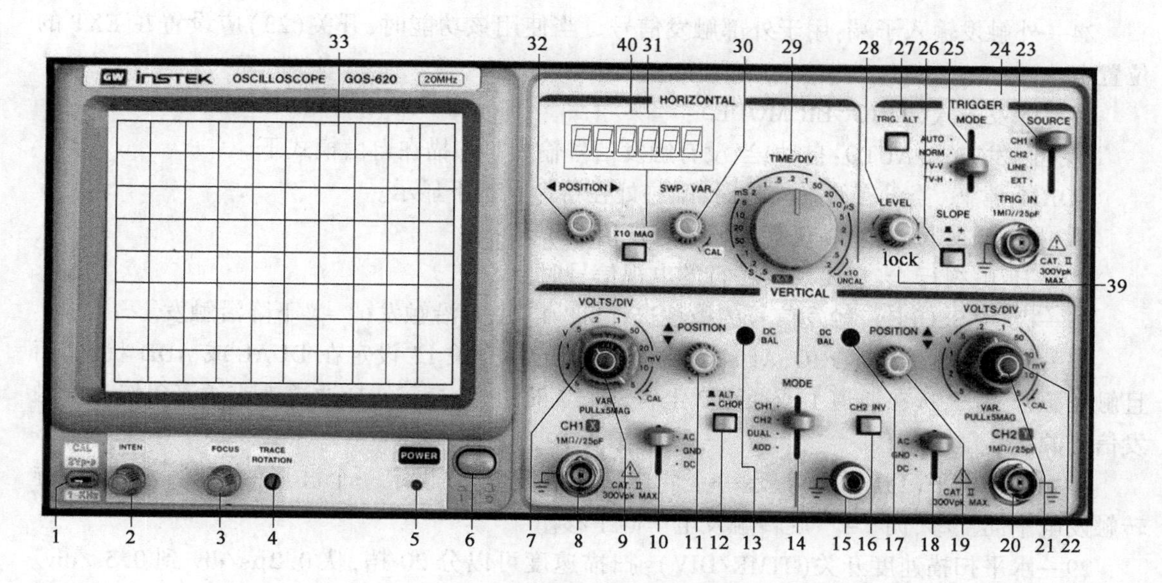

图 S13－6　MOS－620 型双通道通用示波器前面板图

20—CH_2(Y)输入;在 $X-Y$ 模式下,作为 Y 轴输入端。

9,21—垂直微调(VARIABLE):微调灵敏度大于或等于 1/2.5 标示值,在校正位置时,灵敏度校正为标示值。当该旋钮拉出后(×5MAG 状态)放大器的灵敏度乘以 5。

10,18—Y 轴耦合方式(AC－GND－DC):输入信号与垂直放大器连接方式的选择开关,可选择以下几种:(1)AC(交流耦合);(2)DC(直流耦合);(3)GND(输入信号与放大器断开,且放大器输入端接地)。

11,19—垂直位移(POSITION):调节光迹到屏幕上的垂直位置。

12—交替/断续(ALT/CHOP):在双踪显示时,放开此键,表示通道 1 与通道 2 交替显示(通常用在扫描速度较快的情况下);当此键按下时,通道 1 与通道 2 断续显示(通常用在扫描速度较慢的情况下)。

13,17—CH_1 和 CH_2 的 DC　BAL:用于衰减器的平衡调试。

14—垂直方式(MODE):选择放大器的工作模式。

CH_1 或 CH_2:通道 1 或通道 2 单独显示。

DUAL:两个通道同时显示。

ADD:显示两个通道的代数和 CH_1+CH_2。

15—GND:示波器机箱的接地端子。

16—CH_2 INV:通道 2 的信号反向,当此键按下时,通道 2 的信号及通道 2 的触发信号同时反向。

3)触发扫描部分

23—触发源选择(SOURCE):选择内(INT)或外(EXT)触发。

CH_1/CH_2:当垂直开关方式选择开关 14 设定在 DUAL 或 ADD 状态时,选择 1/2 作为内部触发信号源。

LINE:选择交流电源作为触发信号。

EXT:外部触发信号接于 24 作为触发信号。

24—外触发输入子端:用于外部触发信号。当使用该功能时,开关(23)应设置在 EXT 的位置上。

25—触发方式(TRIGGER MODE):

选择触发方式:AUTO:自动,当没有触发信号输入时扫描在自由模式下。

NORM:常态,当没有触发信号时,踪迹处在待命状态下显示。

TV—V:电视场,当想要观察一场的电视信号时。

TV—H:电视行,当想要观察一行的电视信号时。

26—极性(SLOPE):触发信号的极性选择。"+"上升沿触发,"-"下降沿触发。

27—触发交替选择(TRIG. ALT):当垂直方式选择开关 14 设定在 DUAL 或 ADD 状态,而且触发源开关 23 选在通道 1 或通道 2 上,按下 27 时,它会交替选择通道 1 或通道 2 作为内触发信号源。

28—触发电平(LEVEL):显示一个同步稳定的波形,并设定一个波形的起点。向"+"旋转触发电平向上移,向"-"旋转触发电平向下移。

29—水平扫描速度开关(TIME/DIV):扫描速度可以分 20 挡,从 0.2μs/div 到 0.5s/div。当设置到 *X* 位置时可用作 *X* 示波器。

30—水平微调(SWP. VER):微调水平扫描时间,使扫描时间被校正到与面板上 TIME/DIV 指示的一致。TIME/DIV 扫描速度可连续变化,当反时针旋转到底为校正位置。整个延时可达 2.5 倍以上。

31—扫描扩展开关(×10 MAG):按下时扫描速度扩展 10 倍。

32—水平位移(POSITION):调节光迹在屏幕上的水平位置。

33—滤色片:使波形看起来更加清晰。

40—频率数码显示。

2. 示波器后面板介绍

下面以 MOS-620 型示波器后面板(图 S13-7)为例,介绍有关旋钮功能。

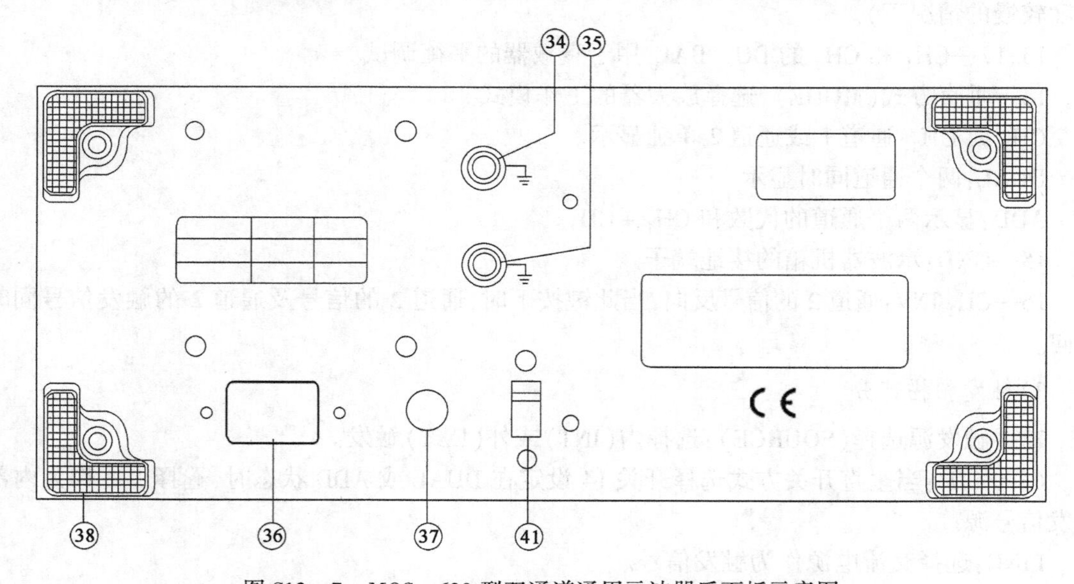

图 S13-7 MOS-620 型双通道通用示波器后面板示意图

34—Z 轴输入:外部亮度调制信号输入端。

35—通道 1 信号输出:提供通道 1 信号(约 20mV/div)去 50Ω 的终端,适合外接频率计或其他仪器。

36—交流电源:交流电源输入插座,交流电源线接于此处。

37—熔断丝。

38—支撑块:当示波器面向上放置时,用于支撑示波器,并且可以引出电源线。

41—电源选择开关:客户可选择 110V/220V 电源输入。(需预先提出)

3. 基本操作

1) 单通道工作

将有关控制元件按表 S13 - 4 设置。

表 S13 - 4　控制元件设置

功　能	序　号	设　置
电源(POWER)	6	关
亮度(INTEN)	2	居中
聚焦(FOCUS)	3	居中
垂直方式(VERT MODE)	4	通道 1
交替/断续(ALT/CHOP)	12	释放(ALT)
通道 2 反响(CH2 INV)	16	释放
垂直位置(POSITION)	11,19	居中
垂直衰减(VOLTS/DIV)	7,22	0.5V/DIV
调节(VARIABLE)	9,21	CAL(校正位置)
AC - GND - DC	10,18	GND
触发源(SOURCE)	23	通道 1
极性(SLOPE)	26	+
触发交替选择(TRIG. ALT)	27	释放
触发方式(TRIGGER MODE)	25	自动
扫描时间(TIME/DIV)	29	0.5mSec/DIV
微调(SWP. VER)	30	校正位置
水平位移(POSITION)	32	居中
扫描扩展(×10 MAG)	31	释放

将开关和控制部分按以上设置后,接上电源线,继续:

(1) 打开电源开关,电源指示灯亮约 20s 后,屏上出现一回扫线,调节"辉度""聚焦"使回扫线亮度适当、清晰。

(2) 调节通道 1 位移旋钮与轨迹旋转电位器,使光迹与水平刻度平行。

(3) 连接测量导线到"Y_1 轴输入端"并将 $2U_{p-p}$ 的"校准信号"加在测量导线的信号端上。

(4) 将"AC - GND - DC"开关置于 AC 上,则屏上出现正方波波形,调节"聚焦"使波形最清晰。

(5) 对于其他信号的观察,可通过调整垂直衰减开关,扫描时间到所需要的位置,从而得到清晰的图形。

(6) 调整垂直和水平位移旋钮,使得波形的幅度与时间容易读出。

(7) 以上是 Y_1(CH_1)方式,Y_2单通道工作方式调节程序与上述相同。

2) 双通道工作

改变垂直方式到 DUAL 状态,于是通道 2 的光迹也会出现在屏幕上(与 CH_1 相同)。这时通道 1 显示一个方波(来自校正信号输出的波形),而通道 2 则仅显示一条直线,因为没有信号接到该通道。现在将校正信号接到 CH_2 的输入端与 CH_1 一致,将"AC - GND - DC"开关置于 AC,调整垂直位移(POSITION),释放(ALT/CHOP)开关,(置于 ALT 方式)。CH_1 和 CH_2 的信号交替地显示在屏幕上,此设定用于观察扫描时间较短的两路信号。按下(ALT/CHOP)开关,(置于 CHOP 方式),二个通道信号以 250kHz 的速度独立地显示在屏幕上,此设定用于观察扫描时间较长的两路信号。在进行双通道操作时(DUAL 或加减方式),必须通过触发信号源的开关来选择通道 1 或通道 2 的信号作为触发信号。如果 CH_1 与 CH_2 的信号同步,则两个波形都会稳定显示出来。反之,则仅有触发源的信号可以稳定地显示出来;如果 TRIG. ALT 开关按下,则两个波形都会同时稳定显示出来。

3) 加减操作

通过设置"垂直方式开关"到"加"的状态,可以显示 CH_1 和 CH_2 信号的代数和,如果 CH_2 INV 开关被按下则为代数减。为了得到加减的精确值,两个通道的衰减设置必须一致。垂直位置可以通过"垂直位置(POSITION)开关"来调整。鉴于垂直放大器的线形变化,最好将该旋钮设置在中间位置。

实验 14　用惠斯通电桥测量中值电阻

[引言]

依据欧姆定律测量电阻是最基本的方法,即用电压表测量电阻两端的电势差,用电流表测量通过电阻的电流强度,这样根据欧姆定律可获得电阻值,然而,实际电压表内阻并非无穷大,而电流表内阻并非无穷小,因此电流表内接时电流表内阻要分压,电流表外接时电压表内阻要分流,这就表明,无论采用电流表内接和外接都会引入系统误差。为了消除此项系统误差,采用不提取电流的思想,设计了电桥电路,若桥路电流为零时,则电桥达到平衡。这就是惠斯通电桥测电阻的基本设计思想。

[实验目的]

(1) 掌握用单臂电桥测量电阻的原理和方法。

(2) 学会自组电桥,并用交换法减小和修正系统误差。

(3) 学习使用箱式电桥测量中值电阻。

(4) 了解电桥灵敏度的概念。

[预习问题]

(1) 熟悉电桥的组成及电桥平衡条件的物理含义。

(2) 掌握自组电桥并思考以下问题:

①稳压电源和检流计接入电桥时为什么不考虑正负极性?

②调节 R 时,为什么应先粗调,再进行细调,次序不能颠倒?

③为什么对不同大小的待测电阻需选取不同的电源电压和相应的比例臂电阻？

(3)如何正确使用检流计？如何选择比率？如何测量灵敏度？

(4)用四转盘箱式电桥测电阻，在选择比例臂时，为什么要求保证 R_s 取满 4 位有效数字？若测一个 500Ω 的电阻，比例臂应选多大？

[实验仪器]

QJ24 型直流单臂电桥(级别为 0.1)、待测电阻、滑线电阻、电阻箱(精度为 0.1Ω)、检流计、学生电源。

[实验原理]

1.用惠斯通电桥测电阻的原理

如图 S14－1 所示。电桥基本组成部分是：桥臂由四个电阻 R_1、R_2、R_s、R_x组成，“桥”平衡指示器(检流计 G)和工作电源 E。一般情况下，接通电源开关 K_1和检流计开关 K_2时，检流计 G 中有电流流过，但当调节四个桥臂为适当阻值时，“桥”上检流计 G 中就无电流流过，这时称为“电桥平衡”。由于检流计中无电流通过，B、D 两点的电势相等，于是流过电阻 R_1和 R_x的电流相同，设为 i_1，流过 R_2和 R_s的电流也相同，设为 i_2，从而可得方程

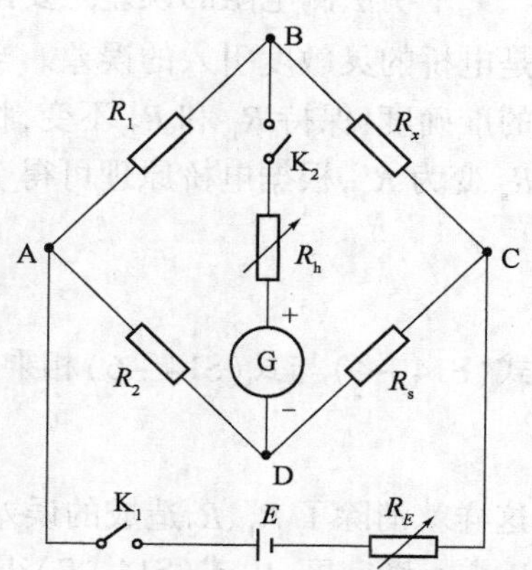

图 S14－1　惠斯通电桥原理图

$$i_1R_1=i_2R_2 \tag{S14－1}$$

$$i_1R_x=i_2R_s \tag{S14－2}$$

由式(S14－1)、式(S14－2)相除得

$$\frac{R_1}{R_x}=\frac{R_2}{R_s} \tag{S14－3}$$

式(S14－3)就是电桥的平衡条件，它说明电桥平衡时，四个桥臂成比例。设 R_x为待测电阻，则它的值为

$$R_x=\frac{R_1}{R_2}R_s=CR_s \tag{S14－4}$$

式中，C 为 R_1与 R_2的比值，称为比率；R_s为比较电阻。若 R_1、R_2(或比率 C)和 R_s已知，则由上式可计算出 R_x。

2.电桥灵敏度

电桥平衡时，流过检流计电流为零，由于检流计灵敏度有限，当 R_s改变 ΔR_s时，电桥虽然已失去平衡，但检流计对于微小的非平衡电流无法分辨，测量者误认为电桥达到了平衡，这就是由于检流计灵敏度不够高，导致分辨率受到限制而带来的测量误差。一般来说，检流计灵敏度越高，电桥灵敏度就越高，除此之外，电桥灵敏度还与工作电压、桥臂电阻、比率、检流计支路中的电阻等因素有关，电桥灵敏度有两种定义方法，多数考虑电桥的相对灵敏度，用 S 表示，其定义为

$$S=\frac{\Delta d}{\Delta R_s/R_s} \tag{S14－5}$$

S 的物理意义是桥臂电阻的单位相对变化所引起的电流计的偏转格数。S 越大，电桥灵敏度越高，引入的测量误差就越小。通常，肉眼能觉察出电流计 $\Delta d=0.2$ 格的偏转，因此，由

电桥灵敏度带来的相对不确定度可按$\frac{\Delta d}{S}$计算。理论上可以证明在电桥平衡条件下，电桥每个桥臂阻值偏离平衡阻值的百分比相等时，电流计有同样大小的偏移，根据这一性质，在测量电桥灵敏度时，可选择任意桥臂作为可变臂，只需电流计有一个明显的偏转即可。

3. 消除系统误差及测量误差

1）交换法（互易法）减小和修正自组电桥系统误差

平衡法测电阻的误差主要由两方面的因素决定：一是电阻 R_1、R_2、R_s 自身带来的误差；二是电桥的灵敏度引入的误差。当电桥灵敏度较高时，被测电阻 R_x 的准确度取决于 R_1、R_2、R_s 的准确度，保持 R_1 和 R_2 不变，将 R_x 与 R_s 的位置互换，再调节 R_s 使电桥平衡，设电桥平衡时 R_s 变为 R'_s，根据电桥原理可得

$$R_x = \frac{R_2}{R_1}R'_s \tag{S14-6}$$

式（S14－4）与式（S14－6）相乘得

$$R_x = \sqrt{R_s R'_s} \tag{S14-7}$$

这样就消除了 R_1、R_2造成的误差，这种方法称为交换法，由于 R_1、R_2均为单次测量，因此只考虑B类不确定度，由式（S14－7）得到交换法测量出 R_x的测量的相对不确定度为

$$\frac{u_{R_x}}{R_x} = \sqrt{\left(\frac{1}{2}\frac{u_{R_s}}{R_s}\right)^2 + \left(\frac{1}{2}\frac{u_{R'_s}}{R'_s}\right)^2} \approx \frac{\sqrt{2}}{2}\frac{u_{R_s}}{R_s} \tag{S14-8}$$

前两项与电阻箱 R_s的仪器误差有关。而 R_s是选用具有一定精度的标准电阻箱，R_x的系统误差就大大减小。实验时 R_s常用十进位转盘直流电阻箱，其仪器的最大允许相对误差为

$$\Delta_{仪} = k\% \cdot R_s \tag{S14-9}$$

式中，k 是电阻箱准确度等级；R_s是电阻箱指示值。

由上述分析可见，平衡电桥交换法测电阻的误差，来源于电桥的灵敏度和电阻箱 R_s自身误差。

2）箱式电桥的误差

箱式电桥的仪器误差可用下式表示：

$$\Delta_{仪} = \pm k\%\left(CR_x + \frac{CR_N}{10}\right) \tag{S14-10}$$

式中，k 为箱式电桥的精度级别；R_x 为测量值；R_N 为基准值，R_N 取 5000。

［实验内容与步骤］

1. 自组惠斯通电桥测中值电阻

（1）按图 S14－1 连接电路，R_1、R_2、R_s 均用 ZX21 型六旋盘电阻箱，取 R_1：R_2＝100：100，“桥”路上滑线变阻器 R_h 的作用是保护检流计，R_E 的作用是保护整个电路。电路连接后应仔细检查无误后才能接通电源。

（2）开始时，电桥一般处于很不平衡的状态，为了保护电路，R_E、R_h 均置于阻值最大，在粗调状态下观察检流计，若不在零位，调节 R_s 使之示零，移动滑线变阻器 R_E 的滑动头，使输出电

压增大，此时电桥的灵敏度增大，电桥若不平衡则检流计会偏转，调节 R_s 使之再次示零，反复调节直至滑线变阻器 R_E 输出电压最大。然后用同样的方法调节 R_h。最后，在 R_E、R_h 的输出电阻均为最小时记录数据。此时电桥的灵敏度最高，误差最小。

(3) 测量待测电阻 R_x，并以交换法进行系统误差研究。

(4) 测量电桥的灵敏度：在电桥平衡的情况下，改变桥臂 R_s 使检流计指针偏转（小于 5 格），记录 ΔR_s 和 Δd，代入式(S14－5)即得电桥灵敏度。

2. 用箱式电桥测电阻

(1) 用万用表欧姆挡粗测电阻 R_x 的大概数值。

(2) 根据 R_x 的数值范围，选择恰当的比率 $\frac{R_b}{R_a}$，为了保证测量数据有四位有效数字，选择比率规定为：千欧级电阻选“1”；百欧级电阻选“0.1”；其他类推。

(3) 将待测电阻 R_x 接到接线柱 X_1 和 X_2 之间，调节转盘电阻 R 的各挡数值到万用电表所示的粗测值，按下开关 B_0 和 G_1，然后依次由大到小逐挡调节，直到检流计指针接近指零为止，再松开 G_1 按钮，并按下 G_0 按钮开关，仔细调节 R，使检流计指零。

(4) 记录转盘电阻 R 的数据，将 R 乘以比率 $\frac{R_b}{R_a}$ 的示值，可得到待测电阻 R_x 的值，即

$$R_x = \frac{R_b}{R_a} R$$

(5) 测量箱式电桥的灵敏度：方法同自组桥法测电桥灵敏度(4)。

[注意事项]

(1) 自组电桥测电阻在合上电键前，必须检查 R_E、R_s 是否在安全位置。

(2) 使用箱式电桥时，通电时间应很短，即不能长时间按下 B_0、G 两键，测量时，应先按 B_0 后按 G；断开时，先松 G，后松 B。

(3) 惠斯通电桥应将“内接”“外接”接线柱上金属片接在一起。

[数据及处理]

1. 数据记录（表 S14－1、表 S14－2）

表 S14－1　自组电桥测中值电阻

R_1	R_2	R_S	R'_S	R_x	ΔR_s	Δd	S

表 S14－2　箱式电桥测中值电阻

k	R_b/R_a	R	R_x	ΔR	Δd	S

2. 数据处理

分别计算自组电桥和箱式电桥的测量不确定度，并给出测量结果的不确定度评价。

3. 结果分析

对实验结果进行分析，并得出实验结论。

视频 S14 用惠斯通电桥测量中值电阻

[问题讨论]

(1)不论如何调节电桥,检流计总是向一边偏,请分析其原因有哪些?

(2)若电桥平衡后改变电源与检流计的位置,电桥是否仍然平衡?试证明之。

(3)总结快速调整电桥的方法。

用惠斯通电桥测量中值电阻如视频 S14 所示。

[补充说明]

本实验分别用自组电桥和 QJ24 型箱式直流单臂电桥来测量电阻。其中 QJ24 型箱式电桥的测量原理和面板图如图 S14－2 和图 S14－3 所示。现结合面板图将它的使用方法介绍如下:

(1)将待测电阻 R_x 接在仪器面板上的 X_1 和 X_2 之间。

(2)电阻 R 实际是由四个可变电阻串联而成。面板图中:在上侧虚线框内的四个转盘就是调节 R 的"转盘电阻箱"。

(3)面板图左上角的转盘为比率转盘,它的指示值表示比率 R_b/R_a 的值,R_b 和 R_a 称为比率臂。为了读数方便,在制作时将比率转盘做成 0.001、0.01、0.1、1、10、100、1000 等七挡。

(4)检流计在面板图的左下方,其上有一个机械调零机构,可通过左右旋转来调节指针的"零点"。

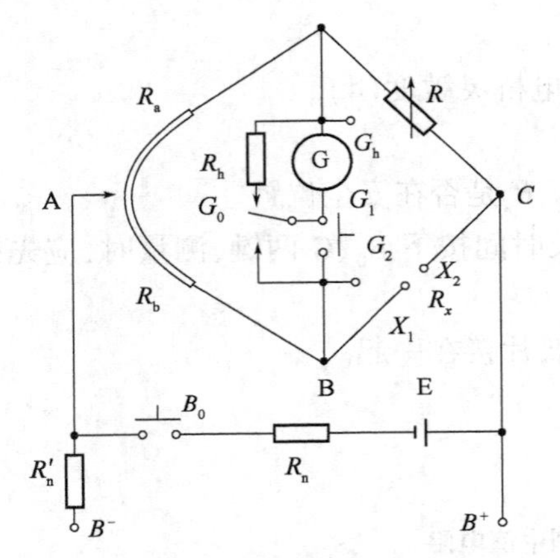

图 S14－2 QJ24 型单臂电桥原理构造图

图 S14－3 QJ24 型单臂电桥面板图

(5)面板图中 B_0 为电源按钮开关,B 为外接电源接线柱。G_1、G_0 是检流计均按钮开关,G 为外接检流计接线柱。测量时,为了保护检流计,开关使用的顺序应先合 B_0,后合 G_1 或 G_0;先松 G_0 或 G_1,后松 B_0。按钮开关不要一直按下,应采用跃接式。为了防止过大的电流通过检流计,一般应先按下 G_1 开关,待电桥调节逐步接近平衡时,再使用 G_0 开关,直至检流计的电流值为零,电桥平衡为止。

实验 15 载流线圈磁场分布的测量

[引言]

当给线圈通有交变电流时,其周围空间可激发出交变磁场,这就使得置于其中的探测线圈所张面积的磁通量发生变化。根据法拉第电磁感应定律,探测线圈闭合回路产生感应电动势,通过测量此感应电动势的大小，就可以测量交变电流所激发的交变磁场的量值。探测线圈相当于一个磁电转换器，它是基于法拉第的磁电效应，将变化的磁场转化为感生电场，从而实现了磁学量的电测技术。

［实验目的］

(1)掌握载流圆线圈和亥姆霍兹线圈的磁场分布特征。

(2)验证亥姆霍兹线圈磁场叠加原理。

(3)学习用感应法测绘磁场分布的方法。

［预习问题］

(1)磁场方向为什么要采用毫伏表读数最小值时所对应的探测线圈法线方向来表示？

(2)本实验是如何测量感应电动势有效值的最大值？

［实验仪器］

JD－HZ－1 型亥姆霍兹线圈磁场描绘仪、探测线圈。

［实验原理］

1. 载流圆线圈轴向和径向磁场的分布

磁场起源于电荷的运动，磁感应强度 $\boldsymbol{B}$ 为定量描述磁场中各点特质的物理量，它是一个矢量，既有大小，又有方向，其大小与介质性质有关。如图 S15－1 所示，取圆线圈的轴线为 Ox 轴，线圈中心 O 为坐标原点，则载流圆线圈轴线上某点 P 的磁感应强度 $\boldsymbol{B}$ 的大小，可根据毕奥——萨伐尔定律求得

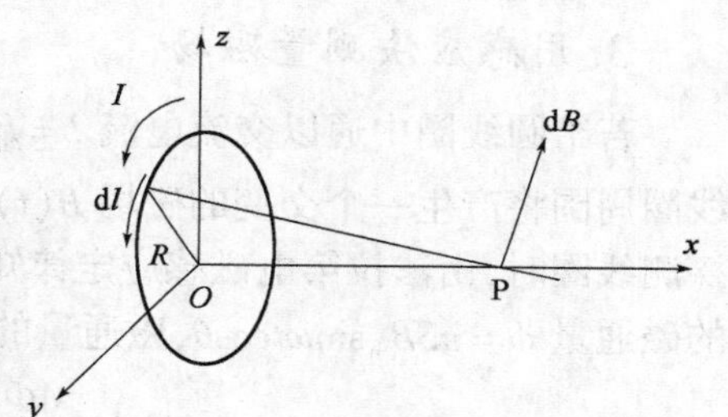

图 S15－1　圆形电流磁场

$$B_x=\frac{\mu NIR^2}{2\left(R^2+x^2\right)^{3/2}} \qquad (S15-1)$$

方向沿 x 轴正方向。式中，x 为 P 点到圆线圈中心 O 的距离；R 为线圈的平均半径，当 $x=0$ 时，圆心处的磁感应强度 $\boldsymbol{B}_0$ 的大小为 $B_0=\frac{\mu_0 NI}{2R}$，还可得圆线圈径向上任意一点 y 处的磁感应强度为

$$B_y=\frac{\mu NI}{2R}\left[\frac{1}{1-\frac{y^2}{R^2}}+\frac{\frac{y^2}{R^2}}{4\left(1-\frac{y^2}{R^2}\right)^{3/2}}+\cdots\right] \qquad (S15-2)$$

磁场方向与轴向磁感应强度方向一致。

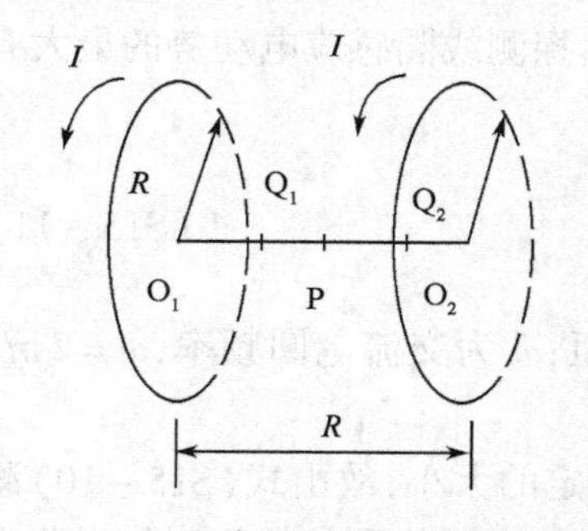

图 S15－2　亥姆霍兹线圈磁场

2. 亥姆霍兹线圈的磁场

亥姆霍兹线圈是由两个半径相同，匝数相同，间距等于其半径的线圈组成的同轴线圈系。

设线圈半径为 R，匝数为 N，两线圈串联，通以同方向电流 I，在轴线上产生的磁场方向相同，线圈轴线上任意一点的磁感应强度的数值等于每个线圈单独在该点所产生的磁感应强度的数值之和。如图 S15－2 所示，O_1、O_2 为两线圈的中心，此处的磁感应强度为

$$B_{O_1}=B_{O_2}=\frac{\mu NI}{2R}+\frac{\mu NIR^2}{2\left(R^{2}+R^2\right)^{3/2}}=0.677\frac{\mu NI}{R} \tag{S15-3}$$

在轴线中点P处产生的磁感应强度为

$$B_P=2\frac{\mu NIR^2}{2\left[R^2+\left(\frac{R}{2}\right)^2\right]^{3/2}}=0.724\frac{\mu NI}{R} \tag{S15-4}$$

在P点两侧各 $R/4$ 处的 Q_1、Q_2 两点的磁场的大小相等,其值为

$$B_{Q_1}=B_{Q_2}=\frac{\mu NIR^2}{2\left[R^2+\left(\frac{R}{4}\right)^2\right]^{3/2}}+\frac{\mu NIR^2}{2\left[R^2+\left(\frac{3}{4}R\right)^2\right]^{3/2}}=0.712\frac{\mu NI}{R} \tag{S15-5}$$

B_P、B_{Q_1} 和 B_{Q_2} 相比,相差不足2%,可以因此认为在P点附近轴线上的磁场基本上是均匀的,即亥姆霍兹线圈所产生的磁场在轴线附近基本上是一个匀强磁场。

3. *用感应法测量磁场*

若给圆线圈中通以交流电流 $I=I_m\sin\omega t$,其中 I_m 为交变电流的峰值,ω 为交流电圆频率,线圈周围将产生一个交变的磁场 $B(t)=B_m\sin\omega t$。给交变磁场中放入截面积为 S、匝数为 n 的探测线圈时,由法拉第电磁感应定律知,探测线圈闭合回路将产生感应电动势。通过探测线圈的磁通量 $\phi=nSB_m\sin\omega t\cos\theta$,磁通量的变化率与感应电动势成正比。

$$|\varepsilon|=\frac{d\Phi}{dt}=n\omega SB_m\cos\theta\cos\omega t=\varepsilon_m\cos\omega t \tag{S15-6}$$

式中,$\varepsilon_m=n\omega sB_m\cos\theta$ 是探测线圈法线和磁场方向成 θ 角时感应电动势的幅值。如果使用毫伏表测量此时线圈的电动势,则毫伏表的示值(有效值)ε_E 应为

$$\varepsilon_E=\frac{\varepsilon_m}{\sqrt{2}}=\frac{1}{\sqrt{2}}n\omega sB_m\cos\theta \tag{S15-7}$$

一般取 $\theta=0$,此时探测线圈的法线方向与磁场方向一致,感应电动势的幅值最大,即

$$\varepsilon_{E_m}=\frac{n\omega sB_m}{\sqrt{2}} \tag{S15-8}$$

则

$$B_m=\frac{\sqrt{2}\varepsilon_{Em}}{n\omega s} \tag{S15-9}$$

有效值为

$$B=\frac{\varepsilon_{Em}}{n\omega s} \tag{S15-10}$$

由以上讨论可以看出,在交变磁场中,用交流电压表测量出探测线圈感应电动势的最大有效值,就可以算出该点的磁感应强度,其大小为

$$B=\frac{108\varepsilon}{13\pi nD^2\omega} \tag{S15-11}$$

式中,探测线圈外径 $D=1.2\times10^{-2}$ m;探测线圈匝数 $n=1200$ 匝;ω 为交流电圆频率,$\omega=2\pi f$,在实验中用 $f=400$Hz 的正弦交流电。

由于圆形线圈的磁场分布是不均匀的,而探测线圈又有一定的大小,故由式(S15-10)测出的数据实际上是所测空间范围内磁感应强度的平均值。如果把探测线圈的体积制作得非常

小则会降低测量灵敏度。为了使平均磁感应强度和探测线圈几何中心磁场一致，一般探测线圈都制成圆柱形且按如下条件制作能够符合要求：

(1)线圈长度 L 与外经 D 之比满足 $\frac{L}{D}\approx\frac{2}{3}$。

(2)线圈内径 d 不大于外径 D 的 $\frac{1}{3}$，$\frac{d}{D}\leqslant\frac{1}{3}$。

(3)线圈体积适当小。

这样，线圈的有效面积 $S=\frac{13\pi}{108}D^2$ 代入式(S15－10)即可计算探测点的磁感应强度。

[实验内容与步骤]

测量载流圆线圈轴向、径向各点的磁感应强度的步骤如下：

(1)将仪器主机和测试装置放置平稳，检查活动线圈已垂直插接于测试装置的插座内，将励磁电压调节旋钮旋到左端。

(2)参考图 S15－3 线圈结构和磁场描绘仪主机面板图 S15－4，将线圈Ⅰ的 1、2 接线柱与磁场描绘仪中“励磁电压” 相连；将探测线圈引线与磁场描绘仪中“探测线圈” 相连；探测线圈置于线圈Ⅰ的中心位置定位孔内。

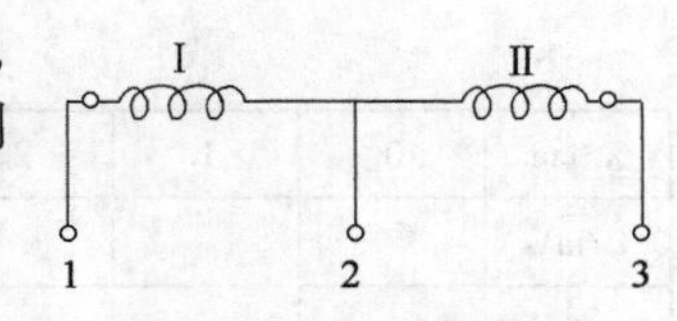

图 S15－3　线圈结构图

(3)将测量转换开关拨到“降压”位置 ，接通主机电源，调节励磁电压调节旋钮，使主机毫伏表显示 10.0mV，此时线圈Ⅰ的电流为 10.0mA，(因测量回路连接有 1Ω 的取样电阻)记录其数值。

(4)将测量转换开关拨至“探测” 位置，此时毫伏表读数为探测线圈回路感应电动势。转动探测线圈方向，找出毫伏表的最大指示值，依次测量线圈Ⅰ径向、轴向各点的最大感应电动势记入表 S15－1、表 S15－2，代入实验公式(S15－11)计算各点磁感应强度。

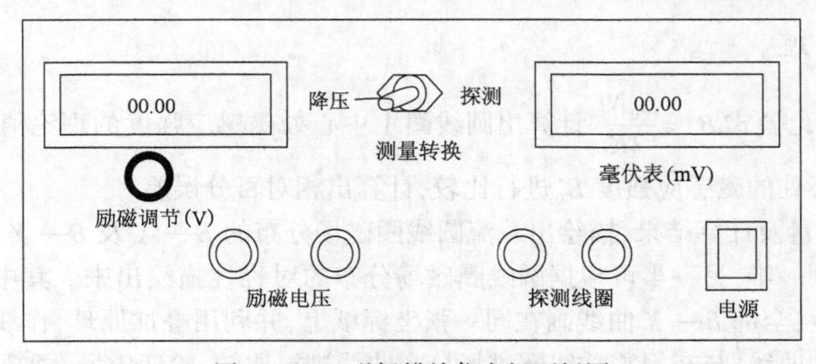

图 S15－4　磁场描绘仪主机面板图

(5)将测量线圈的 2、3 和 1、3 接线柱分别和主机“励磁电压”相连，仍先测量线圈电流值，电流测量方法同步骤(3)线圈Ⅰ电流测量。实验过程保持每个测量线圈回路电流不变，分别测量线圈Ⅱ和亥姆霍兹线圈轴向各点的感应电动势最大值，记入表 S15－3、表 S15－4，代入实验公式(S15－11)计算各点的磁感应强度，并验证磁场的叠加原理。

[注意事项]

(1)励磁电压严禁输出端短路。

(2)磁场测绘期间，应排除较强的电磁场干扰，否则将造成探测线圈感应电动势不稳或者错误。

[数据及处理]

1. 数据记录(表 S15-1~表 S15-4)

表 S15-1　圆线圈Ⅰ磁场的径向分布记录表　　电流 $I=$______mA

Y/cm	0	1	2	3	4	5
ε/mV						
$B/10^{-5}$T						

表 S15-2　圆线圈Ⅰ磁场的轴向分布记录表　　电流 $I=$______mA

X/cm	0	1	2	3	4	5	6	7	8	9	10
ε/mV											
$B/10^{-5}$T											

表 S15-3　圆线圈Ⅱ磁场的轴向分布记录表　　电流 $I=$______mA

X/cm	0	1	2	3	4	5	6	7	8	9	10
ε/mV											
$B/10^{-5}$T											

表 S15-4　亥姆霍兹线圈磁场的轴向分布记录表　　电流 $I=$______mA

X/cm	0	1	2	3	4	5	6	7	8	9	10
ε/mV											
$B/10^{-5}$T											

2. 数据处理

(1)根据理论公式 $B_0=\frac{\mu_0 NI}{2R}$ 计算出圆线圈Ⅰ中心处磁感应强度的理论值 B_0 与实际测出的圆线圈Ⅰ中心处的磁感应强度 B_0' 进行比较,计算出相对百分误差。

(2)根据测量及计算结果,描绘出载流圆线圈磁场分布的 B—X 及 B—Y 曲线。实验中,圆线圈磁场只测一半,另一半可根据圆线圈磁场分布的对称性描绘出来。其中 B—Y 曲线画在一张坐标纸上,全部 B—X 曲线画在同一张坐标纸上,并利用叠加原理,作线圈Ⅰ与线圈Ⅱ轴向磁场的叠加曲线,与亥姆霍兹线圈磁场分布曲线进行比较,验证磁场叠加原理。

3. 结果分析

对实验结果进行分析,并得出实验结论。

视频 S15　载流线圈磁场分布的测量

[问题讨论]

(1)实验中,为什么要求各步骤的电流保持不变?

(2)测量时为什么要旋转探测线圈?

(3)用感应法测磁场的误差来源主要有哪些?

载流线圈磁场分布的测量如视频 S15 所示。

实验16 万用表的使用

【引言】

万用表是一般电工和无线电技术上最常使用的工具，是实验室常用的仪表。万用表把多量程的交直流伏特计、安培计以及欧姆计组合在一起，使用方便。万用表的设计主要是根据电表扩程原理，将一只磁电式电流表（通常称为表头）与多种电学参量的测量电路通过转换开关组合在一起，从而形成的一种多用途的测量电器。

【实验目的】

（1）了解构成万用表的基本电路及欧姆计的基本原理。

（2）掌握万用表的正确使用方法。

【实验仪器】

万用表、测试板及待测电路。

【实验原理】

1. 万用表的基本线路

下面以MF500型万用表为例，介绍其主要的测试电路。

（1）直流电流测试电路：该电路由表头和分流电路构成，如图S16-1所示。显然，由分流电阻的各抽头形成各个量程：0~50μA、0~1mA、0~10mA、0~100mA、0~500mA。下面分析一下50μA挡电路，该挡有两个电流支路：a-b-c和a-d-c。支路a-b-c的总电阻为3.75kΩ，支路a-d-c的总电阻为15kΩ，当有40μA电流流过表头时，支路a-d-c的分流是(3.75kΩ/15kΩ)×40μA=10μA。整个电路可以看成是一个内阻为3kΩ，量程是50μA的微安表。而50μA称为该万用表的极限灵敏度。其他各电流挡，可根据图中的数据自行计算验证。

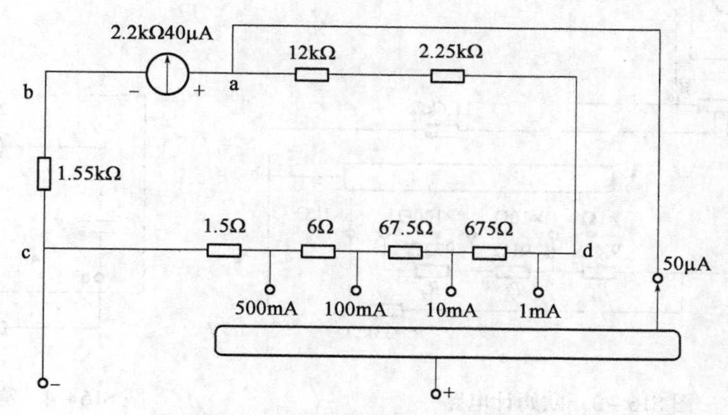

图 S16-1 测量直流电流的原理电路

（2）直流电压测试电路：电路由上述的0~50μA的微安表和一串倍压电阻构成，如图S16-2所示。显然，由倍压电阻的各抽头形成各量程：0~2.5V、0~10V、0~50V、0~250V、0~500V。表征电压表性能的重要参数是电压灵敏度。万用表的电压灵敏度就是极限灵敏度的倒数，即

$$S_V = \frac{1}{50\mu A} = 20k\Omega/V \qquad (S16-1)$$

也称为欧姆每伏，电压表各挡的内阻是：R_V = 量程 × 欧姆/伏。

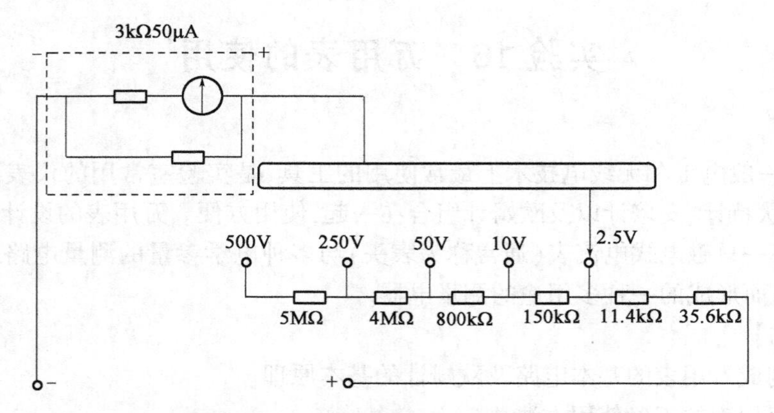

图 S16－2　测量直流电压的原理电路

(3)欧姆计电路：欧姆计电路如图 S16－3 所示，其中 W_0 为"零欧姆"调节电位器，E 为(1.5V和 9V)干电池。测量的挡位在 ×1Ω、×10Ω、×100Ω、×1kΩ 时，$E=1.5V$；在 ×10kΩ(未画出)时，$E=9V$。

2. 欧姆计原理

(1)测量原理：测量电阻的原理电路，如图 S16－4 所示，其中虚线部分为欧姆计，a、b 两端点为表笔插孔。待测电阻 R_x 接在 a、b 上，E 为干电池，R'为限流电阻，此回路的电流为

$$I_x=\frac{E}{R_g+R'+R_x} \qquad (S16-2)$$

可见，对一个欧姆计，如果 E、R'、R_x一定，则 I_x仅由 R_x决定。即 I_x和 R_x是一一对应的，原则上讲，可以在表头上直接标出偏转所对应的 R_x值，形成欧姆计。

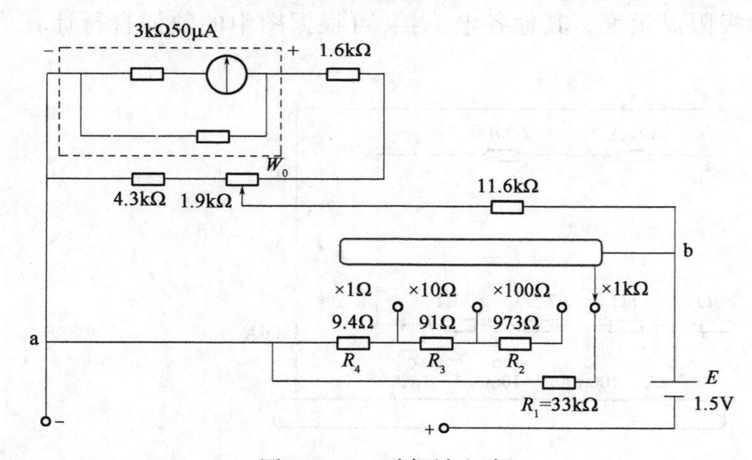

图 S16－3　欧姆计电路

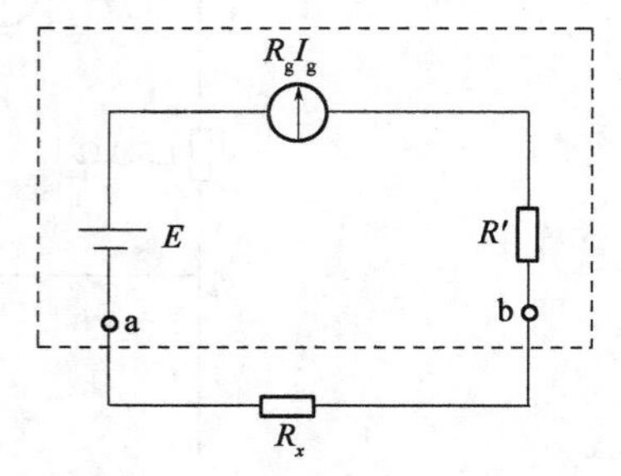

图 S16－4　测量电阻的原理电路

但实际上，由于电池的端电压随着使用时间和负载的不同而变化，因而不能用电流的绝对量表示电阻的绝对量。下面用电流之相对量来表示电阻的相对量。

为此，令 $R_中=R'+R_g$ 代入式(S16－2)，得

$$I_x=\frac{E}{R_中}\frac{R_中}{R_中+R_x}$$

再令

$$I_m=\frac{E}{R_中}$$

代入上式有

$$\frac{I_m}{I_x}-1=\frac{R_x}{R_{中}} \quad (S16-3)$$

由式(S16－3)可以看出,可以用电流之相对大小表示电阻的相对大小,进一步为了便于测量,令 $\alpha I_m=I_g$, $i_x=\alpha I_x$,便有

$$\frac{I_g}{i_x}-1=\frac{R_x}{R_{中}} \quad (S16-4)$$

为欧姆计之刻度公式。

注意这里用 α 的作用,它相当于调零电位器 W_0 的作用,并且 α 随着 W_0 的调节而变化。当 $R_x=0$ 时(表笔短路)调节 W_0 使表头指示满度 I_g。这时就有 $\alpha I_m=I_g$。这一过程称为"欧姆计零点"调节。只有调好欧姆计零点,i_x 才有绝对的意义。即由 i_x 点标出所对应的$\frac{R_x}{R_{中}}$值。

例如,当 $R_x=R_{中}$($R_{中}$即欧姆计的内阻)时,$i_x=\frac{1}{2}I_g$,这时表头指针正指欧姆刻度中央,故此刻度为 i_x 表示中值电阻 $R_{中}$。

由刻度公式(S16－4)可以看出,欧姆计刻度是不均匀的,当 $R_x<<R_{中}$时,$i_x\approx I_g$,指针接近满度,且随 R_x 的变化不明显,因而测量误差大,当 $R_x>>R_{中}$时,$i_x\approx 0$ 测量误差也大,只有当 $R_x=R_{中}$时误差最小。所以,用欧姆计测量电阻,要尽量利用中央附近刻度,如$\frac{1}{5}R_{中}\sim 5R_{中}$这段刻度。

(2)欧姆计测量范围的改变:为适合测量各种阻值,欧姆计需要经常改变测量范围,从式(S16－4)可知,只要改变 $R_{中}$就可以了。如何改变 $R_{中}$呢?参看图 S16－3,只要给 a、b 端上串联不同的电阻就可以了。例如,在 ×1kΩ 挡,a、b 端并接了电阻 $R_1=33\text{k}\Omega$,这时 $R_{中}=10\text{k}\Omega$。在 ×100Ω 挡,a、b 端并接了电阻 $R=R_2+R_3+R_4=1073.4\Omega$,这时 $R_{中}=1\text{k}\Omega$。在 ×10Ω 挡 $R_{中}=100\Omega$,在 ×1Ω 挡 $R_{中}=10\Omega$,这些中值电阻,请同学们自己推算一下。

【实验内容与步骤】

1.准备

认清万用表的面板和刻度,根据待测量的种类及大小将选择开关拨至合适的位置(不知待测量的大小时,一般应选择最大量程进行试测),接好表笔(万用表的"+"端接红色表笔)。

2.测量变压器次级线圈的交流电压

变压器安装在测试板上,测试板电路如图 S16－5 所示。

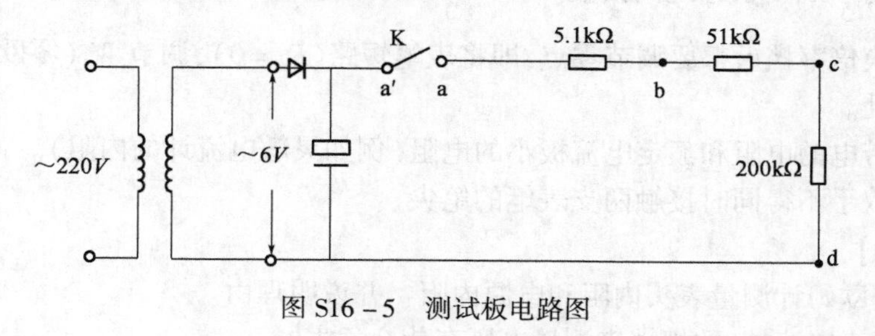

图 S16－5　测试板电路图

3.测量直流电压

参看测试板电路,选择合适的量程(表针应在面板的$\frac{1}{2}$至满刻度之间工作),分别测出

U_{ab}、U_{bc}、U_{cd}、U_{ad}。计算 $U_{ab}+U_{bc}+U_{cd}$是否等于 U_{ad},并解释这些结果。

4. 测量直流电流

断开电键 K,将电流表串接于 a′a 处,选择合适的量程(表针应在面板的$\frac{1}{2}$至满刻度之间工作),(注意极性)测回路电流 $I_{a'a}$。

5. 测量电阻

断开测试板电源,断开电键 K,选择合适的倍率(表针应在面板的$\frac{1}{5}R_{中}\sim 5R_{中}$之间工作),测量测试板上的三只电阻 R_{ab}、R_{bc}、R_{cd}及 R_{ad}。计算 $R_{ab}+R_{bc}+R_{cd}$是否等于 R_{ad},并解释这些结果。

注意:测量时要注意表针的偏转情况,适当调节倍率挡。若测试条件达不到则另当别论。今后在使用数字万用表时也要选择合适的量程才能保证误差最小。

6. 用万用表电压挡和欧姆挡检查一电路故障

电路发生故障时,用万用表检测电路多采用伏特计法,其要点是正确理解电路原理,了解其电压的正常分布,然后用伏特计系统检查各点的电压,由电压不正常点推断出故障点及其原因。若电路中导线可拆卸,也可把导线与电路断开,用万用表"欧姆挡"检查导线是否通。

【注意事项】

(1)使用伏特计或安培计时应注意:

①根据被测量及大小,正确将被测量转换开关及量程转换开关拨至相应位置,并使得指针尽可能在面板的$\frac{1}{2}$至满刻度之间读数。

②测直流电流和电压时,表笔的正、负不能接反。

③执笔时手不能接触任何金属部分。

④测试时采用跃接法,即在用表笔接触测量点的同时,要注意指针偏转的情况,并随时准备在出现异常现象时使表笔跳离测量点。

(2)使用欧姆计时应注意:

①根据被测电阻大小,正确将被测量转换开关及倍率转换开关拨至相应位置,并使得指针尽可能在面板的$\frac{1}{5}R_{中}\sim 5R_{中}$之间读数。

②每次更换倍率挡后都要调节零点,即将表笔短路($R_x=0$),调节 W_0(零欧姆旋钮),使指针指在零欧姆处。

③不得测带电的电阻和额定电流极小的电阻(例如灵敏电流计的内阻)。

④测量时双手不得同时接触两支表笔的笔尖。

【问题讨论】

(1)能否用欧姆计测量表头内阻和电源内阻?并说明理由。

(2)用万用表电流挡、欧姆挡来测量电压有什么危害?

(3)欧姆计的测量范围如何调整?

(4)为什么在欧姆计电路中要设计一个调零电位器 W_0?为什么在欧姆挡有时候调不到零?

实验 17　超声波在空气中传播速度的测量

［引言］

声波是一种在弹性媒质中传播的纵波。声波的波长、频率、强度和传播速度是声波的主要特征量。对超声波传播速度的测量在超声波测距、探伤等方面具有重要意义。在物理实验中，声速的测定可采用驻波法（共轭干涉法）或相位比较法（行波法）。由于超声波具有波长短、容易定向发射等特点，根据入射波在发射器和接收器之间的多次反射相干而形成声波场（即驻波）的物理学规律，选择了空气介质中声波多次反射形成干涉的声压场的物理过程，利用压电陶瓷换能器完成声压和电压之间的转换，再用示波器观测，从而实现对超声波在空气中的传播速度这一非电量的电学测量。

［实验目的］

（1）了解超声波的产生和接收的原理。

（2）学习用驻波法测量超声波在空气中的传播速度。

［预习问题］

（1）什么是行波？什么是驻波？如何由行波获得驻波？

（2）如何调整与判断测量系统处于共振状态？

（3）实验需要测量的物理量（如波长、频率、波节位置等）是如何测量的？

［实验仪器］

SV－DH 系列声速测试仪、SVX－5 型声速测试仪信号源、MOS－620 型示波器等。

［实验原理］

一般来讲，将频率在 20～20000Hz 之间的纵波称为声波；频率大于 20000Hz 的声波称为超声波。声波与超声波在空气中具有相同的传播速度，由波动理论可知波速、频率和波长满足关系式 $v=f\lambda$，所以只要测得频率和波长就可以求得波速。

1．超声波在空气中的传播速度

在理想气体中声波的传播速度为

$$v=\sqrt{\frac{\gamma RT}{M}} \tag{S17－1}$$

式中，$\gamma=c_p/c_V$ 称为比热容比，即气体比定压热容与比定容热容的比值；M 是气体的摩尔质量；T 是热力学温度；$R=8.31441\mathrm{J/(mol\cdot K)}$ 为普适气体常量。由于空气实际上并不是理想气体，总含有一些水蒸气，经过对空气平均摩尔质量 M 和比热容比 γ 的修正，在温度为 t、相对湿度为 r 的空气中，声速的修正计算公式可表示为

$$v=v_0\sqrt{\left(1+\frac{t}{T_0}\right)\left(1+0.31\frac{rp_s}{p}\right)} \tag{S17－2}$$

式中，p_s 为摄氏温度 t 时空气的饱和蒸气压，可从饱和蒸气压和温度的关系表中查出；p 为大气压，取 $p=1.013\times10^5\mathrm{Pa}$，相对湿度 r 可从温度计读出。由这些气体参量可以计算出声速，故式（S17－2）可作为实验室环境下声速的理论计算公式。

2．压电换能器

压电换能器是由压电陶瓷片和两种金属组成。压电陶瓷片是由一种多晶结构的压电材料

(如石英或锆钛酸铅压电陶瓷等),在一定温度下经极化处理制成。它具有压电效应,即受到与极化方向一致的应力 T 时,在极化方向上产生一定的电场强度 E 且具有线性关系:$E = gT$,称为正压电效应;当与极化方向一致的外加电压 U 加在压电材料上时,材料的伸缩形变 S 与 U 之间有简单的线性关系:$S = d \cdot U$,称为逆压电效应。其中 g 为比例系数,d 为压电常数,与材料的性质有关。由于 E 与 T,S 与 U 之间有简单的线性关系,因此可以将正弦交流电信号变成压电材料纵向的长度伸缩,使压电陶瓷片成为超声波的波源,即压电换能器可以把电能转换为超声能量作为超声波发生器;反过来也可以使声压变化转化为电压变化,即用压电陶瓷片作为声频信号接收器。因此,压电换能器可以把电能转换为超声能量作为声波发生器,也可把超声能量转换为电能作为声波接收器之用。压电陶瓷换能器根据它的工作方式,可分为纵向(振动)换能器、径向(振动)换能器及弯曲振动换能器。图 S17 - 1 所示为纵向换能器的结构简图。

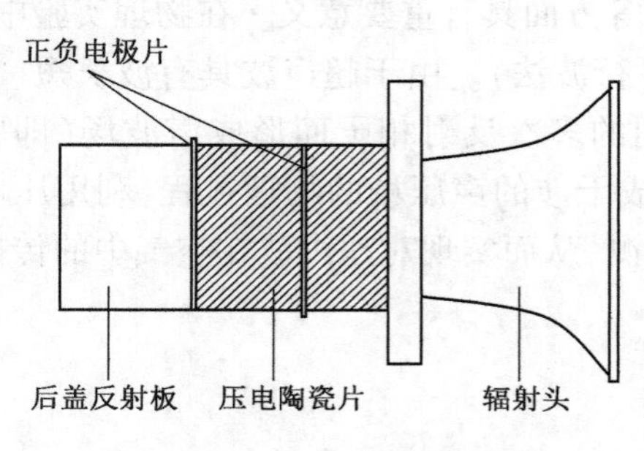

图 S17 - 1　纵向换能器的结构

实验中,需要使用两个完全相同的换能器,一个换能器用来产生超声波称为发射换能器,另一个用来接收超声波称为接收换能器。其中发射换能器把交变的电压信号转化成超声信号并由发射平面发射出去;接收换能器除了反射一部分超声波能量外,还通过压电陶瓷将超声信号转化成电信号。而且只有当输入电信号的频率与换能器的固有谐振频率相同时,换能器进行电声或者声电转换的效率才最高,此种现象称为驻波系统的共振。

3. 驻波的形成

设发射换能器发出的平面波经前方反射后,入射波和反射波叠加,当满足驻波条件时形成驻波。设两列平面波频率、振幅和振动方向相同,在 x 轴上沿相反方向传播,振动方程可写为

入射波:
$$y_1 = A\cos\left(\omega t - \frac{2\pi}{\lambda}x\right)$$

反射波:
$$y_1 = A\cos\left(\omega t + \frac{2\pi}{\lambda}x\right)$$

两平面波进行叠加,合成波为

$$y = y_1 + y_2 = A\cos\left(\omega t - \frac{2\pi}{\lambda}x\right) + A\cos\left(\omega t + \frac{2\pi}{\lambda}x\right) = \left(2A\cos\frac{2\pi}{\lambda}x\right)\cos\omega t \qquad (S17-3)$$

上式表明合成波为驻波。驻波场中各点都在作同频率的振动,各点的振幅 $2A\cos\frac{2\pi}{\lambda}x$ 是位置 x 的余弦函数。对应于 $\left|\cos\frac{2\pi}{\lambda}x\right| = 1$ 的各点振幅最大,称为波腹,即在 $x = \pm n\frac{\lambda}{2}(n = 1,2,3,\cdots)$ 处是波腹位置;对应于 $\left|\cos\frac{2\pi}{\lambda}x\right| = 0$ 的各点始终静止不动,称为波节,即在 $x = \pm(2n+1)\frac{\lambda}{4}(n = 1,2,3,\cdots)$ 处是波节位置。可见,相邻波腹或者波节之间的距离差为 $\frac{\lambda}{2}$, 然而,在驻波场中,波腹处声压最小,经接收换能器转化成的电压信号应最弱;波节处声压最大,经接收换能器转化成的电压信号则最强。因此,可以通过观察接收换能器输出信号的强弱来判断换能器处在波节还是波腹位置。所以,当改变两个换能器的间距时,同时观察接收换能器输出电压的变化,若能

测得相邻两次最大电压值的位置读数,则两次读数值之差便为$\frac{\lambda}{2}$,即可求得超声波的波长 λ。

声速、波长和频率之间的关系为

$$v=f\lambda \tag{S17-4}$$

式中,λ 为波长;f 为输入电信号的频率;f 可用示波器测出。因此,根据式(S17-4)就可以得到声速 v。

[**实验内容与步骤**]

(1)通过开机预热,使其自动工作在连续波方式。

(2)用驻波法测量声速。

①测量装置的连接 如图 S17-2 所示,信号源面板上的发射端换能器接口(S_1),用于输出一定频率的电压信号,接至测试架的发射端(S_1);信号源面板上的发射端的发射波形 Y1,接至双踪示波器的 CH_1(Y_1),用于观察发射波形;接收换能器(S_2)的输出端接至示波器的 CH_2(Y_2)。

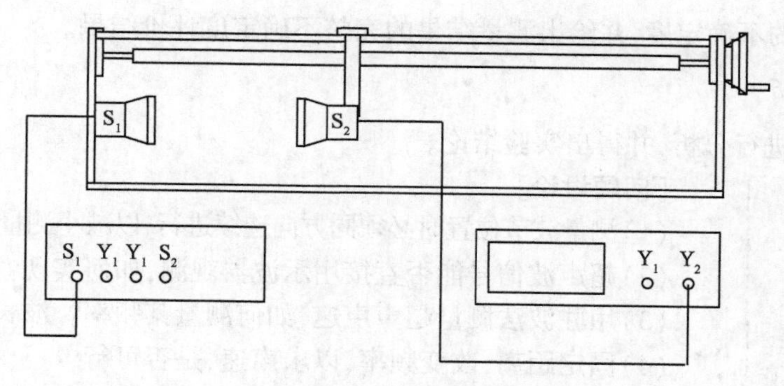

图 S17-2 实验接线图

②测定压电换能器的最佳工作点。

只有当换能器 S_1 和 S_2 发射面与接收面保持平行时才有较好的接收效果;为了得到较清晰的接收波形,应将外加的驱动信号频率调节到换能器 S_1、S_2 谐振频率点处,才能较好地进行声能与电能的相互转换从而提高测量精度,以得到较好的实验效果。超声换能器工作状态的调节方法如下:各仪器都正常工作以后,首先调节声速测试仪信号源输出电压(100~500mV),调节信号频率(25~45kHz),观察频率调整时示波器显示电压幅度的变化,在某一频率点处电压幅度最大时,此频率即是压电换能器 S_1、S_2 相匹配的频率点,记录信号源的频率 f。

③测量步骤。

a. 按图 S17-2 所示连接好电路。

b. 将测试方法设置到连续波方式,将声速测试仪信号源频率调至换能器的谐振频率。

c. 在谐振频率下,单方向移动 S_2,依次记下各振幅最大时,接收换能器在读数标尺上的位置 $L_1, L_2, \cdots$ 共 16 个值。

d. 记下室温 t。

e. 用逐差法处理数据。

[**注意事项**]

(1)防止将信号线和接地线接错或将信号输出短路。

(2)防止过力扭动示波器及信号源的各个旋钮,导致损坏各调节旋钮。

(3)确保信号强度应与示波器衰减挡适配,以免损坏仪器。

[数据及处理]

1. 数据记录(表 S17-1)

表 S17-1 声速测量数据记录表

$t=$________℃, $f=$________Hz, 相对湿度 $r=$________ 单位:mm

次数	1	2	3	4	5	6	7	8
卡尺读数 L_i								
次数	9	10	11	12	13	14	15	16
卡尺读数 L_i								
$L_{i+8}-L_i$								

2. 数据处理

计算声速 v 的不确定度,并给出测量结果的完整不确定度评价结果。

3. 结果分析

对实验结果进行分析,并得出实验结论。

视频 S17 超声波在空气中传播速度的测量

[问题讨论]

(1)测量波节位置时必须同方向连续进行,以减小测量误差,为什么?

(2)超声波信号能否直接用示波器观测,如何实现?

(3)用驻波法测量超声声速,如何测量其频率?波长又如何测量?

(4)固定距离,改变频率,以求声速,是否可行?

超声波在空气中传播速度的测量如视频 S17 所示。

[补充说明]

图 17-3 是声速测试仪的结构示意图,其中发射换能器固定在支座上,接收换能器固定在游标卡尺副尺上,当接收换能器发生移动时,其距离的变化可由游标卡尺读出。

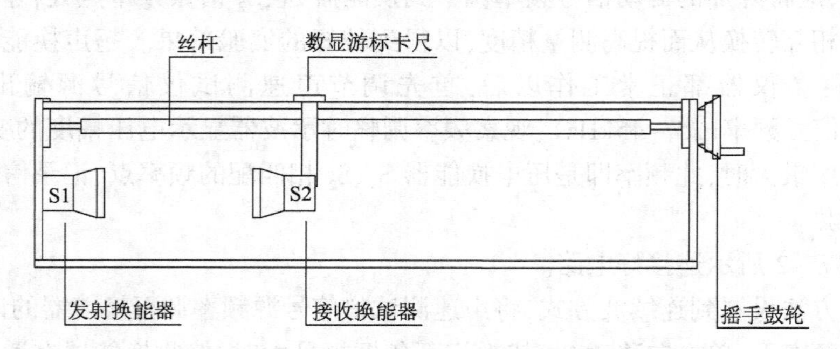

图 17-3 声速测试仪结构示意图

实验 18 用双臂电桥测量低值电阻

[引言]

通常,电阻按照其阻值大小可分为高值电阻($R>1\text{M}\Omega$)、中值电阻($1\Omega<R<1\text{M}\Omega$)和低值电阻($R<1\Omega$)。当使用单臂电桥测量低值电阻时,导线本身的电阻和接点处接触电阻等附加

电阻的影响，已成为影响低值电阻测量不可忽视的主要因素。消除这些附加电阻影响的基本方法是采用“四端钮接入法”，将附加电阻部分转移到电源回路，部分转移到比其大许多量级的电桥桥臂电阻的电路中。这种“四端钮接入法”只能靠双臂电桥来实现，这就是设计双臂电桥的基本思想。

[实验目的]

(1)了解双臂电桥的组成及其工作原理。

(2)掌握双臂电桥测量低值电阻的原理和方法。

(3)学会用双臂电桥测量低值电阻。

[预习问题]

(1)为什么不能用单臂电桥测量低值电阻？

(2)双臂电桥与单臂电桥有哪些异同？

(3)双臂电桥中的一根粗短导线应该接在何处？

[实验仪器]

QJ19 型单双臂两用电桥、AC15 型直流复射式检流计、电流表、标准电阻、滑线变阻器、直流电源、待测电阻、电键、导线。

[实验原理]

1. 双臂电桥测低值电阻的原理

如图 S18－1 所示用惠斯通电桥(单臂电桥)测量中值电阻，在此电路中有若干导线和 A、B、C、D 四个接触点。电桥平衡时待测电阻 $R_x=\frac{R_2}{R_1}R_N$。由 A、C 两点到电源的导线电阻可计入电源的内阻中，D、B 两点到检流计的导线电阻并入检流计的内阻中，均对测量没有影响。比例臂 R_1 和 R_2 一般选用阻值较大的电阻，因此和这两个电阻连接的导线电阻和 D 点的接触电阻都可忽略不计。测量中值电阻时与 R_x 和 R_N 相连的导线电阻以及接触电阻也是可以忽略的，不会对测量产生影响。但是如果被测电阻 R_x 是低值电阻，与 R_x 和 R_N 相连的导线电阻以及接触电阻就不能忽略，这附加电阻值都已超过或大大超过被测电阻的阻值，这样会造成很大误差，甚至完全无法得出测量结果。所以，用单臂电桥来测量低值电阻是不可能精确的，必须在测量线路上采取措施，避免附加电阻对低值电阻测量的影响。

双臂电桥也称开尔文电桥，是在惠斯通电桥的基础上发展起来的，它采用了“四端钮接入法”，如图 S18－2 所示。待测电阻 R_x 是 c、d 之间的电阻值，c、d 称为电压端，a、b 称为电流端。如图 S18－3 所示，将待测电阻 R_x 接入惠斯通电桥电路。图中 R_x 的附加电阻 r_1、r_2、r_1'、r_2' 的值都比较小。r_1 与电源内阻串联，r_1' 与桥臂电阻 R_2 串联，因为 R_2 和电源内阻远大于和它们相串联的附加电阻，故附加电阻 r_1、r'_1 的影响可以忽略。而 r_2 和检流计 G 相串联，r'_2 与标准电阻 R_N 串联，R_N 的阻值本身较小，故附加电阻 r_2 的影响不能忽略。那么，如何使得附加电阻 r_2 和 r_2' 的影响也能忽略呢？仿照待测电阻 R_x，将标准电阻 R_N 也采取“四端钮接入法”，如图 S18－4所示。标准电阻两端的附加电阻 r_{N1}、r_{N2}、r_{N1}'、r_{N2}'，采用“四端钮接入法”以后，r_{N1}' 和桥臂电阻 R_1 串联，r_{N1} 与电源内阻串联，为了消除 r_{N2}' 和 r_2' 的影响，分别串联比它们阻值大得多的电阻 R_4、R_3，构成一对新的桥臂。r_{N2} 和 r_2 串联，设其阻值总和为 r，其影响可以通过调节桥臂电阻 R_1、R_2、R_3、R_4 的阻值消去。

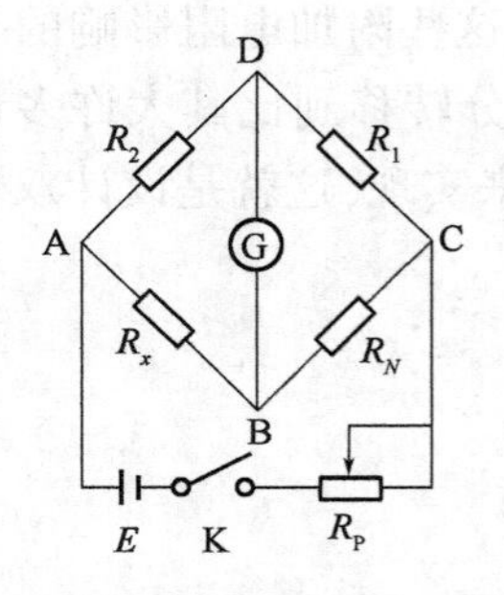

图 S18－1　单臂电桥工作原理

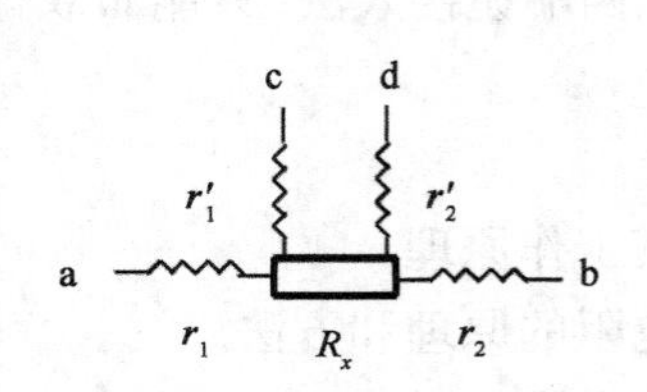

图 S18－2　四端钮接入法

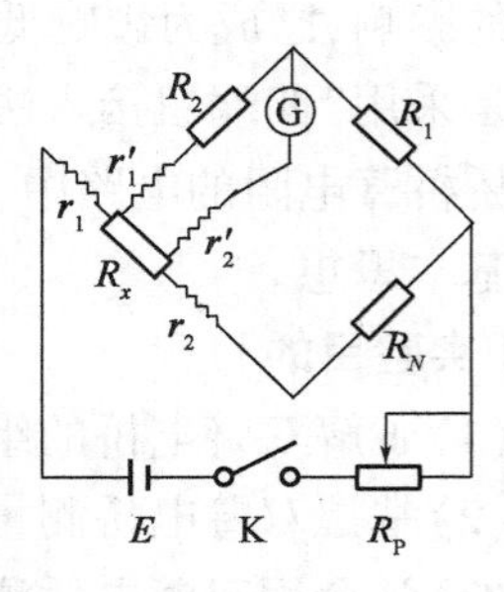

图 S18－3　电阻的四端钮接入法

双臂电桥的电路如图 S18－5 所示。由于双臂电桥具备 $r_2' \ll R_4, r_{N1}' \ll R_1, r_{N2}' \ll R_3$ 的条件，调节 R_1、R_2、R_3、R_4，使电桥达到平衡（检流计指零）时流经 R_1 和 R_2 上的电流相等设为 I_1，R_3 和 R_4 上的电流相等设为 I_2，流经 R_x 和 R_N 的电流也相等设为 I_3，且有 $I_1 \ll I_3, I_2 \ll I_3$，故根据基尔霍夫第二定律有

$$I_1R_1 = I_3R_x + I_2R_2 \tag{S18－1}$$

$$I_1R_2 = I_3R_N + I_2R_4 \tag{S18－2}$$

$$I_2(R_3 + R_4) = (I_3 - I_2)r \tag{S18－3}$$

联立求解可得

$$R_x = \frac{R_2}{R_1}R_N + \frac{rR_3}{R_3 + R_4 + r}\left(\frac{R_2}{R_1} - \frac{R_4}{R_3}\right) \tag{S18－4}$$

若取$\frac{R_2}{R_1} = \frac{R_4}{R_3}$或 $R_1 = R_3$、$R_2 = R_4$，则式（S18－4）右边的第二项为零，于是有

$$R_x = \frac{R_2}{R_1}R_N \tag{S18－5}$$

这样就消除了 r 对测量的影响。将$\frac{R_2}{R_1} = \frac{R_4}{R_3}$或 $R_1 = R_3$、$R_2 = R_4$ 称为双臂电桥平衡的辅助条件。

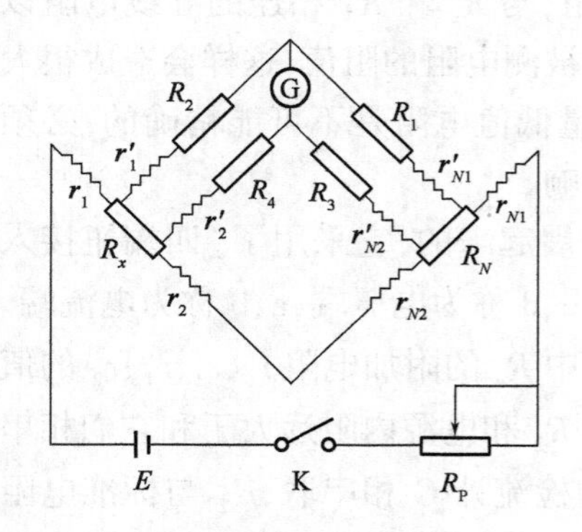

图 S18－4　双臂电桥原理图

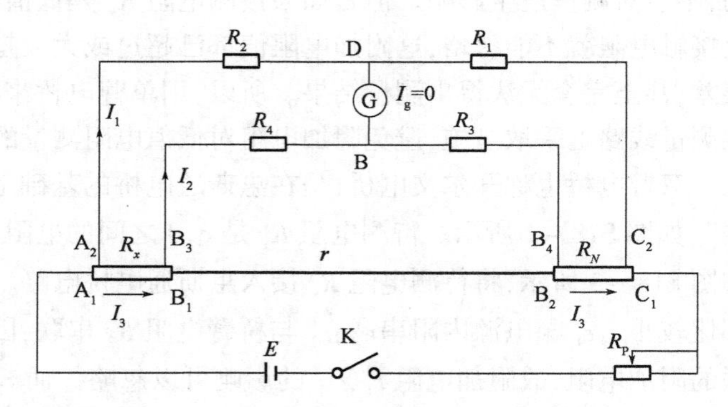

图 S18－5　双臂电桥电路图

实验时，始终保持$\frac{R_2}{R_1} = \frac{R_4}{R_3}$或 $R_1 = R_3$、$R_2 = R_4$，并且四个电阻臂 R_1、R_2、R_3、R_4 的阻值应远大于 R_x 和 R_N 的接线电阻和接触电阻。这些要求都可以用电桥的结构来保证，通常把电桥做成一种特殊的结构，将两对比率臂$\frac{R_2}{R_1}$和$\frac{R_4}{R_3}$都采用比率相同的双十进电阻箱，并且将它们的转臂

连接在同一转轴上,这样在臂的任一位置都能满足平衡电桥的辅助条件,四个电阻臂也可以取到足够大的阻值。当然,制造仪器时不可能将 R_1、R_2、R_3、R_4 做到绝对相等,为了消除 r 对待测电阻 R_x 的影响,要求连接 R_x 和 R_N 两电流端的导线尽可能短而粗,使 r 尽可能小。

双臂电桥内部结构原理与思想原理设计的实质一致,但是为了和 QJ19 型电桥面板图上的符号一致,稍做了点改动,使 R_2 和 R_3 的编号对换,这样双臂电桥平衡必须满足的辅助条件变为:

$$\frac{R_3}{R_1}=\frac{R_4}{R_2} \quad 或 \quad R_3=R_4、R_1=R_2$$

当电桥达到平衡时,有

$$R_x=\frac{R_3}{R_1}R_N=\frac{R_4}{R_2}R_N$$

通常在双臂电桥中规定 $R_3=R_4=R$,则

$$R_x=\frac{R}{R_1}R_N=\frac{R}{R_2}R_N \tag{S18-6}$$

这就是双臂电桥测量低值电阻的计算公式,式中 R 就是电桥平衡时面板上的指示值。

双臂电桥在测量时,由于 R_x 很小,R_x 两端的电压也很小,而工作电流又不会很小,此时热电势的存在使 R_x 两端的电压有增大或减小的可能,这将破坏电桥的平衡条件,在有热电势的情况下检流计指零,电桥桥臂之间并不严格满足式(S18-6)的关系。但由于热电势与电流的方向有关,故可采用改变电源极性(电流方向)的方法予以消除,在实验中通过换向开关来达到此目的。在不同极性下测得不同的 R,求其平均值。则式(S18-6)可写成

$$R_x=\frac{1}{2}(R_+ + R_-)\frac{R_N}{R_1} \tag{S18-7}$$

2. 电桥的相对灵敏度

电桥达到平衡时,通过检流计的电流 $I_g=0$,然而检流计是否真正指零,是凭借人的判断,由于视差的存在,必然会给测量结果带来一定的误差,误差的大小取决于电桥的相对灵敏度。

电桥的相对灵敏度定义为:电桥达到平衡时,电桥面板指示值为 R,此时再调节测量旋钮使面板指示值改变某值 ΔR,若检流计偏离平衡位置 Δn,则定义电桥的相对灵敏度为

$$S=\frac{\Delta n}{\frac{\Delta R}{R}} \tag{S18-8}$$

S 的物理意义是桥臂电阻的单位相对变化所引起的电流计的偏转格数。S 越大,电桥灵敏度就越高,引入的测量误差也就越小。

[实验内容与步骤]

1. 熟悉 QJ19 型单双臂两用电桥的功能,掌握测量的基本要领

参见本实验补充说明。

2. 测量金属丝的电阻值

(1)按图 S18-6 接好线路,根据 R_x 的估计值和已知标准电阻 R_N 值,在保证 R 取满 5 位有效数字的前提下,合理选择 $R_1=R_2$ 的值,并且估计 R 的值,将 R 置于估计之值;

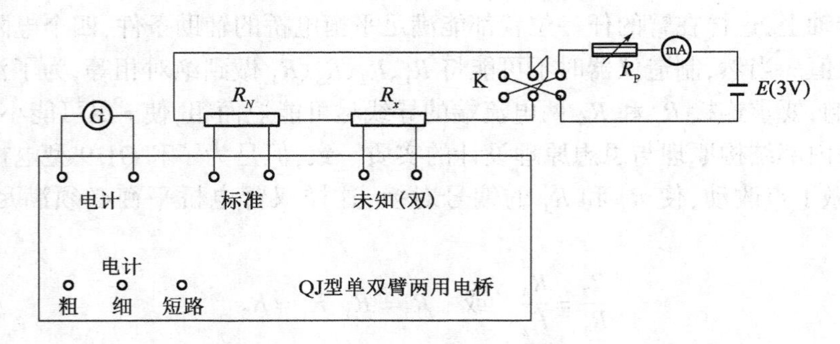

图 S18－6　双臂电桥测低值电阻实验线路

(2)将灵敏检流计接上220V的交流电源,把分流旋钮置于“直接”挡,调节零点;

(3)把检流计分流旋钮置0.01挡,闭合电源开关K到任一边,按先“粗”后“细”的步骤,用逐步逼近法调节R,使检流计指零即电桥平衡;

(4)再将分流旋钮分别置于“0.1”挡、“1”挡及“直接”挡,每一挡都按照步骤(3)中的方法调节R使电桥平衡,记录在“直接”挡时电桥平衡时的R示值R_+;

(5)将换向开关K反向,再观察检流计还是否指零,即电桥是否平衡,若平衡被破坏,再调节R使其指零,记录下电桥平衡时R的示值R_-,根据式(S18－7)计算出待测电阻R_x的值。

3. 测量电桥的相对灵敏度S

在平衡电桥的基础上,让R改变某一值ΔR,记录检流计偏离平衡位置的格数Δn,由式(S18－8)计算电桥的相对灵敏度S。

[注意事项]

(1)实验中应保持桌面平稳,以免检流计不能正常工作。

(2)勿将“粗”或“细”调按钮长期按下并锁住。

(3)旋动各旋钮切勿用力过大而损坏仪器。

(4)实验中检流计应随时调零。

(5)采用“四端钮接入法”时电阻的电压端与电流端不能接混。

[数据及处理]

1. 数据记录(表S18－1、表S18－2)

表 S18－1　测量低值电阻数据记录

R_x估计值________　R_N = ________　$R_1 = R_2$ = ________　k = ________

比较臂R的读数/Ω		测量值/Ω
正向读数R_+	反向读数R_-	$R_x=\frac{1}{2}(R_+ + R_-)\frac{R_N}{R_1}$

表 S18－2　测量电桥灵敏度数据记录

平衡电桥R/Ω	改变量ΔR/Ω	偏离格数Δn	相对灵敏度$S=\frac{\Delta n}{\Delta R/R}$

2. 数据处理

计算测量电阻的不确定度,给出测量结果的不确定度的评价。

3. 结果分析

对实验结果进行分析,并得出实验结论。

[问题讨论]

(1)双臂电桥中是如何消除引线、接触电阻影响的?

(2)如何消除热电势对测量结果的影响?

(3)四端电阻的电流端和电压端是如何区分的?

(4)如果低值电阻的电流端和电压端互相接错,有什么不好影响?

用双臂电桥测量低值电阻如视频 S18 所示。

视频 S18 用双臂电桥测量低值电阻

[补充说明]

QJ19 型单双臂两用电桥准确度等级 0.05,当双桥使用时测量范围为$10^{-5} \sim 10^{2}\Omega$。其各有效量程误差如表 S18-3 所示。图 S18-7 是 QJ19 型单双臂两用电桥的面板结构,包括“标准(双)”外接标准电阻、“未知(双)”外接待测电阻(双桥与单桥的接法略有不同)、“检流计”外接灵敏检流计、静电屏蔽端 11。R 是由五个十进位转盘电阻箱组成,可使 R 读取五位有效数字,其值分别为×100、×10、×1、×0.1、×0.01。R_1 和 R_2 也是两个转盘电阻箱。测量时先按下电计“粗”按钮,根据检流计的光点偏转方向,采用逐步逼近法调节 R 使检流计指零,然后松开“粗”按钮,按下电计“细”按钮,再调节 R 使检流计指零。在调节过程中,如果遇到检流计的光点偏转很快,应立即松开电计按钮,按下短路按钮,这种情况说明电桥面板上的 R 值距平衡时的阻值相差甚远,应再仔细调节;如果遇到检流计的光点始终向一个方向偏转,则说明工作电路不正常,应认真检查各接点是否接触良好;如果遇到电流表无电流指示,应认真检查待测电阻丝是否断路。实验完毕,各按钮均应放开。

表 S18-3 QJ19 型单双臂两用电桥有效量程误差(双桥)

被测电阻 R_x/Ω	标准电阻 R_N/Ω	比例臂电阻 $R_1=R_2\,\Omega$	电源电压/V	误差/%
$10 \sim 10^{2}$	100	10^{3}	2~6	≤0.05
$1 \sim 10$	10			
$0.1 \sim 1$	1			
$10^{-2} \sim 10^{-1}$	10^{-1}			
$10^{-3} \sim 10^{-2}$	10^{-2}			

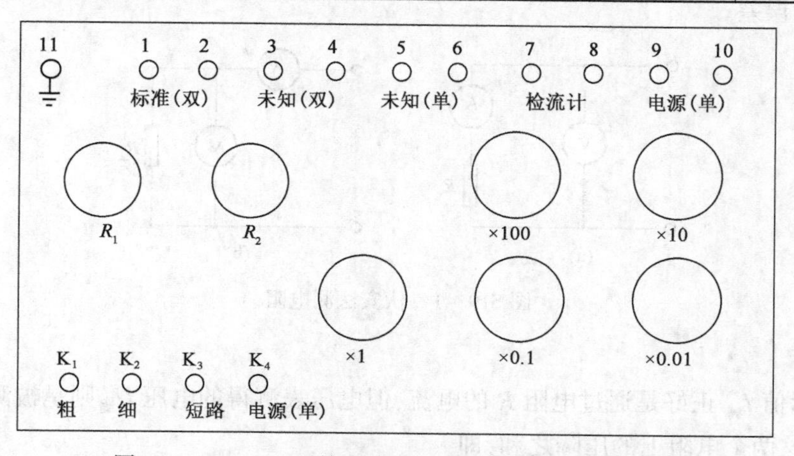

图 S18-7 QJ19 型单双臂两用电桥的面板结构示意图

实验19　非线性元件伏安特性的研究

［引言］

伏安法常用于测量材料的导电特性，在应用中，很多电子元件的电阻特性是用其伏安特性表征的，即由元件两端加上一定范围的电压，其中有一定范围的电流流过，这种一一对应关系再现了元件电阻的重要特性，即伏安特性。各种电子元件按其呈现的伏安特性曲线性质不同分为线性电阻元件和非线性电阻元件。例如，金属膜电阻、碳膜电阻及各种线绕电阻属于线性电阻元件，而各类半导体二极管、半导体三极管、光电二极管及光电三极管等属于非线性电阻元件。本实验利用伏安法研究半导体二极管的伏安特性。

［实验目的］

(1)掌握电学元件伏安特性测量的基本方法。

(2)学会分析伏安法的电表接入误差，能正确选择测量电路。

(3)学会用作图法处理数据的方法。

［预习提示］

(1)电流表内接法与外接法适用的条件是什么？

(2)测二极管正向特性曲线、反向特性曲线时，为什么一个用外接，一个用内接？

(3)本实验使用的半导体二极管正向导通阈值电压是多大？反向击穿电压是多大？如何避免二极管二次击穿？

［实验仪器］

直流电源、直流电压表(C30－V)、直流电流表(C30－A)(毫安表、微安表)、电阻箱(ZX21)、滑线变阻器(BX7D)、金属膜电阻、二极管。

［实验原理］

1. 伏安法测电阻及其误差

根据欧姆定律，如果测出电阻两端的电压和通过该电阻的电流，就可以算出被测电阻的阻值，这种方法称为伏安法测电阻。用伏安法测电阻，有两种接法：在图 S19－1(a)中电流表接在电压表的内侧，简称内接法；在图 S19－1(b)电流表接在电压表外侧，简称外接法。以下分别讨论其测量误差。

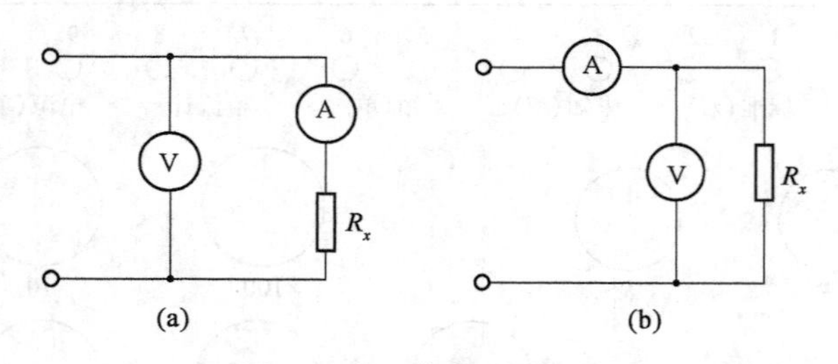

图 S19－1　伏安法测电阻

1)内接法

电流表示值 $I_{测}$ 正好是通过电阻 R 的电流，但电压表测得的电压 $U_{测}$ 则是被测电阻 R_x 与电流表内阻 R_A 这两个电阻上的压降之和，即

$$U_{测} = I_{测}(R_x + R_A) \tag{S19-1}$$

也可以写成

$$R_x = \frac{U_{测}}{I_{测}} - R_A = R_{测} - R_A \tag{S19-2}$$

由此看出，若以 $U_{测}/I_{测}$ 作为实验结果，则存在一个修正值 S_I，$S_I = R_A$。只有当被测电阻远大于电流表内阻 R_A 时，内接法测得的结果才是近似正确的，所以此方法适用于测量较大阻值的电阻。

如果知道电流表的内阻 R_A，则可以对测量结果按式（S19－2）进行修正，从而获得准确结果。

2）外接法

电压表读数 $U_{测}$ 正好反映了 R_x 上的压降，而电流表的读数 $I_{测}$ 却是 R_x 与电压表两部分通过的电流之和。设电压表内阻为 R_V，则有

$$I_{测} = U_{测}\left(\frac{1}{R_x} + \frac{1}{R_V}\right) \tag{S19-3}$$

$$R_{测} = \frac{U_{测}}{I_{测}} = R_x \frac{R_V}{R_x + R_V} \tag{S19-4}$$

显然，$R_{测}$ 实际上小于被测电阻的真实阻值 R_x，若以它作为结果，就存在偏差 S_U，且满足

$$S_U = \frac{U_{测}}{I_{测}} - \frac{U_{测} R_V}{R_V I_{测} - U_{测}} \tag{S19-5}$$

因此，只有当 $R_x \ll R_V$ 时，外接法才是近似准确的。如果已知电压表的内阻 R_V，对测量结果可根据式（S19－4）修正为

$$R_x = \frac{R_V R_{测}}{R_V - R_{测}} \tag{S19-6}$$

综上所述，在实际测量中，必须根据被测电阻与电表内阻的相对大小，适当选择测量电路，使电表内阻对测量的影响降到最小。如果电表内阻的影响不可忽略时，则必须对测量结果进行修正，从而取得正确的结果。

2. 不确定度的计算

设电阻测量中的不确定度为 u_R，根据不确定度的传递公式可得

$$u_R = \sqrt{\left(\frac{\partial R}{\partial U}\right)^2 u_1^2 + \left(\frac{\partial R}{\partial I}\right)^2 u_2^2} \tag{S19-7}$$

$$u_1 = \sqrt{S_U^2 + u_U^2}; \quad u_2 = \sqrt{S_I^2 + u_I^2}$$

式中，S、u 为电压和电流测量中的 A 类和 B 类不确定度。对于 A 类不确定度，按第 2 章测量不确定度的方法进行计算。B 类不确定度可以由电表的级别和量程计算，即

$$\Delta_{仪} = 量程 \times 级别\%, \quad u = \frac{\Delta_{仪}}{\sqrt{3}} \tag{S19-8}$$

伏安法测量电阻的最后结果可表示为

内接法 $$R_x = (R_{测} - R_A) \pm u_R, \quad E = \frac{\sigma_R}{R_x} \times 100\% \tag{S19-9}$$

外接法 $$R_x=\frac{R_V R_{测}}{R_V-R_{测}}\pm u_R,\quad E=\frac{u_R}{R_x}\times 100\% \tag{S19-10}$$

3. 电子元件

通常把电子元件伏安特性曲线某点切线的斜率的倒数，称为该电子元件在该点（工作状态下）的动态电阻，即作

$$R=\lim\Delta U_R/\Delta I_R=\mathrm{d}U_R/\mathrm{d}I_R \tag{S19-11}$$

显然，动态电阻若是常数，对应的伏安特性曲线呈直线型，如图 S19－2 所示。这类电阻元件是严格服从欧姆定律的，称为线性电阻，如线绕电阻、碳膜电阻及金属膜电阻等。动态电阻若是变量，是状态函数，则对应的伏安特性曲线呈曲线形，如图 S19－3 所示。这类加在其上的电压与通过的电路没有线性关系的元件均称为非线性元件，如二极管、三极管、光敏电阻、热敏电阻等。

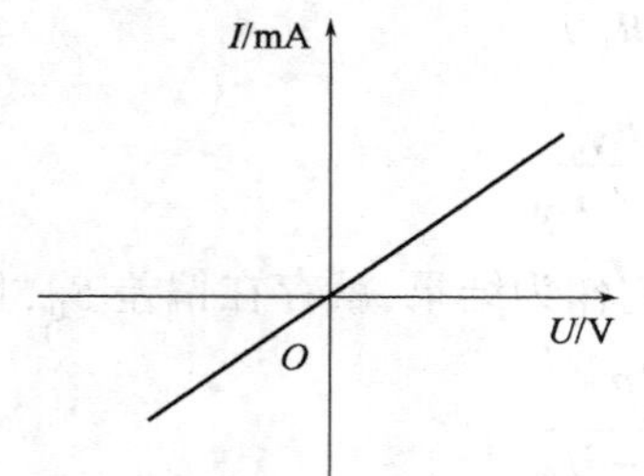

图 S19－2　线性元件的伏安特性

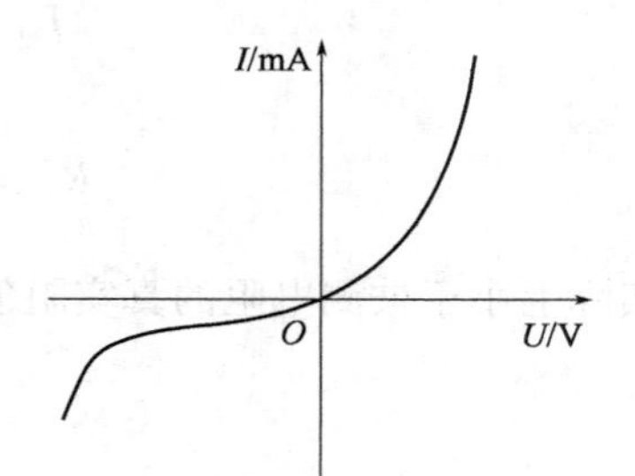

图 S19－3　非线性元件的伏安特性

非线性电阻元件的伏安特性所反映的规律，总是与一定的物理过程相联系的。例如半导体整流二极管，正向导通电流为 mA 数量级，而反向电流仅为 μA 数量级（设计测量电路时应注意两者的区别），这与二极管的 p—n 结阻挡层结构的物理过程相关联，又如发光二极管伏安特性曲线上出现的拐点，是与发光二极管未发光和发光两种工作状态的转变相对应的，这除了共同的 p—n 结阻挡层的作用影响了阻值的变化以外，还表明由于额外的光能量输出而影响二极管内部的能量分配。所以，对非线性电阻特性和规律的研究，有助于对有关物理过程的理解。

[实验内容与步骤]

1. 伏安法测电阻

（1）按图 S19－4 接好线路，图中 K_2 为单刀双掷开关，倒向 A 为电流表内接，倒向 B 为电流表外接，R_x 为待测电阻。

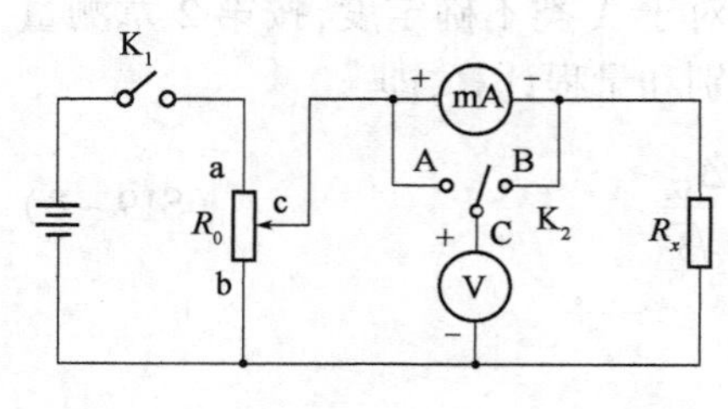

图 S19－4　伏安法测电阻的电路图

（2）取待测电阻为 250Ω（$R_x>100R_A$），将开关 K_2 掷向 A，滑线变阻器滑动头 C 从分压最小位置开始移动，移动到某一位置，使电压表、毫安表指针为一合适读数，并记录之。在滑线变阻滑动头位置不变的情况下，再使开关 K_2 掷向 B，记录其电压值、电流值。

（3）取待测电阻为 30Ω（$100\ R_x<R_V$），用步骤（2）同样的方法记录电流表内接、外接两次测量的电压值、电流值。

2. 二极管的伏安特性曲线的测量

1）二极管的正向特性曲线的测量

按图 S19－5 接好线，（因二极管正向电阻很小，远小于电压表内阻，所以用电流表外接法）滑线变阻器滑动头置最小分压处，R 为保护电阻。合上开关 K，滑动滑线变阻器滑动头使电压为 0.10～0.70V 的范围内变化，先定性观察现象，再定量测量。注意，电压的测试值不应等间隔取，而是在电流变化缓慢区电压间隔取得疏一些，在电流变化迅速区，电压间隔取得密一些。记录相应电压及电流值，填入表 S19－1 中。

2）二极管的反向特性曲线的测量

按图 S19－6 接好线，（因二极管反向电阻很大，远大于电流表内阻，所以用电流表内接法）使滑线变阻器滑动头置最小分压处。合上开关 K，滑动滑线变阻器的滑动头，使电压在 0.00～3.00V 的范围内变化，同样先定性观察现象，再定量测量。电压的测试值依然本着电流变化缓慢区则疏，电流变化迅速区则密的原则。记录相应电压及电流值，填入表 S19－2 中。

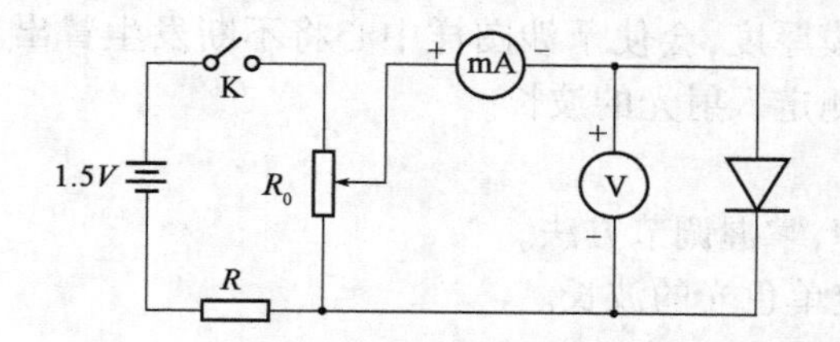

图 S19－5　测晶体二极管正向伏安特性的电路

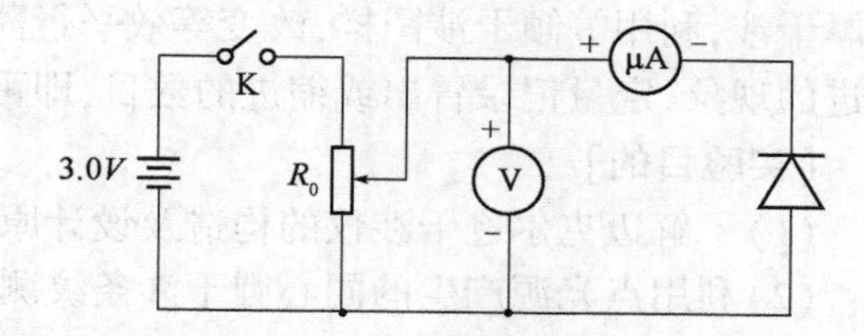

图 S19－6　测晶体二极管反向伏安特性的电路

［注意事项］

（1）更换测量内容前，必须先将滑线变阻器置于安全位置，然后逐渐增加至需要值。

（2）电源不得短路。

（3）接线时电表正负极不能接反；注意正确选择电表量程。

（4）测二极管伏安特性时，注意最大正向电流值和最大反向电压值。

［数据及处理］

1. 数据记录（表 S19－1、表 S19－2）

表 S19－1　二极管正向特性曲线数据记录表

U/V									
I/mA									

表 S19－2　二极管反向特性曲线数据记录表

U/V									
I/μA									

2. 数据处理

把表 S19－1 和表 S19－2 中的数据作在一张坐标图上，反映二极管的伏安特性。

3. 结果分析

对实验结果进行分析，并得出实验结论。

[问题讨论]

(1)试比较滑线电阻器分压用法和限流用法中电压的变化范围。

(2)伏安特性曲线的斜率代表什么?分别讨论直线和曲线的情况。

(3)测量电压表(电流表)的内阻时,用什么接法测量结果比较准确?

实验20 用迈克尔逊干涉仪测量激光的波长

[引言]

迈克尔逊干涉仪是由迈克尔逊和莫雷合作,为研究"以太"漂移实验而设计的精密光学仪器,实验结果否定了"以太"的存在,促进了相对论的建立,此后迈克尔逊又用它做了两个重要的实验,首次系统研究了光谱线的精细结构,以及直接将光谱线的波长与标准米尺进行比较。由于干涉仪的设计思想丰富,启迪后人相继研制出各种专用的干涉仪。迈克尔逊干涉仪基于一种最典型的分振幅双光束干涉原理。由其光路决定,可灵活实现分振幅双光束等倾干涉和等厚干涉,利用等倾干涉图样,改变等效平行薄膜厚度,会使干涉图样中心将不断发生冒出或缩进的现象,定量记录冒出或缩进的数目,即可测定入射光的波长。

[实验目的]

(1)了解迈克尔逊干涉仪的构造及设计原理,掌握调节方法。

(2)利用点光源产生的同心圆干涉条纹测定单色光的波长。

[预习问题]

(1)迈克尔逊干涉仪由哪几部分组成?各部分的作用是什么?

(2)简述迈克尔逊干涉仪的调节方法、读数方法及使用注意事项。

(3)如何利用点光源产生的非定义域干涉条纹测定激光的波长?

[实验仪器]

WSM-100型迈克尔逊干涉仪、HN-ⅡQ型激光器、扩束透镜。

[实验原理]

1.产生干涉的等效光路

光路如图S20-1所示,G_1、G_2为平行平面玻璃板,G_1为分光板,它的一个表面镀有半反射金属膜,G_2为补偿板。当平面镜M_1、M_2互垂直时,G_1、G_2与M_1、M_2均成45°角。

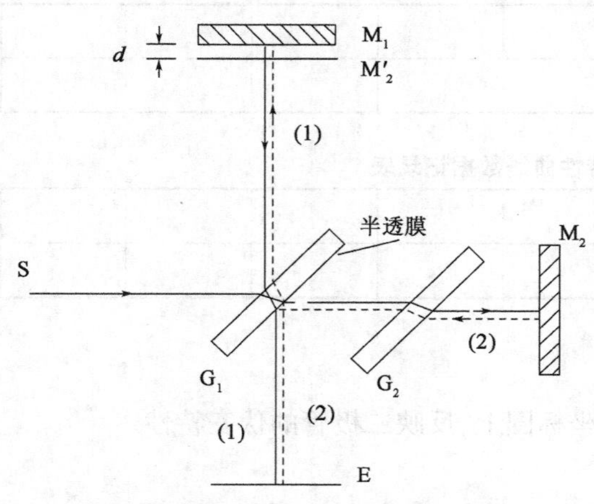

图S20-1 迈克尔逊干涉仪光路示意图

光源S发出的光在G_1的半反射面上被分成反射光(1)和透射光(2)。两束光光强近似相等。光束(1)投向M_1,经M_1反射后穿过G_1。光束(2)经过G_2投向M_2,经M_2反射后再经过G_2,在G_1半透膜面上反射。由于这两束光是来自同一光源S的同一束光,因此它们是两列相干光束,在光屏E处相遇有干涉图样形成,即使白光也能产生干涉。

补偿板G_2的作用是补偿光束(1),因在G_1板中往返两次所多走的光程,使光束(1)、(2)在G_1、

G_2 板中满足等光程。因此 G_1、G_2 板的折射率和厚度都应相同,而且要互相平行。在制作工艺上将同一块平行平面玻璃板切割成两块,一块镀上金属膜作为分光板,一块作为补偿板。

2. 单色光波长的测量

当 $M_2 \perp M_1$ 时,即 M_2' 平行于 M_1,若光以同一倾角 θ 入射在 M_2' 和 M_1 上,反射后形成如图 S20-2所示(1)和(2)两束相互平行的相干光,过 C 作 CN 垂直于光线(1),因 M_2' 和 M_1 之间为空气层,$n \approx 1$,两束光的光程差为

$$\Delta = (AB + BC) - AN = \frac{2d}{\cos\theta} - 2d\tan\theta\sin\theta = 2d\left(\frac{1}{\cos\theta} - \frac{\sin^2\theta}{\cos\theta}\right) = 2d\cos\theta \qquad \text{(S20-1)}$$

d 固定时,可看出倾角 θ 相同方向上两相干光光程差 Δ 均相等。具有相等 θ 的各方向光束形成一圆锥面,因此在无穷远处形成等倾干涉条纹呈圆环形,这时眼睛对无穷远调焦就可以看到一系列同心圆。θ 越小,干涉圆环直径越小,它的级次 k 越高。圆心处 $\theta = 0$,$\cos\theta$ 值最大,这时有

$$\Delta = 2d = k\lambda \qquad \text{(S20-2)}$$

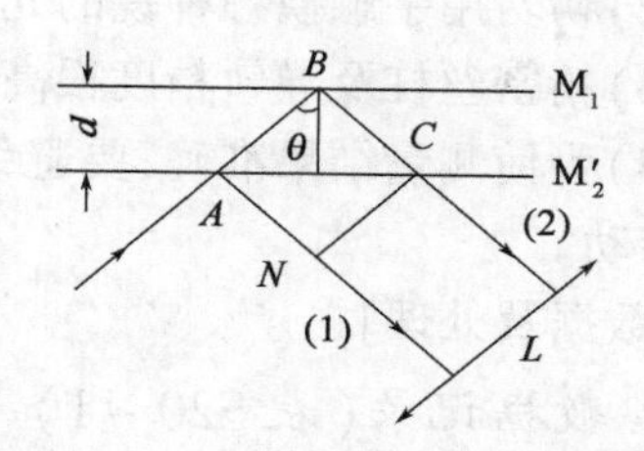

图 S20-2　两相干光的光程差计算参考图

所以圆心处级次最高。

当 d 增加时,圆心的干涉级次越来越高,就看到圆环一个一个从中心"冒"出来;反之,d 减小时,就看到圆环一个一个从中心"缩"进去;每当 d 增加或减少 $\frac{\lambda}{2}$,就会冒出或缩进一个圆环。因此,若测出移动的距离 Δd 和冒出(或缩进)圆环数 N,就可以求出波长:

$$\lambda = \frac{2\Delta d}{N} \qquad \text{(S20-3)}$$

反之,若已知 λ 和冒出(或缩进)圆环数就可以求出 M_1 镜移动的距离,这就是测量长度的原理。

[实验内容与步骤]

1. 调节仪器

根据式(S20-3)测量波长,要求 M_1 和 M_2 的像 M_2' 平行,即 M_1 和 M_2 相互垂直。调节方法如下:

(1)打开激光器,粗略调节迈克尔逊干涉仪与激光器大致处于同一水平高度,让激光束通过分光板和补偿板中心垂直入射到平面镜 M_2 的中心(目测)。

(2)将平面镜 M_1 和 M_2 背后的倾度粗调螺钉置于松紧合适的中间位置(留有可调余地)。

(3)转动手轮,尽量使 M_1、M_2 同分光板的距离相等(目测)。

(4)遮住 M_1,调节 M_2 背后的倾度粗调螺钉,使其反射的光点正好射回激光器的发射孔中。

(5)遮住 M_2,调节 M_1 背后的倾度调节螺钉,使 M_1 反射的三个光点中间的一个(最亮)回到激光器的发射孔中,此时达到了粗调的要求。

(6)将扩束透镜放置在激光器与分光板之间的适当位置,让透射光照射到分光板上,在观

察屏 E 的背面就可以观察到等倾干涉圆环条纹,这时的条纹可能不够圆或者中心偏移,再微调节 M_2 的倾度微调螺钉,使条纹变圆且居中,此时,已满足 $M_1 \perp M_2$ 的要求。

2. 测量

由测量波长关系式(S20-3)可知,λ 是一定值,单方向平移 M_1 来改变 d,观察等倾圆环条纹的变化规律并记录。每"冒出"或"缩进"50 个圆环(中央亮斑最大)记录一次 M_1 镜的位置,连续测 9 次,用逐差法处理实验数据。

[注意事项]

(1)切勿用眼睛直视激光。

(2)切勿用手触摸各种镜的光学表面。

(3)精密丝杠及导轨精度很高,受损会影响仪器精度,操作时动作要慢,严禁粗鲁、急躁。

(4)为使测量结果准确,要避免引入空程误差,记录 M_1 镜在各位置读数时,微动手轮应单方向转动。

[数据及处理]

1. 数据记录(表 S20-1)

表 S20-1 平面镜 M_1 位置变化测量值

条纹的吞吐数 N_1	0	50	100	150	200
d_i/mm					
条纹的吞吐数 N_2	250	300	350	400	450
d_{i+5}/mm					
$\Delta N = N_2 - N_1$	250	250	250	250	250
$\Delta d_i = \frac{d_{i+5} - d_i}{5}$/mm					
$\Delta(\Delta d_i) = \Delta d_i - \overline{\Delta d}$/mm					
$\overline{\lambda} = \frac{2\overline{\Delta d}}{N}$/m					$N=50$

2. 数据处理

用逐差法计算 λ,并与标准值 λ_0 进行比较,计算其百分差。

3. 结果分析

对实验结果进行分析,并得出实验结论。

[问题讨论]

(1)实验中观察的干涉条纹有何特点及规律?

(2)测量时,若多数或少数了一个条纹,会产生多大的误差?

(3)什么是空程? 测量中如何操作才能避免空程误差?

(4)迈克尔逊干涉仪产生的环形干涉条纹与牛顿环干涉条纹有何不同?

视频 S20 用迈克尔逊干涉仪测量激光的波长

用迈克尔逊干涉仪测量激光的波长如视频 S20 所示。

[补充说明]

迈克尔逊干涉仪的结构如图 S20－3 所示。

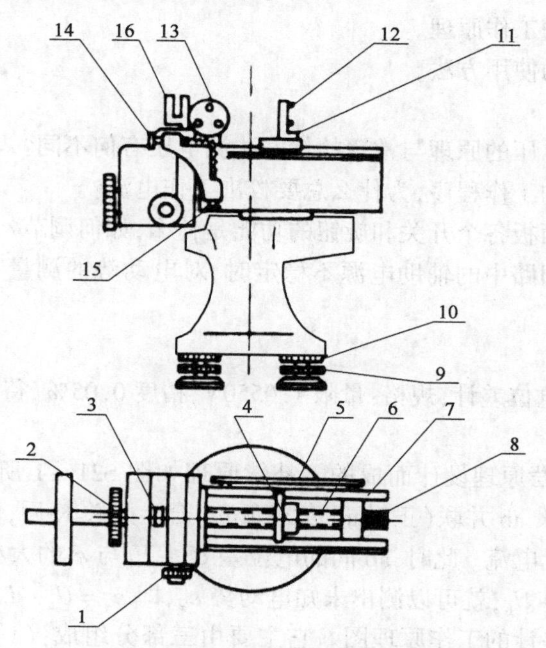

图 S20－3　迈克尔逊干涉仪结构图

1—微动手轮；2—转动手轮；3—读数窗口；4—蜗杆齿合螺母；5—主尺；6—蜗杆；7—导轨；8—顶进螺母；9—水平调节螺母；10—锁紧圈；11—可动镜 M_1；12—粗调螺钉；13—固定镜 M_2；14，15—水平、竖直微调螺钉；16—分光板 G_1、补偿板 G_2

平面镜 M_1 可通过转动手轮 2 或微动手轮 1 沿导轨 7 移动。M_1 的位置可由干涉仪直接读出，读数由主尺 5，转动手轮 3 的读数窗口，微动手轮 1 三部分组成，从 5 上可以读到整毫米数，在 3 上可以读到 0.01mm，从 1 上可以准确读到 10^{-4}mm，这样最小读数可以估读到10^{-5}mm。底座有三个水平调节螺钉 9，用来调节台面水平。M_1 和 M_2 是两片精密磨光的平面反射镜，M_2 是固定不动的，M_1 用螺旋控制，可前后移动。M_1 和 M_2 的背面各有粗调螺钉 12 可调节镜面倾斜度。M_2 下方还有一个水平方向的微调螺钉 14 和垂直方向的微调螺钉 15，其松紧可使 M_2 镜产生微小倾斜，从而对 M_2 镜倾斜度作更精细调节。16 为分光板 G_1 和补偿板 G_2。

实验 21　低压电位差计的使用

[引言]

若用伏特表直接并接在电源输出端，则伏特表的读数并不等于该电源电动势。因为电源存在内阻，电源向外输出电流引起内阻产生压降。因此，不能用伏特表测量电池的电动势。然而，测量回路中使电流为零的设想，为设计电位差计提供了重要的理论依据。如何实现测量回路中电流为零？设想将一个连续可调的标准电源和待测电池同极性相接，并接入检流计作为回路中有无电流的判据，对于任何待测电池，通过调节标准电源的输出电压大小，总能实现使待测电池电动势与标准电源输出电压相等（即检流计为零）。此时，称回路中待测电池向外提供电流已被可调的标准电源向外提供电流“补偿”，因此这种方法称为补偿法。电位差计就是根据这种补偿原理设计而成的。

[实验目的]

(1)了解电位差计的工作原理。

(2)掌握电位差计的使用方法。

[预习问题]

(1)补偿原理测量电压的原理与常用的伏安法测电压有何不同?

(2)简述电位差计的工作程序,为什么总要校准工作电流?

(3)箱式电位差计面板各个开关和旋钮的功能是什么,如何调节?

(4)电位差计工作回路中的辅助电源不稳定时,对电动势的测量有何影响?拟采用稳压电源还是恒流电源?

[实验仪器]

UJ33a 型低压直流电位差计,规格:量限 1.0550V,精度 0.05%,待测件及导线若干。

[实验原理]

电位差计是依据补偿原理设计而成的。补偿原理如图 S21-1 所示,将电动势为 ε_x 的电源和电阻丝 AB 的某一段 ab 并联(正对正,负对负)。合上开关 K 后,调节滑动 a(或 b)头,直至检流计指零,回路中无电流。此时,ab 间的电位差 U_a-U_b 与 ε_x 的大小相等、方向相同。只要测得 ab 间的电位差 U_a-U_b,就可以测出未知电动势 ε_x,即 $\varepsilon_x=U_a-U_b$。

图 S21-2 是电位差计的工作原理图。它主要由三部分组成:(1)调节工作电流回路,由 E、r、K_0、R_N、R_0 组成;(2)校正回路,由 ε_N、K(接 N 端)、G、R_N 组成;(3)测量回路,由 ε_x、R、G、K(接 X 端)组成。图 S21-2 中标准电阻 R_N 是标准电池 ε_N 的补偿电阻;可变电阻 R_0 是待测电动势 ε_x 的补偿电阻。

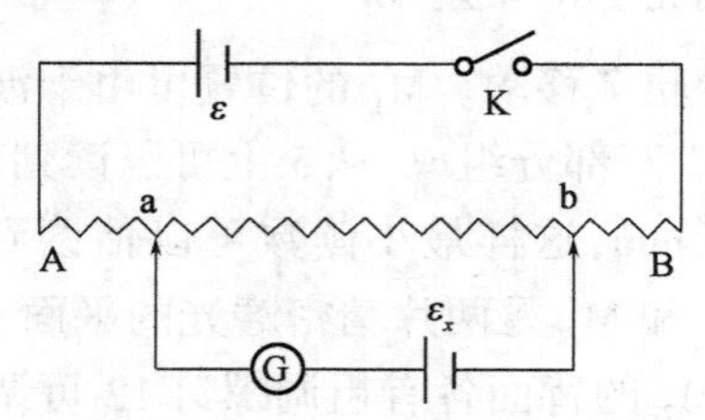

图 S21-1　补偿原理图

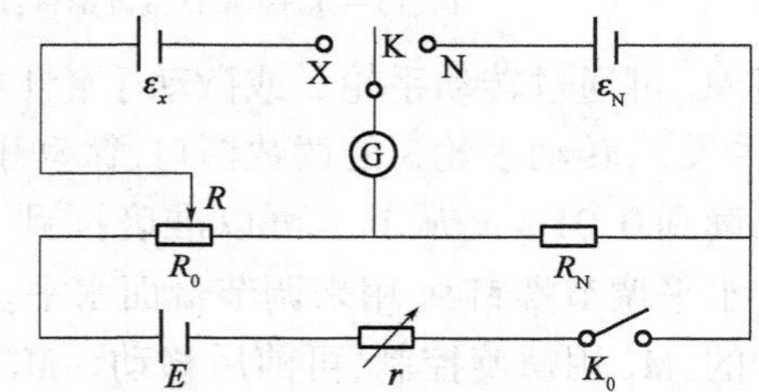

图 S21-2　电位差计工作原理图

测量前,先调节工作电流,实现工作电流标准化(对于给定型号的电位差计来说,工作电流回路中的电流是事先设计好的)。将开关 K 倒向 N 端,合上电源开关 K_0,调节 r 使检流计 G 指零。这时,标准电池的电动势等于标准电阻 R_N 上的电压降,即

$$\varepsilon_N=IR_N \quad 或 \quad I_0=\varepsilon_N/R_N \tag{S21-1}$$

进行测量时,将 K 倒向 X 端,调节 R 使检流计 G 示零。此时,未知电动势 ε_x 等于 R 上的压降,即 $\varepsilon_x=IR$,此式中的电流 I,就是式(S21-1)中已调好的工作电流 I_0。故将式(S21-1),代入 $\varepsilon_x=IR$ 式中,得

$$\varepsilon_x=\frac{\varepsilon_N}{R_N}R \tag{S21-2}$$

可见,用电位差计测量电动势,测量结果主要取决于标准电池 ε_N、标准电阻 R_N 和补偿电阻 R 的精度以及检流计的灵敏度。

工作电流 I_0 是仪器设计的重要参数。仪器面板上的标准电池在各种温度下的电动势示值

和 R_0旋钮旁的示数也都是以电压 I_0R_N、I_0R 标定好的。因此,在测量之前必须对工作电流标准化,在测量中,必须保持已标准化了的工作电流 I_0不变。

[实验内容与步骤]

(1)准备。根据仪器的使用说明,连接电路,调节电位差计使之处于使用状态。以 UJ33a 型电位差计为例,将倍率开关从"断"旋到所需倍率,此时内部电源接通。调节"调零"旋钮,使检流计指针示零。将被测电压(电势)按极性接入"未知"端钮,"测量-输出"开关置于"测量"位置。

(2)校准。根据仪器的使用说明,校准电位差计的工作电流。以 UJ33a 型电位差计为例,扳键开关 K_2扳向"标准",调节"粗""微"旋钮,直到检流计示零。此时,电位差计工作电流回路中的电流已被标准化了,其值为 I_0。

(3)测量未知电动势。以 UJ33a 型电位差计为例:扳键开关 K_2扳向"未知"。调节Ⅰ、Ⅱ、Ⅲ测量转盘,使检流计示零,被测电动势等于测量盘读数之和与倍率的乘积。

[注意事项]

(1)测量电压不得超过电位差计的最大量程。

(2)测量过程中,注意检查工作电流是否发生变化,如发生变化应随时校准。

(3)校准电池严禁短路。

(4)测量完成后,务必关上电位差计的电源。

[数据及处理]

1. 数据记录(表 S21-1)

表 S21-1 电位差计测量结果记录表

倍率 K	×10	×1	×0.1	U_x
×0.1				
×1				
×5				

2. 数据处理

计算测量结果的不确定度,并给出测量结果的不确定度评价。

3. 结果分析

对实验结果进行分析,并得出实验结论。

[问题讨论]

(1)为什么用电压表不能精确测量电源的电动势?

(2)对电位差计进行校准或测量时,如果发现检流计指针总是向一边偏,可能有哪些原因?(提示:从电位差计的工作原理图分析。)

(3)电位差计由哪几部分组成?试简要说明它们之间的关系。

低压电位差计的使用如视频 S21 所示。

视频 S21 低压电位差计的使用

[补充说明]

UJ33a 型电位差计准确度等级为 0.05,测量上限为 1.0550V。仪器有内附微电流放大电路,以提高检流计的灵敏度,标准电池及工业电池,不需外加附件便可进行测量。仪器内附标准电池为 BC5 型不饱和标准电池,温度系数小,不必对它进行温度补偿即可测量。图 S21-5 是 UJ33a 型直流电位差计的面板示意图。当倍率开关 K_1置于"×0.1"时,量程为 21.10mV,基

准值 R_0为 0.01V;置于“×1”时,量程为 211.0mV,基准值 R_0为 0.1V;置于“×5”时,量程为 1.0550V,基准值 R_0为 1V。图中开关 K_2为检流计转换开关,置于“标准”时,进行工作电流调节,置于“未知”时,进行测量(或电压输出)。面板左上方为内附微电流放大式检流计,在其下方有检流计“调零”旋钮。该型号电位差计只有“粗”“细”两个转盘进行工作电流的调节。图中标有“×10”“×1”“×0.1”的转盘组成测量转盘。开关 K_3作为用途转换开关,置于“测量”时,用此电位差计测量电压、电流等,被测电压为测量转盘读数之和与倍率的乘积;置于“输出”时,此电位差计可用于电压信号输出,输出电压值为测量转盘读数之和与倍率的乘积。图中标有“未知”的接线柱,当 K_3置于“测量”时,用于连接待测电路的输入端,当 K_3置于“输出”时(检流计被短路),用于电压输出端。电位差计使用完毕,“倍率”开关应置于“断”位置。

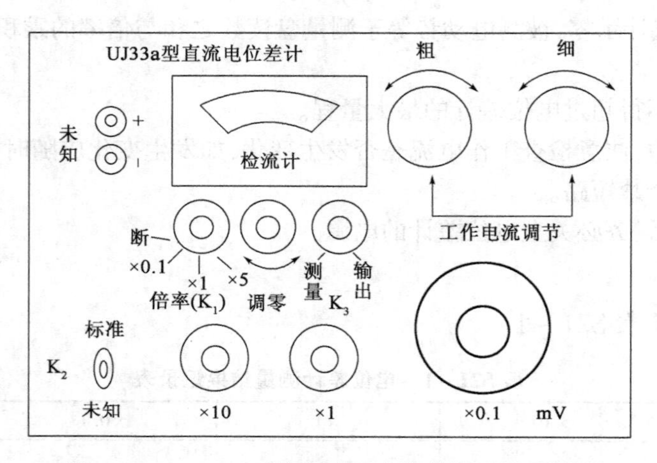

图 S21-3　UJ33a 型低压直流电位差计面板示意图

实验 22　电流表和电压表的改装与校准

[引言]

电表是用来测量电流、电压的仪表,实验室使用的电表大部分是磁电式仪表,它具有灵敏度高、功率消耗小、受磁场影响小、刻度均匀、读数方便等优点。未经改装的电表,由于灵敏度高,满度电流(或电压)很小,只允许通过微安级或毫安级的电流,一般只能测量很小的电流和电压,如果要想测量较大的电流或电压,就必须进行改装。在生产和实验中,常常选用量程比较小的电表,并联一个电阻扩程为较大量程的电流表,或串联一个电阻改装成为较大量程的电压表。

[实验目的]

(1)了解电流表、电压表的结构及规格。

(2)掌握电表的扩程原理及校准方法。

(3)学会确定改装表的级别的方法。

(4)学会变阻器、电表、电阻箱等基本电学仪器的正确使用方法。

[实验仪器]

50μA 微安表头、标准电流表、标准电压表、电阻箱、滑线变阻器、电源、电键等。

[实验原理]

电流表、电压表等电表一般都是由一个微安表配以不同的电路和元件后改装而成的。本实验将分别阐述怎样将微安表改装成电流表和电压表。其改装原理如下：

1. 将微安表改装成电流表

用来改装的微安表通常称为“表头”，它有两个主要参数，即量程和内阻。量程是指表头指针偏转到满刻度的电流 I_g，超过这个限度就无法读数，还会损坏电表。内阻是指表头线圈所具有的电阻 R_g，表头能够测量的电流很小，若要测量较大的电流，则需要扩大量程。利用并联电阻的分流作用，只要在表头两端并联一个分流电阻 R_s，使被测电流大部分从 R_s 流过，而表头仍保持原来允许通过的最大电流 I_g，就可把微安表改装成电流表。其原理电路图如图 S22－1 所示。

分流电阻 R_s 的大小根据待扩量程 I 的大小而定。设表头改装后的量程 $I=nI_g$，由并联电路两端电压相等得

$$I_gR_g=(I-I_g)R_s=(n-1)I_gR_s$$

则

$$R_s=\frac{R_g}{n-1} \tag{S22-1}$$

上式表明，若将微安表的量程扩大 n 倍，只需在表头两端并联一个阻值为 $R_g/(n-1)$ 的分流电阻，扩程倍数 n 越高，分流电阻值越小。在表头上并联不同阻值的分流电阻，便可制成多量程的电流表。图 S22－1 所示接法称为开路置换式，实际使用的多量程电流表，分流电路的接法一般采用闭路抽头式，即将分流电阻 R_s 分成若干个电阻串联起来，并进行抽头，不同抽头可得不同的分流电阻，从而获得不同量程的电流表，如图 S22－2 所示。

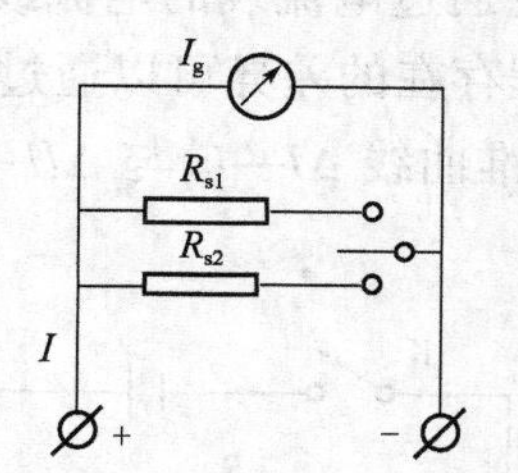

图 S22－1　电流表扩程原理

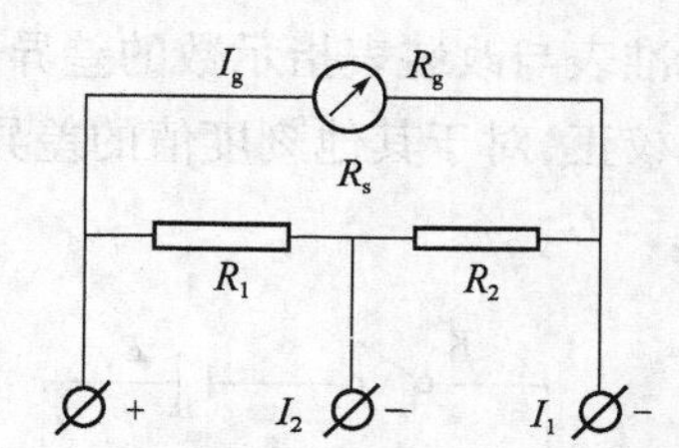

图 S22－2　两量程电流表结构

例如，将 $I_g=1\text{mA}$，$R_g=50\Omega$ 的表头改装成具有两个量程 $I_1=50.0\text{mA}$，$I_2=100.0\text{mA}$ 的毫安表，分流电阻 R_1、R_2 的计算方法是：首先根据最小电流量程 $I_1=50.0\text{mA}$ 求出总的分流电阻 R_s，因 $n=I/I_g=500$，由式（S22－1）得 $R_s=R_g/49\approx10.200\Omega$。再计算电流量程 $I_2=100.0\text{mA}$ 的分流电阻 R_1，由 $I_g(R_g+R_2)=(I_2-I_g)R_1$，$R_2=R_s-R_1$ 和 $I_2=100I_g$ 可以求得 $R_1=R_g/98\approx5.102\Omega$，$R_2=5.098\Omega$。

2. 将微安表改装成电压表

当电流通过量程为 I_g、内阻为 R_g 的表头时，根据欧姆定律，表头上的电压降为 $U_g=I_gR_g$，因此，表头也可以用来测量电压，只是它的量程 U_g 很小，一般不高于一二百毫伏，若要测量较高的电压，则需要利用串联电阻的分压作用，使被测电压大部分降落在分压电阻 R_p 上，而表头上仍保持量程电压 U_g，这样就可以把微安表改装成电压表，其原理电路如图 S22－3 所示。

设欲改装电压表的量程 $U=nU_g=nI_gR_g$，则根据欧姆定律得

$$I_g(R_g+R_p)=U=nU_g$$

$$R_p = \frac{nU_g}{I_g} - R_g = (n-1)R_g \tag{S22-2}$$

式(S22－2)表明:要将表头改装成量程为 U 的电压表,只需在表头上串联一个阻值为$(n-1)R_g$的分压电阻即可。

在表头上串联不同阻值的分压电阻 R_p,就可以得到不同量程的电压表,图 S22－4 是两个量程电压表结构示意图。

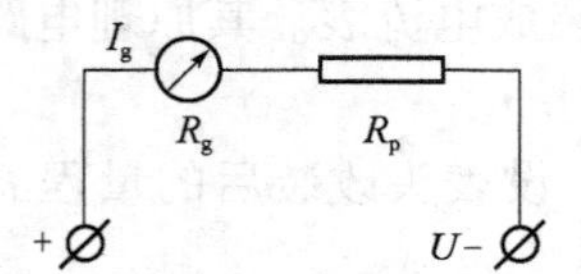

图 S22－3　电压表扩程原理

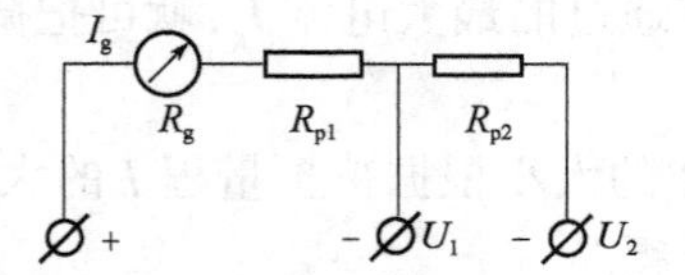

图 S22－4　两量程电压表结构

3. 改装电流表与改装电压表的校正

根据计算得出的分流电阻 R_s或分压电阻 R_p,与表头组成了扩程后的电流表或电压表,由于种种原因,它们指示的读数不能保证与规定的计量标准完全相符。因此必须与标准表进行比较,将可校正的偏差(例如量限不准)予以校正外,无法校正的偏差则通过作校正曲线予以反映。

图 S22－5 与图 S22－6 分别为电流表与电压表校正电路,其中 A_{st} 及 v_{st} 是标准电流表及电压表,虚线框内的为改装表,R 为分压电阻,用来改变通过电流表的电流或加到电压表两端的电压值。比较标准表与改装表指示数的差异,对量程存在的差异可以通过调节 R_s、R_p,按照标准表指示数予以校正,对于其他刻度值的差异,作校准曲线 $\Delta I—I_{改}$ 与 $\Delta U—U_{改}$ 予以修正,$\Delta I = I_{标} - I_{改}$,$\Delta U = U_{标} - U_{改}$。

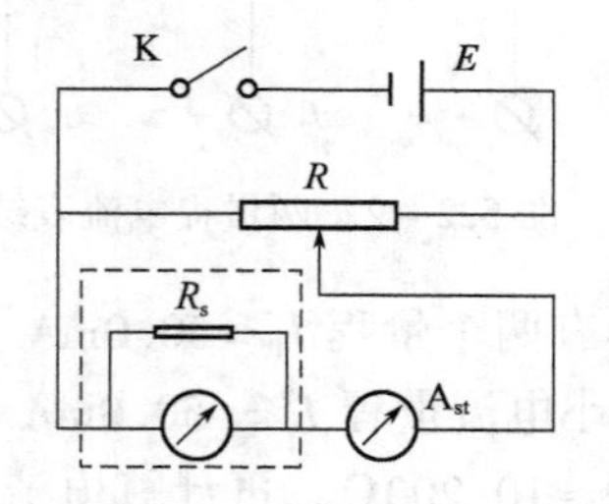

图 S22－5　电流表校正电路

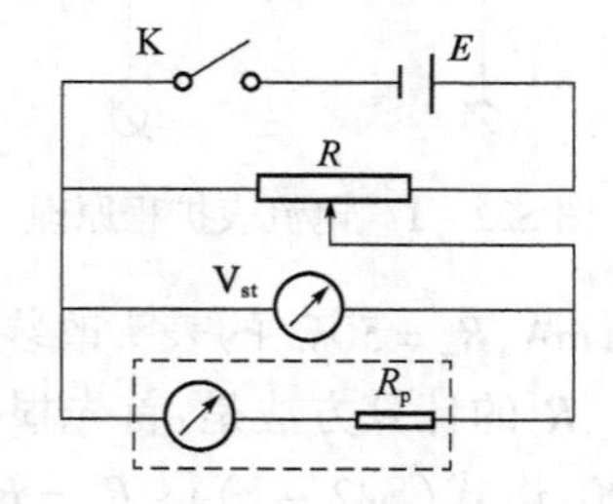

图 S22－6　电压表校正电路

4. 改装表的标称误差及准确度级别

改装表与标准表进行比较,得到一系列对应标准表刻度的绝对误差,把这些误差中的最大误差和改装表的量程之比的绝对值称为该电表的标称误差,即

$$\text{标称误差} = \left| \frac{\text{最大绝对误差}}{\text{改装表量程}} \right| \times 100\%$$

电表准确度等级,就是根据标称误差大小确定的,例如,某改装电流表的 $\Delta I_{最大} = 0.05\text{mA}$,量程为 5mA,则

$$标称误差 = \left|\frac{最大绝对误差}{改装表量程}\right| \times 100\% = \left|\frac{0.05}{5}\right| \times 100\% = 1.0\%$$

所以该改装电流表相对于实验用的标准表为 1.0 级电表。

[设计举例]

1. 将量程为 50μA 的表头改装成量程为 5mA 的电流表

(1)表头的内阻由实验室给定。根据实验设计提示,计算出扩程倍数 n。

(2)按式(S22-1)计算出分流电阻 R_s 的数值,从电阻箱上取相应的电阻值与表头并联,组成改装的毫安表(以下简称改装表)。

(3)将改装表按图 S22-5 接好校准电路。注意:线路未接通前,要调节两表指针指零,滑线变阻器应取最小值,标准表量程应选择等于或略大于改装表量程。

(4)校准量程:调节滑线变阻器 R 使标准表的指针指向 5mA,此时观察改装表指针是否指向满刻度,若有差异,通过调节 R_s 值和 R 值,使改装表指针指向满刻度,标准表指针指向 5mA 刻度值。

(5)校准其他刻度:在校准满刻度值 5mA 后,调节 R 使电流从大到小(对改装表取整刻度,即为 4.00mA、3.00mA、2.00mA、1.00mA、0.00mA)依次校准。记录改装表、标准表的相应的各刻度值,然后使电流从小到大依相反次序复测一遍。作出改装电流表的 $\Delta I—I$ 的校正曲线,求出改装表的准确度等级。

实验过程中填写数据记录(表 S22-1)。

数据记录

标准表量程:________ 标准表级别:________

表头量程:________ 表头内阻:________

表头级别:________ 扩程倍数:________

R_s 计算值:________ R_s 实测值:________

表 S22-1 校正改装电流表数据记录表

改装表		标准表读数 $I_{标}$/mA			$\Delta I = \overline{I}_{标} - I_{改}$ mA
表面读数	$I_{改}$/mA	电流从大到小	电流从小到大	$\overline{I}_{标}$	
0	0.00				
10	1.00				
20	2.00				
30	3.00				
40	4.00				
50	5.00				

(6)作出改装电流表 $\Delta I—I_{改}$ 校准曲线。

(7)计算出改装表的标称误差,并确定其准确度级别。

2. 将量程为 50μA 的表头改装成量程为 5V 的电压表

(1)根据实验设计提示,计算出扩程倍数 n。

(2)按式(S22-2)计算出分压电阻 R_p 的数值,从电阻箱上取相应的电阻值与表头串联,组成改装的毫伏表(以下简称改装表)。

(3)将改装表按图 S22－6 接好校准电路。注意:线路未接通前,要调节两表指针指零,滑线变阻器应取最小值,标准表量程应选择等于或略大于改装表量程。

(4)校准量程:调节滑线变阻器 R 使标准表的指针指向 5V,此时观察改装表指针是否指向满刻度,若有差异,通过调节 R_p 值和 R 值,使改装表指针指向满刻度,标准表指针指向 5V 刻度值。

(5)校准其他刻度:在校准满刻度值 5V 后,调节 R 使电压从大到小(对改装表取整刻度,即为 4.00V、3.00V、2.00V、1.00V、0.00V)依次校准。记录改装表、标准表的相应的各刻度值,然后使电压从小到大依相反次序复测一遍。作出改装电压表的 $\Delta U—U$ 的校正曲线,求出改装表的准确度等级。

实验过程中填写数据记录(表 S22－2)。

数据记录

标准表量程:__________　标准表级别:__________

表头量程:__________　表头内阻:__________

表头级别:__________　扩程倍数:__________

R_p 计算值:__________　R_p 实测值:__________

表 S22－2　校正改装电压表数据记录表

改装表		标准表读数 $U_{标}$/V			$\Delta U=\overline{U}_{标}-U_{改}$ V
表面读数	$U_{改}$/V	电压从大到小	电压从小到大	$\overline{U}_{标}$	
0	0.00				
10	1.00				
20	2.00				
30	3.00				
40	4.00				
50	5.00				

(6)作出改装电压表 $\Delta U—U_{改}$ 校准曲线。

(7)计算出改装表的标称误差,并确定其准确度级别。

[注意事项]

(1)接通线路之前,先调整好表头及标准表的零点,并使滑线变阻器置最小值。

(2)校准改装表时,应取改装表为整刻度,记录标准表读数,并注意取够有效数字。

(3)做校准曲线时,两点间连线应为折线。

视频 S22　电流表和电压表的改装与校准

电流表和电压表的改装与校准如视频 S22 所示。

实验 23　夫兰克—赫兹实验

[引言]

1914 年,德国物理学家夫兰克(J. Franck)和赫兹(G. Hertz)对勒纳用来测量电离电位的实验装置做了改进,他们同样采取慢电子(几个到几十个电子伏特)与单元素气体原子碰撞的办法，但着重观察碰撞后电子发生什么变化(勒纳则观察碰撞后离子流的情况)。通过实验测量，电子和原子碰撞时会交换某一定值的能量，且可以使原子从低能级激发到高能级。这直接证明了原子发生跃变时吸收和发射的能量是分立的、不连续的，证明了原子能级的存在，从而证明了玻尔理论的正确，因此获得了1925年诺贝尔物理学奖金。夫兰克—赫兹实验至今仍是探索原子结构的重要手段之一，实验中用的"拒斥电压"筛去小能量电子的方法，已成为广泛应用的实验技术。

[实验目的]

学会测量氩原子的第一激发电位，加深对原子能级的理解。

[预习问题]

实验在正常工作范围内可得到几个波峰?

[实验仪器]

FH-2智能夫兰克—赫兹实验仪、MOS-620型示波器等。

[实验原理]

1. 激发电位

玻尔的原子模型认为，原子是由原子核和以原子核为中心、沿不同轨道运动的电子构成的。对于不同的原子，这些轨道上的电子数分布各不相同。一定轨道上的电子具有一定的能量。当同一原子的电子从低能量的轨道跃迁到较高能量的轨道时，原子就处于受激状态。若轨道Ⅰ为正常状态(即基态)，则较高能量的Ⅱ和Ⅲ依次称为第一受激态和第二受激态。但是原子所处的能量状态并不是任意的，而是受玻尔理论的两个基本假设制约。

1)定态假说

原子只能处于稳定态中，其中每一状态相应于一定的能量值E_i($i=1,2,3,\cdots$)，这些能量值是彼此分立的，不连续的。

2)跃迁假设

当原子从一个稳定态过渡到另一个稳定态时，就吸收或放出一定频率的电磁辐射，频率的大小取决于原子所处两定态之间的能量差，并满足如下关系

$$h\nu = E_m - E_n \tag{S23-1}$$

式中，普朗克常数$h=6.63\times10^{-34}\,\mathrm{J\cdot s}$；$E_m$和$E_n$分别为跃迁前后原子的能量，$E_m>E_n$为发射辐射，反之为吸收辐射。

通常两种情况下会改变原子状态：一是当原子本身吸收或发出电磁辐射；二是当原子与其他粒子相互作用而交换能量。本实验就是利用具有一定能量的电子与氩原子相互作用而发生能量交换实现氩原子状态的改变。

设初速度为零的电子在电位差为U_0的加速电场作用下，获得能量eU_0。当具有这种能量的电子与稀薄气体的原子发生相互作用时就会发生能量交换。如以E_1代表氩原子的基态能量、E_2代表氩原子的第一激发态能量，当氩原子从基态跃迁到第一激发态时，氩原子吸收从电子传递来的能量恰好为

$$eU_0 = E_2 - E_1 \tag{S23-2}$$

相应的电位差称为氩原子的第一激发电位。测定出这个电位差U_0，就可以根据式(S23-2)求

出氩原子的基态和第一激发态之间的能量差。

2. 夫兰克—赫兹实验原理

夫兰克—赫兹实验原理如图 S23 - 1 所示。在充氩的夫兰克—赫兹管中，电子由于阴极加热而发出，阴极 K 和第二栅极G_2之间的加速电压U_{G_2K}使电子加速。在板极 A 和第二栅极G_2之间加有反向拒斥电压U_{G_2A}。当电子通过 KG_2 空间进入G_2A空间时，如果有较大的能量（$\geqslant eU_{G_2A}$），就能反抗反向拒斥电场而到达板极形成板流，为电流计 μA 表检出。如果电子在KG_2空间与氩原子相互作用，按照玻尔原子理论，氩原子只能吸收确定的能量。电子本身所剩余的能量就很小，通过第二栅极后已不足以克服拒斥电场而到达极板 A，这时，通过电流计 μA 表的电流将显著减小。

实验时，使U_{G_2K}电压逐渐增加并仔细观察电流变化，如果原子能级确实存在，且基态和第一激发态之间有确定的能量差，就能观察到如图 S23 - 2 所示的I_A—U_{G_2K}曲线。图 S23 - 2 所示的曲线反映了氩原子在 KG_2空间与电子进行能量交换的情况。当 KG_2空间电压逐渐增加时，电流计检测到的电流随之变化的关系如图 S23 - 2 所示，称为充氩的夫兰克—赫兹管伏安特性曲线，即I_A—U_{G_2K}曲线。电子在 KG_2空间被加速而取得越来越大的能量。但起始阶段，由于电压较低，电子的能量较少，即使在运动过程中它与原子相碰撞也只有微小的能量交换。穿过第二栅极的电子所形成的板流I_A将随第二栅极电压的U_{G_2K}增加而增大（如图 S23 - 2 的 OA 段）。当 KG_2间的电压达到氩原子的第一激发电位U_0时，电子在第二栅极附近与氩原子相互作用，将自己从加速电场中获得的全部能量交给后者，并且使后者从基态激发到第一激发态。而电子本身由于把全部能量给了氩原子，即使穿过了第二栅极也不能克服反向拒斥电场而不能到达极板 A（被筛选掉），所以板极电流将显著减小（图 S23 - 2 所示 ab 段）。随着第二栅极电压的增加，电子的能量也随之增加，在与氩原子相互作用后还留下足够的能量，可以克服反向拒斥电场而达到板极 A，这时电流又开始上升（bc 段）。直到 KG_2间电压是二倍氩原子的第一激发电位时，电子在 KG_2间又会因二次碰撞而失去能量，因而又会造成第二次板极电流的下降（cd）段，故有

$$U_{G_2K} = n\,U_0 \qquad (n = 1,\ 2,\ 3,\cdots) \tag{S23 - 3}$$

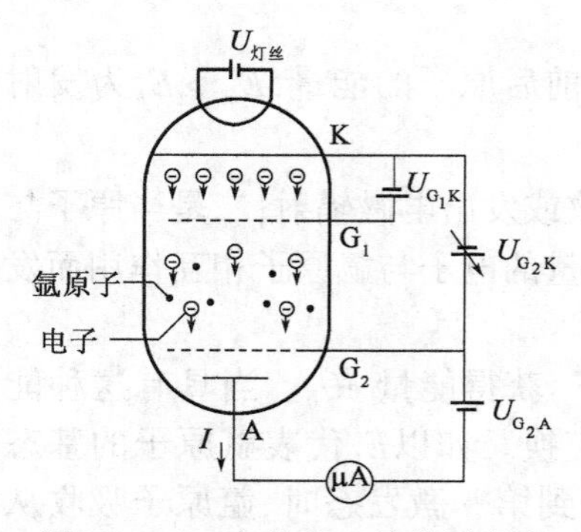

图 S23 - 1　夫兰克—赫兹原理图

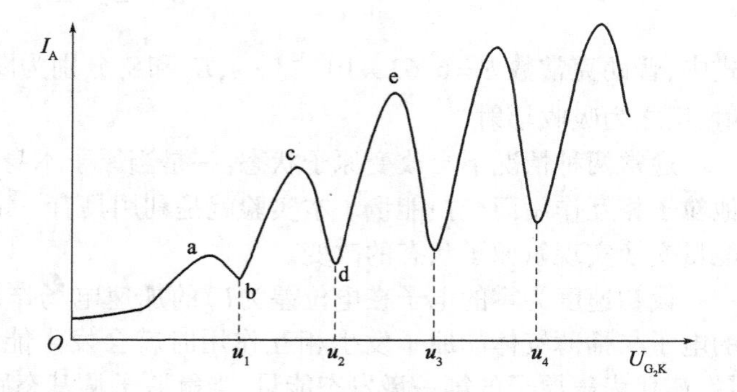

图 S23 - 2　夫兰克—赫兹管的 I_A—U_{G_2K}曲线

凡在满足式（S23 - 3）的地方板极电流I_A都会相应下跌的，形成规则起伏变化的I_A—U_{G_2K}曲线。而各次板极电流I_A下降相对应的阴、栅极电位差$U_{n+1} - U_n$应该是氩原子的第一激发电位U_0。

本实验就是要测定氩原子的第一激发电位（公认值为$U_0 = 11.5\text{V}$）。原子处于激发态是不稳定的。在实验中被慢电子轰击到第一激发态的原子要跳回基态，进行这种反跃迁时，就应该有eU_0电子伏特的能量发射出来。反跃迁时，原子是以放出光量子的形式向外辐射能量。这种光辐射的波长为

$$eU_0 = hv = h\frac{c\lambda}{\lambda} \tag{S23-4}$$

对于氩原子 $$\lambda = \frac{hc}{eU_0} = \frac{6.63 \times 10^{-34} \times 3.00 \times 10^{8}}{1.6 \times 10^{-19} \times 11.5} = 1081(\text{Å})$$

如果夫兰克—赫兹管中充有其他元素，则也可类似地得到它们的第一激发电位，见表S23－1。

表S23－1　几种元素的第一激发电位

元素	钠(Na)	钾(K)	锂(Li)	镁(Mg)	汞(Hg)	氦(He)	氖(Ne)
U_0/V	2.12	1.63	1.84	3.2	4.9	21.2	18.6
λ/Å	5898 5896	7664 7699	6707.8	4571	2500	584.3	640.2

[实验内容与步骤]

1. 实验内容

测定氩原子的第一激发电位。

2. 实验步骤

1）准备

（1）熟悉实验仪的使用方法（参考说明书）。

（2）按要求（参考说明书）连接夫兰克—赫兹管各组工作电源线，检查无误后开机。开机后的初始状态如下：

①实验仪的“1mA”电流挡位指示灯亮，表明此时电流的量程为1mA挡；电流显示值为000.0mA。

②实验仪的“灯丝电压”挡位指示灯亮，表明此时修改的电压为灯丝电压；电压显示值为000.0V；最后一位在闪动，表明现在修改位为最后一位。

③“手动”指示灯亮。表明仪器工作正常。

2）氩原子第一激发电位的测定

（1）手动测量。

①设置仪器为“手动”工作状态，按“手动/自动”键，“手动”指示灯亮。

②设定电流量程（电流量程可参考机箱盖上提供的数据），按下相应电流量程键，对应的量程指示灯点亮。

③设定电压源的电压值（设定值可参考机箱盖上提供的数据），用↓/↑，←/→键完成，需设定的电压源有：灯丝电压U_F、第一加速电压U_{G_1K}、拒斥电压U_{G_2A}。

④按下“启动”键，实验开始。用↓/↑，←/→键完成电压值的调节，从0.0V起，按步长1V（或0.5V）的电压值调节电压源，同步记录值和对应的值，同时仔细观察夫兰克—赫兹管的板极电流值的变化（可用示波器观察）。注意：为保证实验数据的唯一性，电压必须从小到大单向调节，记录完最后一组数据后，立即将电压快速归零。

⑤重新启动。在手动测量过程中，按下启动按键，U_{G_2K}的电压值将被设置为零，内部存储的测量数据被清除，示波器上显示的波形被清除，但U_F、U_{G_1K}、U_{G_2A}电流挡等的状态不发生改变。这时，操作者可以在该状态下重新进行测试，或修改状态后再进行测量。

建议：手动测量，每修改一次U值，进行一次对应的测量。

（2）自动测量。

智能夫兰克—赫兹实验仪除可以进行手动测量外，还可以进行自动测量。进行自动测量时，实验仪将自动产生U_{G_2K}扫描电压，完成整个测量过程，将示波器与实验仪相连接，在示波器上可看到夫兰克—赫兹管板极电流I_A随U_{G_2K}电压变化的波形。

①自动测量状态的设置。自动测量时，U_F、U_{G_1K}、U_{G_2A}电流挡位等状态设置的操作过程，以及与夫兰克—赫兹管的连线过程与手动测量操作过程一样。

②U_{G_2A}扫描终止电压的设定。进行自动测量时，实验仪将自动产生U_{G_2K}扫描电压。实验仪默认U_{G_2K}扫描电压的初始值为零，U_{G_2K}扫描电压大约每0.4s递增0.2V。直到扫描终止电压。要进行自动测量，必须设置电压U_{G_2K}的扫描终止电压。首先，将“手动/自动”测试键按下，自动测试指示灯亮，按下U_{G_2K}电压源选择键，U_{G_2K}电压源选择指示灯亮。用↓/↑，←/→键完成U_{G_2K}电压值的具体设定。U_{G_2K}设定终止值建议不超过80V为好。

③自动测量的启动。将“电压源选择”选为U_{G_2K}后再按面板上的“启动”键，自动测量开始。在自动测量过程中，观察U_{G_2K}与夫兰克—赫兹管板极电流的相关变化情况。（可通过示波器观察夫兰克—赫兹管板极电流I_A随扫描电压U_{G_2K}变化的输出波形）在自动测量过程中，为避免面板按键误操作，导致自动测量失败，面板上除“手动/自动”按键外的所有按键都被屏蔽禁止。

④自动测量过程正常结束。当扫描电压U_{G_2K}的电压值大于设定的测试终止电压值后，实验仪将自动结束本次自动测量过程，进入数据查询工作状态。测量数据保留在实验仪主机的存储器中，供数据查询过程使用，所以，示波器仍可观测到本次测量数据所形成的波形。直到下次测量开始时才刷新存储器的内容。

⑤自动测量后的数据查询。自动测量过程正常结束后，实验仪进入数据查询工作状态。这时面板按键除测量电流指示区外，其他都已开启。自动测量指示灯亮，电流量程指示灯指示于本次测量的电流量程选择挡位。各电压源选择按键可选择各电压源的电压值指示，其中U_F、U_{G_1K}、u_{G_2A}三电压源只能显示设定电压值，不能通过按键改变相应的电压值。用↓/↑，←/→键改变电压源U_{G_2K}的指示值，就可查阅到在本次测量过程中，电压源U_{G_2K}的扫描电压值为当前显示值时，对应的夫兰克—赫兹管板极电流值I_A的大小，记录I_A的峰值、谷值和对应U_{G_2K}的值（为便于作图，在I_A的峰、谷值附近需多取点数）。

⑥中断自动测试过程。在自动测量过程中，只要按下“手动/自动”键，手动测量指示灯亮，实验仪就中断了自动测量过程，恢复到开机初始状态。所有按键都被再次开启工作。这时可进行下一次的测量准备工作。本次测量的数据依然保留在实验仪主机的存储器中，直到下

次测量开始时才被清除。所以,示波器仍会观测到部分波形。

⑦结束查询过程恢复初始状态。当需要结束查询过程时,只要按下“手动/自动”键,手动测量指示灯亮,查询过程结束,面板按键再次全部开启。原设置的电压状态被清除,实验仪存储的测量数据被清除,实验仪恢复到初始状态。

建议:自动测量应变化两次值,测量两组I_A—U_{G_2K}数据。若实验时间允许,还可进行多次I_A—U_{G_2K}测量。

[注意事项]

(1)线路连好后应反复检查有无错误,方可开启电源。

(2)各参考电压一定要在给定的数据范围之内,否则将造成管子的老化或击穿。

(3)调节灯丝电压U_F时,电流I_A在1min后达到稳定,才能读数测量。

[数据及处理]

1. 数据记录(表S23-2)

表S23-2 实验数据记录表格

	峰1	谷1	峰2	谷2	峰3	谷3	峰4	谷4
U_{G_2K}/V								
I_A/μA								

2. 数据处理

(1)作出I_A—U_{G_2K}曲线,曲线上相邻两峰值之间的电位差就是氩原子的第一激发电位。

(2)用逐差法求得氩原子的第一激发电位。

3. 结果分析

对实验结果进行分析,并得出实验结论。

[问题讨论]

(1)试讨论温度对实验曲线(如曲线形状、峰谷值以及峰的数目等)的影响。

(2)I_A—U_{G_2K}曲线中的第一峰值对应的横坐标(即电压值)是否就是氩原子的第一激发电位?请说明原因。

实验24 用光电效应法测定普朗克常数

[引言]

1887年赫兹在用两套电极做电磁波的发射与接收的实验中,发现当紫外光照射到接收电极的负极时,接收电极间更易于产生放电,赫兹的发现吸引许多人去做这方面的研究工作。斯托列托夫发现负电极在光的照射下会放出带负电的粒子,形成光电流,光电流的大小与入射光强度成正比,光电流实际是在照射开始时立即产生,无需时间上的积累。

普朗克常数是在辐射定律研究过程中,1900年由普朗克(1858—1947)引入并与黑体发射和吸收相关的普适常量。普朗克公式与实验符合得很好。他的这一研究成果发表后不久,他又通过解释提出了与经典理论相悖的假设,认为能量不能连续变化,只能取一些分立值,这些值是最小能量的整数倍。1905年爱因斯坦(1879—1955)把这一观点推广到光辐射,提出了光量子概念,由他建立的方程即爱因斯坦方程成功地解释了光电效应,并验证了光量子理论的正确性。普朗克的理论解释和公式推导是量子论诞生的标志,具有里程碑的意义。本实验就是根据光电效应的物理学规律,选择光电二极管阴极材料在大于其红限条件下的不同频率光照射下发生光电效应的物理过程,利用截止电压与照射光频率成正比的关系,测定普朗克常数。

[实验目的]

(1)了解光电效应的规律,加深对光的量子性的理解。

(2)掌握测量普朗克常数的方法。

[预习问题]

(1)如何测量不同入射光照射下的截止电压?

(2)影响截止电压测量结果的主要因素有哪些?

[实验仪器]

ZKY-GD-4智能光电效应(普朗克常数)实验仪、MOS-620型示波器等。

[实验原理]

光电效应的实验原理如图S24-1所示。入射光照射到光电管阴极K上,产生的光电子在电场的作用下向阳极A迁移构成光电流,改变外加电压U_{AK},测量出光电流I的大小,即可得出光电管的伏安特性曲线。

1. 光电效应的基本实验现象

(1)当某一频率的光照射时,光电效应的I—U_{AK}关系如图S24-2所示。从图中可见,对一定的频率,有一电压U_0,当$U_{AK} \leqslant U_0$时,电流为零,这个相对于阴极的负值的阳极电压U_0,称为截止电压。

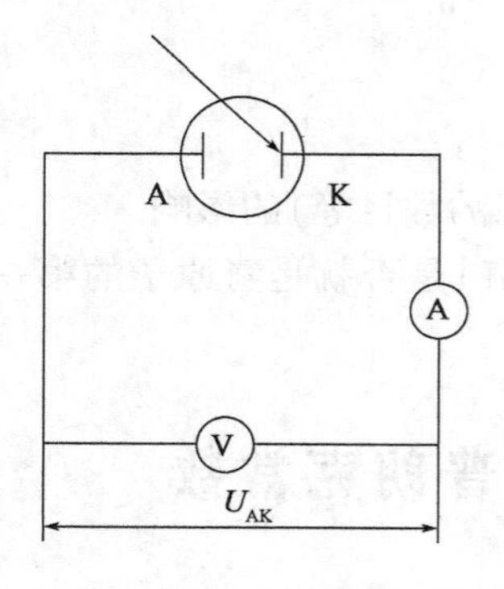

图S24-1 实验原理图

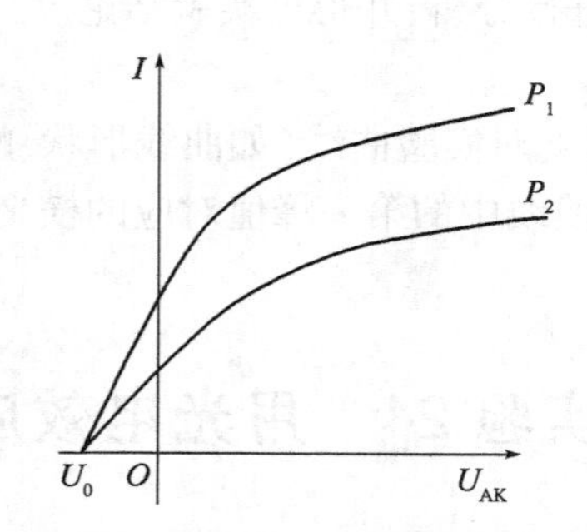

图S24-2 同一频率,不同光强时的伏安特性曲线

(2)当$U_{AK} > U_0$后,I迅速增加,然后趋于饱和,饱和光电流I_M的大小与入射光的强度P成正比。

(3)对于不同频率的入射光,其截止电压的值不同,如图S24-3所示。

(4)截止电压U_0与频率ν的关系如图S24-4所示。U_0与ν成正比关系。当入射光频率低于某极限值ν_0(ν_0随不同金属而异)时,不论光的强度如何,照射时间多长,都没有光电流产生。

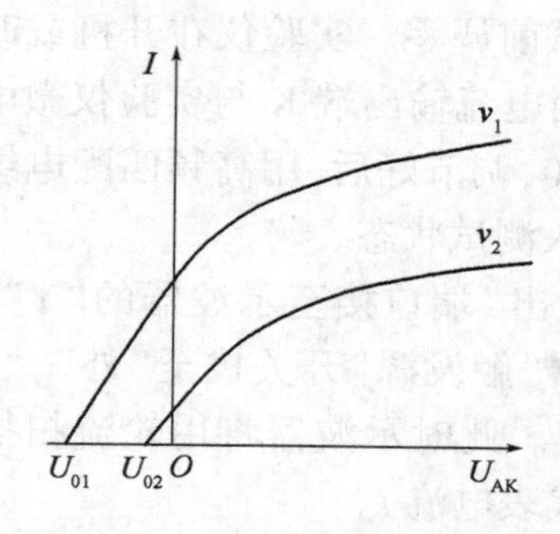

图 S24-3 不同频率时光电管的伏安特性曲线

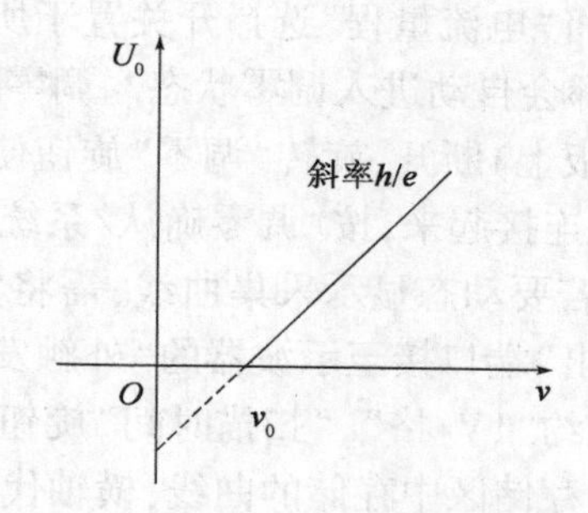

图 S24-4 截止电压 U 与入射光频率的关系图

(5)光电效应是瞬时效应。即使入射光的强度非常微弱，只要入射光频率 ν 大于ν_0，在开始照射后立即就有光电子产生，所经历的时间仅为 10^{-9} s 的量级。

2. 爱因斯坦光量子理论

按照爱因斯坦的光量子理论，光子具有波粒二象性。频率为 ν 的光子具有能量 $E=h\nu$，h 为普朗克常数。当光子照射到金属表面上时，光子能量被金属中的电子全部吸收，电子把吸收光子的能量的一部分用来克服金属表面对它的吸引力，余下的就变为电子离开金属表面后的动能，按照能量守恒原理，爱因斯坦提出了著名的光电效应方程

$$h\nu=\frac{1}{2}mv_0^2+A \tag{S24-1}$$

式中，A 为金属的逸出功，$\frac{1}{2}mv_0^2$为光电子获得的初始动能。由该式可见，入射到金属表面的光频率越高，逸出的电子动能越大，所以即使阳极电位比阴极电位低时也会有电子落入阳极形成光电流，直至阳极电位低于截止电压，光电流才为零，此时有关系

$$eU_0=\frac{1}{2}mv_0^2 \tag{S24-2}$$

阳极电位高于截止电压后，随着阳极电位的升高，阳极对阴极发射的电子的收集作用越强，光电流随之上升：当阳极电位高到一定程度，已把阴极发射的光电子几乎全收集到阳极，再增加U_{AK}时 I 不再变化，光电流出现饱和，饱和光电流I_M的大小与入射光的强度 P 成正比。

光子的能量 $h\nu_0<A$ 时，吸收的光子能量不足以使电子脱离金属表面，对光电流做出贡献，因而没有光电流产生。产生光电效应的最低频率（截止频率）是 $\nu_0=A/h_0$，将式(S24-2)代入式(S24-1)可得

$$eU_0=h\nu-A \tag{S24-3}$$

此式表明截止电压U_0是频率 ν 的线性函数，直线斜率 $K=h/e$，只要用实验方法测出不同的频率对应的截止电压，求出直线斜率，就可算得普朗克常数 h。

［实验内容与步骤］

1. 利用光电效应测定普朗克常数 h

(1)将实验仪及汞灯电源接通（汞灯及光电管暗箱遮光盖盖上），预热 20min。调整光电管与汞灯距离约为 40cm 并保持不变。

(2)用专用连接线将光电管暗箱电压输入端与实验仪电压输出端（在后面板上）连接起来（红—红，蓝—蓝）。

(3)将“电流量程”选择开关置于所选挡位，进行测试前调零。实验仪在开机或改变电流量程后，都会自动进入调零状态。调零时应将光电管暗箱电流输出端 K 与实验仪微电流输入端（后面板上）断开，旋转“调零”旋钮使电流指示为 000.0，调节好后，用高频匹配电缆将电流输入端子连接起来，按“调零确认/系统清零”键，系统进入测试状态。

(4)若要动态显示采集曲线，需将实验仪的“信号输出”端口接至示波器的“Y”输入端，“同步输出”端口接至示波器的“外触发”输入端。示波器“触发源”开关拨至“外”，“Y 衰减”旋钮拨至约“1V/格”，“扫描时间”旋钮拨至约“20*us*/格”。此时示波器将用轮流扫描的方式显示 5 个存储区中存储的曲线，横轴代表电压U_{AK}，纵轴代表电流 I。

(5)问题讨论及测量方法。

理论上，测出各频率的光照射下阴极电流为零时对应的U_{AK}，其绝对值即为该频率对应的截止电压，然而实际上由于光电管的阳极反向电流、暗电流、本底电流及极间接触电位差的影响，实测电流并非阴极电流，实测电流为零时对应的U_{AK}也并非截止电压。其原因是光电管制作过程中阳极往往被污染，沾上少许阴极材料，入射光照射阳极或入射光从阴极反射到阳极之后造成阳极光电子发射，U_{AK}为负值时，阳极发射的电子向阴极迁移构成了阳极反向电流。

暗电流和本底电流是热激发产生的光电流与杂散光照射光电管产生的光电流，可以在光电管制作或测量过程中采取适当措施减少它们的影响。

极间接触电位差与入射光频率无关，只影响U_0准确性，不影响 U_0—v 直线斜率，对测定 h 无大影响。

由于本实验仪的电流放大器灵敏度高，稳定性好，光电管阳极反向电流、暗电流也较小。在测量各谱线的截止电压U_0时，可采用零电流法，即直接将各谱线照射下测得的电流为零时对应U_{AK}的绝对值作为截止U_0。此法的前提是阳极反向电流、暗电流和本底电流都很小，用零电流法测得的截止电压与真实值相差较小。且各谱线的截止电压都相差 ΔU 对U_0—v 曲线的斜率无大的影响，因此对 h 的测量不会产生大的影响。

2. 测量截止电压

测量截止电压时，“伏安特性测试/截止电压测试”状态键应为截止电压测试状态。“电流量程”开关处于 10^{-13}A 挡。

1)手动测量

使“手动/自动”模式键入手动模式。将直径 4mm 的光阑及 365.0nm 的滤色片装在光电管暗箱光输入窗口上，打开汞灯遮光盖。此时电压表显示U_{AK}的值，单位为伏；电流表显示与U_{AK}对应的电流值 I，单位为所选择的“电流量程”。用电压调节键→、←、↑、↓可调节U_{AK}的值，→、←键用于选择调节位↑、↓键用于调节值的大小。

从低到高调节电压（绝对值减小），观察电流值的变化，寻找电流为零时对应的U_{AK}，以其绝对值作为该波长对应的U_0的值，并记录数据。为尽快找到U_0的值，调节时应从高电位到低电位，先确定高位的值，再依次往低电位调节。

依次换上 404.7nm、435.8nm、546.1nm、577.0nm 的滤色片，重复以上测量步骤。

2)自动测量

按“手动/自动”模式键切换到自动模式。此时电流表左边的指示灯闪烁，表示系统处于自动测量扫描范围设置状态，用电压调节键可设置扫描起始电压和终止电压。

对各条谱线，建议扫描范围大致设置为：(1)365nm：1.90 ~ −1.50V；(2)405nm：−1.60 ~

−1.20V；(3)436nm：−1.35 ~ −0.95V；(4)546nm：−0.80 ~ −0.40V；(5)577nm：−0.65 ~ −0.25V。实验仪设有5个数据存储区，每个存储区可存储500组数据，并有指示灯表示其状态。灯亮表示该存储区已存有数据，灯不亮为空存储区，灯闪烁表示系统预选存储区或正在存储数据的存储区。设置好扫描起始和终止电压后，按动相应的存储区按键，仪器将先清除存储区原有数据，等待约30s，然后按“4mV”的步长自动扫描，并显示，存储相应的电压和电流值。

扫描完成后，仪器自动进入数据查询状态，此时查询指示灯亮，显示区显示扫描起始电压和相应的电流值。用电压调节键改变电压值，就可查阅到在测试过程中扫描电压为当前显示值时相应的电流值。读取电流为零时对应的U_{AK}值，以其绝对值作为该波长对应的U_0的值，并记录数据。

按“查询”键，查询指示灯灭，系统回复到扫描范围设置状态，可进行下一次测量。

在自动测量过程中或测量完成后，按“手动/自动”键，系统恢复到手动测量模式，模式转换前工作的存储区内的数据将被清除。若仪器与示波器连接，则可观察到U_{AK}为负值时各谱线在选定的扫描范围内的伏安特性曲线。

[注意事项]

(1)汞灯打开后直至实验全部完成后再关闭。一旦中途关闭电源，至少等5min后再启动。

(2)注意勿使电源输出端与地短路，以免烧毁电源。

(3)注意保持滤色片的清洁，但不要随意擦拭滤色片。

(4)实验后用遮光罩罩住光电管暗盒，以保护光电管。

[数据及处理]

1. 数据记录(表S24−1)

表S24−1　U_0—ν关系数据记录表　　光阑孔 ϕ = ______ mm

波长 λ_i/nm		365.0	404.7	435.8	546.1	577.0
频率ν_i/$\times 10^{14}$Hz		8.214	7.408	6.879	5.490	5.196
截止电压 U_{0i}/V	手动					
	自动					

2. 数据处理

(1)根据表S24−1的数据在坐标纸上作U_0—ν图线，由直线斜率求得普朗克常数。

(2)与公认值比较给出百分差。

3. 结果分析

对实验结果进行分析，并得出实验结论。

[问题讨论]

(1)光电流是否随光源的强度变化？截止电压是否因光源强度不同而改变？

(2)在实验过程中若改变了光源与光电管之间距离，会产生什么影响？

(3)光电管的阴极和阳极之间存在接触电位差，试分析这对本实验结果有无影响？

(4)光电管的阳极、阴极材料选用应考虑哪些因素？

[补充说明]

ZKY−GD−4智能光电效应(普朗克常数)实验仪由汞灯及电源、滤色片、光阑、光电管、测试仪构成，如图 24－5 所示。实验仪有手动和自动两种工作模式，具有数据自动采集、存储、实时显示采集数据、动态显示采集曲线(连接普通示波器，可同时显示 5 个存储区中存储的曲线)，以及采集完成后查询数据等功能。

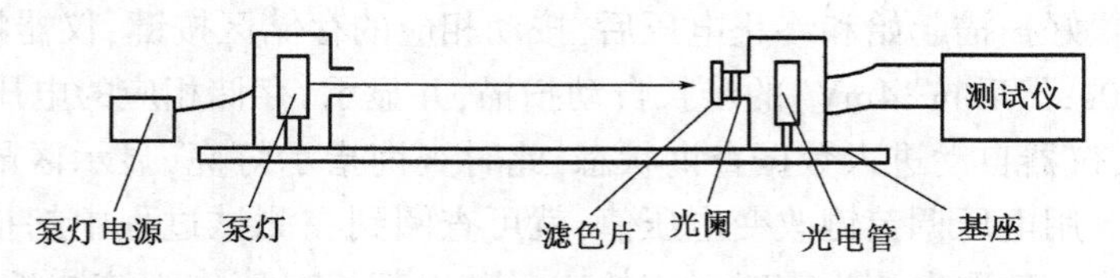

图 24－5　ZKY－GD－4 智能光电效应(普朗克常数)实验仪

实验 25　霍尔效应及其应用研究

[引言]

霍尔效应是导电材料中的电流与外磁场相互作用而在垂直于磁场和电流方向的两个端面之间产生电动势的效应。1879 年美国霍普金斯大学研究生霍尔在研究金属导电机构时发现了这种电磁现象，故称霍尔效应。后来曾有人利用霍尔效应制成测量磁场的磁传感器，但因金属的霍尔效应太弱而未能得到实际应用。随着半导体材料和制造工艺的发展，人们又利用半导体材料制成霍尔元件，由于它的霍尔效应显著而得到实用和发展，现在广泛用于非电量检测、电动控制、电磁测量和计算装置等方面。在电流体中的霍尔效应也是目前在研究中的“磁流体发电”的理论基础。1980 年物理学家冯·克利青(K. Von Klitzing)研究二维电子气系统的输运特性，在低温和强磁场下发现了量子霍尔效应，这是凝聚态物理领域最重要的发现之一。目前对量子霍尔效应正在进行深入研究，并取得了重要应用，例如用于确定电阻的自然基准、可以极为精确地测量光谱精细结构常数等。

[实验目的]

(1)学习用霍尔效应仪测量磁场，加深对霍尔效应产生机理的理解。

(2)学习用“对称测量法”消除霍尔效应实验中的副效应影响。

(3)计算霍尔元件灵敏度、载流子的浓度和迁移率，并判断其载流子的类型。

(4)学习利用霍尔效应测量磁感应强度 B 及磁场分布。

[预习问题]

(1)如何使用数字式特斯拉计？

(2)何谓对称测量法？本实验采用对称测量法的目的是什么？

(3)选择半导体材料做霍尔器件时，主要考虑哪些参数？

[实验仪器]

霍尔效应仪、可调直流稳压电源(0～500mA)、直流稳流电源(0～5mA)、直流数字电压表，数字式特斯拉计、直流电阻(取样电阻)电磁铁、霍尔元件(砷化镓霍尔元件)、双刀双向开关、导线等。

[实验原理]

1. 霍尔效应

霍尔效应从本质上讲是运动的带电粒子在磁场中受洛仑兹力的作用而引起带电粒子的偏转。当带电粒子(电子或空穴)被约束在固体材料中，这种偏转就导致在垂直于电流和磁场方向的两个端面产生正负电荷的聚积，从而形成附加的横向电场。

如图 S25－1 所示，沿 z 轴的正向加以磁场 B，与 z 轴垂直的半导体薄片上沿 x 正向通以电流 I_s（称为工作电流或控制电流），假设载流子为电子［如 N 型半导体材料，图 S25－1(a)］，它沿着与电流 I_s 相反的 x 负向运动。由于洛伦兹力 F_m 的作用，电子即向图中的 D 侧偏转，并使 D 侧形成电子积累，而相对的 C 侧形成正电荷积累。与此同时，运动的电子还受到由于两侧积累的异种电荷形成的反向电场力 F_e 的作用。随着电荷的积累，F_e 逐渐增大，当两力大小相等，方向相反时，电子积累便达到动态平衡。这时在 C、D 两端面之间建立的电场称为霍尔电场 E_H，相应的电势差称为霍尔电压 U_H。

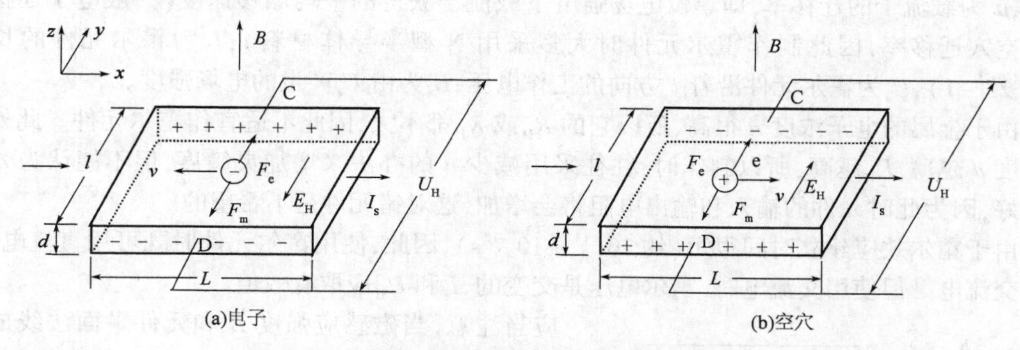

图 S25－1　霍尔元件中载流子在外磁场下的运动情况

设电子按相同平均漂移速率 $\boldsymbol{v}$ 向图 S25－1 中的 x 轴负方向运动，在磁场 B 作用下，所受洛伦兹力为：

$$\boldsymbol{F}_m = -e\boldsymbol{v} \times \boldsymbol{B} \tag{S25-1}$$

式中，e 为电子电量，取 1.6×10^{-19} C；$\boldsymbol{v}$ 为电子漂移平均速度；$\boldsymbol{B}$ 为磁感应强度。

同时，电场作用于电子的力为：

$$\boldsymbol{F}_e = -e\boldsymbol{E}_H = -e\frac{\boldsymbol{U}_H}{l} \tag{S25-2}$$

式中，$\boldsymbol{E}_H$ 为霍尔电场强度；$\boldsymbol{U}_H$ 为霍尔电压；l 为霍尔元件宽度。

当达到动态平衡时，$\boldsymbol{F}_m = -\boldsymbol{F}_e$，从而得到

$$vB = \frac{U_H}{l} \tag{S25-3}$$

霍尔元件宽度为 l，厚度为 d，载流子浓度为 n，则霍尔元件的工作电流为

$$I_s = nevld \tag{S25-4}$$

由式（S25－3）、式（S25－4）可得

$$U_H = \frac{1}{ne}\frac{I_s B}{d} = R_H \frac{I_s B}{d} = K_H I_s B \tag{S25-5}$$

即霍尔电压 U_H（此时为 C、D 间电压）与 I_s、B 成正比，与霍尔元件的厚度 d 成反比。式中：比例系数 $R_H = \frac{1}{ne}$ 称为霍尔系数，它是反映材料霍尔效应强弱的重要参数；比例系数 $K_H = \frac{1}{ned}$ 称为霍尔元件的灵敏度，它表示霍尔元件在单位磁感应强度和单位工作电流下的霍尔电势大小，其单位是 mV/(mA·T)，一般要求 K_H 越大越好。

当霍尔元件的材料和厚度确定时，根据霍尔系数或灵敏度可以得到载流子的浓度 n：

$$n = \frac{1}{eR_{\mathrm{H}}} = \frac{1}{edK_{\mathrm{H}}} \qquad (S25-6)$$

霍尔元件中载流子迁移率μ:

$$\mu = \frac{v}{E_{\mathrm{s}}} = \frac{vL}{U_{\mathrm{s}}} \qquad (S25-7)$$

将式(S25-4)、式(S25-5)、式(S25-7)联立求得

$$\mu = K_{\mathrm{H}} \frac{L}{l} \cdot \frac{I_{\mathrm{s}}}{U_{\mathrm{s}}} \qquad (S25-8)$$

式中,μ为载流子的迁移率,即单位电场强度下载流子获得的平均漂移速度(一般电子迁移率大于空穴迁移率,因此制作霍尔元件时大多采用N型半导体材料);L为霍尔元件的长度(图S25-1),U_{s}为霍尔元件沿着I_{s}方向的工作电压;E_{s}为由U_{s}产生的电场强度。

由于金属的电子浓度n很高,所以它的R_{H}或K_{H}都不大,因此不适宜作霍尔元件。此外元件厚度d越薄,K_{H}越高,所以制作时,往往采用减少d的办法来增加灵敏度,但不能认为d越薄越好,因为此时元件的输入和输出电阻将会增加,这对锗元件是不希望的。

由于霍尔效应建立时间很短(约$10^{-14} \sim 10^{-12}$s),因此,使用霍尔元件时既可用直流电,也可用交流电。但使用交流电时,霍尔电压是交变的,I_{s}和U_{H}应取有效值。

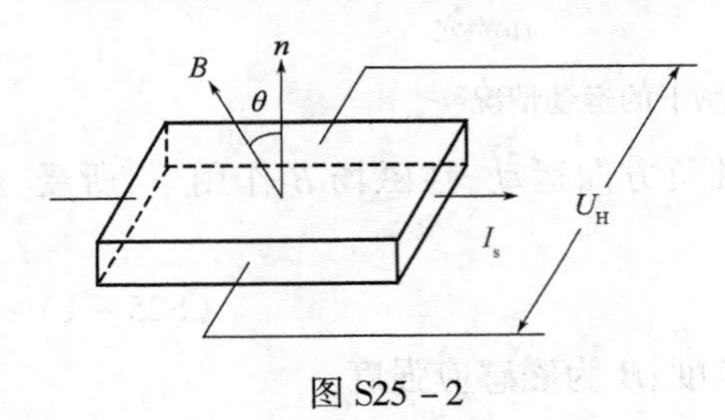

图 S25-2

应当注意,当磁感应强度B和元件平面法线成一角度时(图S25-2),作用在元件上的有效磁场是其法线方向上的分量$B\cos\theta$,此时

$$U_{\mathrm{H}} = K_{\mathrm{H}} I_{\mathrm{s}} B\cos\theta \qquad (S25-9)$$

所以,一般在使用时应调整元件平面朝向,使U_{H}达到最大,即$\theta=0$, $U_{\mathrm{H}} = K_{\mathrm{H}} I_{\mathrm{s}} B\cos\theta = K_{\mathrm{H}} I_{\mathrm{s}} B$。

由式(S25-9)可知,当工作电流I_{s}或磁感应强度B,两者之一改变方向时,霍尔电压U_{H}的方向随之改变;若两者方向同时改变,则霍尔电压U_{H}极性不变。

2. 霍尔效应的副效应及其消除

测量霍尔电势U_{H}时,不可避免地会产生一些副效应,由此而产生的附加电势叠加在霍尔电势上,形成测量系统误差。

1)不等位电势U_0

由于制作时,两个霍尔电极不可能绝对对称地焊在霍尔元件两侧[图S25-3(a)]、霍尔元件电阻率不均匀、工作电流极的端面接触不良[图S25-3(b)]都可能造成C、D两极不处在同一等位面上,此时虽未加磁场,但C、D间存在电势差U_0,称为不等位电势,$U_0 = I_{\mathrm{s}}R_0$,R_0是C、D两极间的不等位电阻。由此可见,在R_0确定的情况下,U_0与I_{s}的大小成正比,且其正负随I_{s}的方向改变而改变。

2)爱廷豪森(Ettingshausen)效应

当霍尔元件的x方向通以工作电流I_{s},z方向加磁场B时,由于霍尔元件内的载流子速度服从统计分布,有快有慢。在达到动态平衡时,在磁场的作用下慢速与快速的载流子将在洛伦兹力和霍尔电场的共同作用下,沿y轴分别向相反的两侧偏转,这些载流子的动能将转化为热能,使两侧的温度不同,因而造成y方向上两侧出现温差($\Delta T = T_{\mathrm{C}} - T_{\mathrm{D}}$)。

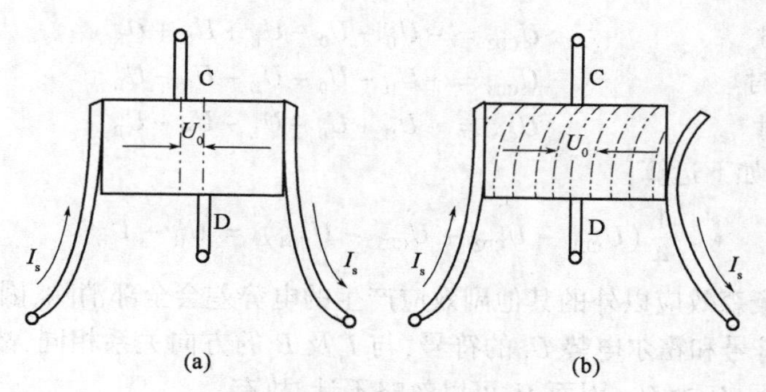

图 S25－3　不等位电势 U_0。

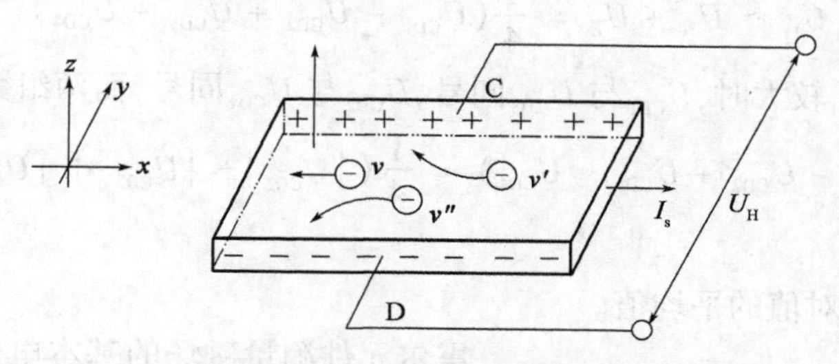

图 S25－4　霍尔元件中电子实际运动情况（图中 $v'<v, v''>v$）

因为霍尔电极和元件两者材料不同，电极和元件之间形成温差电偶，这一温差在 C、D 间产生温差电动势 U_E，$U_E \propto I_s B$。

这一效应称爱廷豪森效应，U_E的大小及正负符号与 I_s、B 的大小和方向有关，跟 U_H与 I_s、B 的关系相同，所以不能在测量中消除。

3）伦斯脱（Nernst）效应

由于工作电流的两个电极与霍尔元件的接触电阻不同，工作电流在两电极处将产生不同的焦耳热，引起工作电流两极间的温差电动势，此电动势又产生温差电流（称为热电流）I_Q，热电流在磁场作用下将发生偏转，结果在 y 方向上产生附加的电势差 U_N且 $U_N \propto I_Q B$，这一效应称为伦斯脱效应，由上式可知 U_N的符号只与 B 的方向有关。

4）里纪—勒杜克（Righi-Leduc）效应

如伦斯脱效应所述霍尔元件在 x 方向有温度梯度，引起载流子沿梯度方向扩散而有热电流 I_Q通过霍尔元件，在此过程中载流子受 z 方向的磁场 B 作用，在 y 方向引起类似爱廷豪森效应的温差 $\Delta T = T_C - T_D$，由此产生的电势差 $U_R \propto I_Q B$，其符号与 B 的方向有关，与 I_s的方向无关。

在确定的磁场 B 和工作电流 I_s下，实际测出的电压是 U_H、U_0、U_E、U_N和 U_R这 5 种电势差的代数和。上述 5 种电势差与 B 和 I_s方向的关系见表 S25－1。

表 S25－1　电势差与 B 和 I_s方向的关系表

U_H		U_0		U_E		U_N		U_R	
B	I_s	B	I_s	B	I_s	B	I_s	B	I_s
有关	有关	无关	有关	有关	有关	有关	无关	有关	无关

为了减少和消除以上效应引起的附加电势差，利用这些附加电势差与霍尔元件工作电流 I_s、磁场 B（即相应的励磁电流 I_M）的关系，采用对称（交换）测量法测量 C、D 间电势差：

当 $+I_M$，$+I_s$时　　　$U_{CD1} = +U_H + U_0 + U_E + U_N + U_R$

当 $+I_M$，$-I_s$ 时　　$U_{CD2}=-U_H-U_0-U_E+U_N+U_R$

当 $-I_M$，$-I_s$ 时　　$U_{CD3}=+U_H-U_0+U_E-U_N-U_R$

当 $-I_M$，$+I_s$ 时　　$U_{CD4}=-U_H+U_0-U_E-U_N-U_R$

对以上四式做如下运算：

$$\frac{1}{4}(U_{CD1}-U_{CD2}+U_{CD3}-U_{CD4})=U_H+U_E \tag{S25-10}$$

可见，除爱廷豪森效应以外的其他副效应产生的电势差会全部消除，因爱廷豪森效应所产生的电势差 U_E 的符号和霍尔电势 U_H 的符号，与 I_s 及 B 的方向关系相同，故无法消除，但在非大电流、非强磁场下，$U_H \gg U_E$，因而 U_E 可以忽略不计，故有

$$U_H \approx U_H+U_E=\frac{1}{4}(U_{CD1}-U_{CD2}+U_{CD3}-U_{CD4}) \tag{S25-11}$$

一般情况下当 U_H 较大时，U_{CD1} 与 U_{CD3} 同号，U_{CD2} 与 U_{CD4} 同号，而两组数据反号，故

$$U_H=\frac{1}{4}(U_{CD1}-U_{CD2}+U_{CD3}-U_{CD4})=\frac{1}{4}(|U_{CD1}|+|U_{CD2}|+|U_{CD3}|+|U_{CD4}|) \tag{S25-12}$$

即用四次测量值的绝对值的平均值。

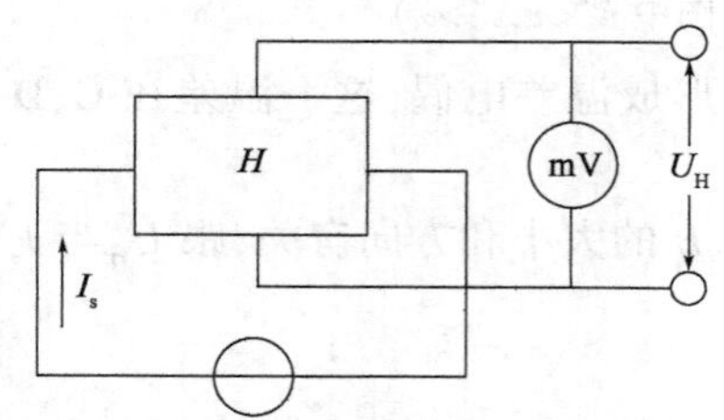

图 S25－5　霍尔元件测量磁场的电路图

霍尔元件测量磁场的基本电路如图 S25－5 所示，将霍尔元件置于待测磁场的相应位置，并使元件平面与磁感应强度 B 垂直，在其控制端输入恒定的工作电流 I_s，霍尔元件的霍尔电压输出端接毫伏表，测量霍尔电压 U_H 的值。

[实验内容与步骤]

（1）按仪器面板上的文字和符号提示将 ZKY－HS 霍尔效应实验仪（以下简称“实验仪”）与 ZKY－H/L 霍尔效应螺线管磁场测试仪（以下简称“测试仪”）正确连接。

①将工作电流、励磁电流调节旋钮逆时针旋转到底，使电流最小。

②将测试仪的电压量程调至高量程。

③测试仪面板右下方为提供励磁电流 I_M 的恒流源输出端，接实验仪上励磁电流的输入端（将接线叉口与接线柱连接）。

④测试仪左下方为提供霍尔元件工作电流 I_s 的恒流源输出端接实验仪工作电流输入端（将插头插入插孔）。

⑤实验仪上的霍尔电压输出端接测试仪中部下方的霍尔电压输入端。

⑥将测试仪与 220V 交流电源相连，按下开机键。

注：为了提高霍尔元件测量的准确性，实验前霍尔元件应至少预热 5min，具体操作如下：断开励磁电流开关，闭合工作电流开关，通入工作电流 5mA，待至少 5min 可以开始实验。

（2）测量霍尔元件灵敏度 K_H，计算载流子浓度 n。

①移动二维移动尺，使霍尔元件处于电磁铁气隙中心位置（其法线方向已调至平行于磁场方向），闭合励磁电流开关，调节励磁电流 $I_M=300\text{mA}$，通过公式：$B=C\cdot I_M$ 求得并记录此时电磁铁气隙中的磁感应强度 B（C 为电磁铁的线圈常数，C 的值见面板标示牌）。

②调节工作电流 $I_s=1.00\text{mA}, 2.00\text{mA}, \cdots, 10.00\text{mA}$（间隔 1.00mA），通过变换实验仪各

换向开关，在（$+I_M$，$+I_s$）、（$-I_M$，$+I_s$）、（$-I_M$，$-I_s$）、（$+I_M$，$-I_s$）四种测量条件下，分别测出对应的C、D间电压值U_i（$i=1$、2、3、4），根据式（S25－12）计算霍尔电压U_H填入表S25－1，并绘制U_H-I_s关系曲线，求得斜率K_1（$K_1=U_H/I_s$）。

③根据式（S25－5）可知$K_H=K_1/B$；据式（S25－6）可计算载流子浓度n。

（3）测量霍尔元件的载流子迁移率μ（选做）。

注：该实验需自备一只电压表，用于测量工作电压U_s，电压表挡位选为“直流20V挡”。电压表的正负极分别接测试仪上工作电流输出端的红、黑插孔。

①断开励磁电流开关，使$I_M=0$（电磁铁剩磁很小，约零点几毫特，可忽略不计）。调节$I_s=$0.50mA，1.00mA，…，5.00mA（间隔0.50mA），记录对应的工作电压U_s填入表S25－2，绘制I_s-U_s关系曲线，求得斜率K_2（$K_2=I_s/U_s$）。

②根据上面求得的K_H，结合式（S25－8）可以求得载流子迁移率μ（霍尔元件长度L、宽度l已知，见面板标示牌）。

（4）判定霍尔元件半导体类型（P型或N型）或者反推磁感应强度B的方向。

①根据电磁铁导线绕向及励磁电流I_M的流向，可判定气隙中磁感应强度B的方向。

②根据闸刀开关接线以及霍尔测试仪I_s输出端引线，可判定I_s在霍尔元件中流向。

③根据换向闸刀开关接线以及霍尔测试仪U_H输入端引线，可以得出U_H的正负与霍尔元件上正负电荷积累的对应关系。

④由B的方向、I_s流向以及U_H的正负并结合霍尔元件的引脚位置可以判定霍尔元件半导体的类型（P型或N型）。反之，若已知I_s流向、U_H的正负以及霍尔元件半导体的类型，可以判定磁感应强度B的方向。

（5）研究霍尔电压U_H与励磁电流I_M之间的关系。

霍尔元件仍位于电磁铁气隙中心，调定$I_s=3.00$mA，分别调节$I_M=100$mA，200mA，…，1000mA（间隔为100mA），分别测量C、D间电压值U_i，计算霍尔电压U_H填入表S25－3，并绘出U_H-I_M曲线，分析磁感应强度B与励磁电流I_M之间的关系。

（6）测量一定I_M条件下电磁铁气隙中磁感应强度B的大小及分布情况。

①调节$I_M=600$mA，$I_s=5.00$mA，调节二位移动尺的垂直标尺，使霍尔元件处于电磁铁气隙垂直方向的中心位置。调节水平标尺至0刻度位置，测量相应的U_i。

②调节水平标尺按表4中给出的位置测量U_i，填入表S25－4。

③根据以上测得的U_i，计算霍尔电压U_H值，根据式（S25－5）计算出各点的磁感应强度B，并绘出$B-X$图，描述电磁铁气隙内X方向上B的分布状态。

［数据及处理］

将数据记录到表S25－2至表S25－5中。

表S25－2　霍尔电压U_H与工作电流I_s的关系

$I_M=300$mA，$C=$______ mT/A

I_s/mA	U_1/mV	U_2/mV	U_3/mV	U_4/mV	$U_H=\frac{1}{4}(\lvert U_1\rvert+\lvert U_2\rvert+\lvert U_3\rvert+\lvert U_4\rvert)$ mV
	$+I_M$，$+I_s$	$-I_M$，$+I_s$	$-I_M$，$-I_s$	$+I_M$，$-I_s$	
1.00					
2.00					
3.00					

续表

I_s/mA	U_1/mV	U_2/mV	U_3/mV	U_4/mV	$U_H=\frac{1}{4}(\lvert U_1\rvert+\lvert U_2\rvert+\lvert U_3\rvert+\lvert U_4\rvert)$ mV
	$+I_M,+I_s$	$-I_M,+I_s$	$-I_M,-I_s$	$+I_M,-I_s$	
4.00					
5.00					
6.00					
7.00					
8.00					
9.00					
10.00					

表 S25－3　工作电流 I_s 与工作电压 U_s 的关系（$I_M=0$mA）

I_s/mA	0.50	1.00	1.50	2.00	2.50	3.00	3.50	4.00	4.50	5.00
U_s/mV										

表 S25－4　霍尔电压 U_H 与励磁电流 I_M 之间的关系（$I_s=3.00$mA）

I_M/mA	U_1/mV	U_2/mV	U_3/mV	U_4/mV	$U_H=\frac{1}{4}(\lvert U_1\rvert+\lvert U_2\rvert+\lvert U_3\rvert+\lvert U_4\rvert)$ mV	B/mT
	$+I_M,+I_s$	$-I_M,+I_s$	$-I_M,-I_s$	$+I_M,-I_s$		
100						
200						
300						
400						
500						
600						
700						
800						
900						
1000						

表 S25－5　电磁铁气隙中磁感应强度 B 的分布（$I_M=600$mA，$I_s=5.00$mA

x/mm	U_1/mV	U_2/mV	U_3/mV	U_4/mV	$U_H=\frac{1}{4}(\lvert U_1\rvert+\lvert U_2\rvert+\lvert U_3\rvert+\lvert U_4\rvert)$ mV	B/mT
	$+I_M,+I_s$	$-I_M,+I_s$	$-I_M,-I_s$	$+I_M,-I_s$		
0						
2						
4						
6						
8						
10						
12						
15						
20						

续表

x/mm	U_1/mV $+I_M$, $+I_s$	U_2/mV $-I_M$, $+I_s$	U_3/mV $-I_M$, $-I_s$	U_4/mV $+I_M$, $-I_s$	$U_H=\frac{1}{4}(\lvert U_1\rvert+\lvert U_2\rvert+\lvert U_3\rvert+\lvert U_4\rvert)$ mV	B/mT
25						
30						
35						
40						
45						
48						
50						

[注意事项]

(1)由于励磁电流较大,所以千万不能将 I_M 和 I_s 接错,否则励磁电流将烧坏霍尔元件。

(2)霍尔元件及二维移动尺容易折断、变形,应注意避免受挤压、碰撞等。实验前应检查两者及电磁铁是否松动、移位,并加以调整。

(3)为了不使电磁铁因过热而受到损害,或影响测量精度,除在短时间内读取有关数据,通以励磁电流 I_M 外,其余时间最好断开励磁电流开关。

(4)仪器不宜在强光、高温、强磁场和有腐蚀性气体的环境下工作和存放。

[问题讨论]

(1)实验中要求霍尔元件与磁场垂直,若不垂直会产生什么影响?

(2)霍尔电压的极性随工作电流及磁感应强度的方向改变如何变化?工作电流与励磁电流同时换向时霍尔电压的极性为什么不变?

(3)实验中如何消除附加电势差的影响?

(4)利用霍尔效应法能测交变磁场吗?为什么?

[补充说明]

本套仪器由 ZKY - HS 霍尔效应实验仪和 ZKY - H/L 霍尔效应螺线管磁场测试仪两大部分组成,如图 S25 - 6 所示。

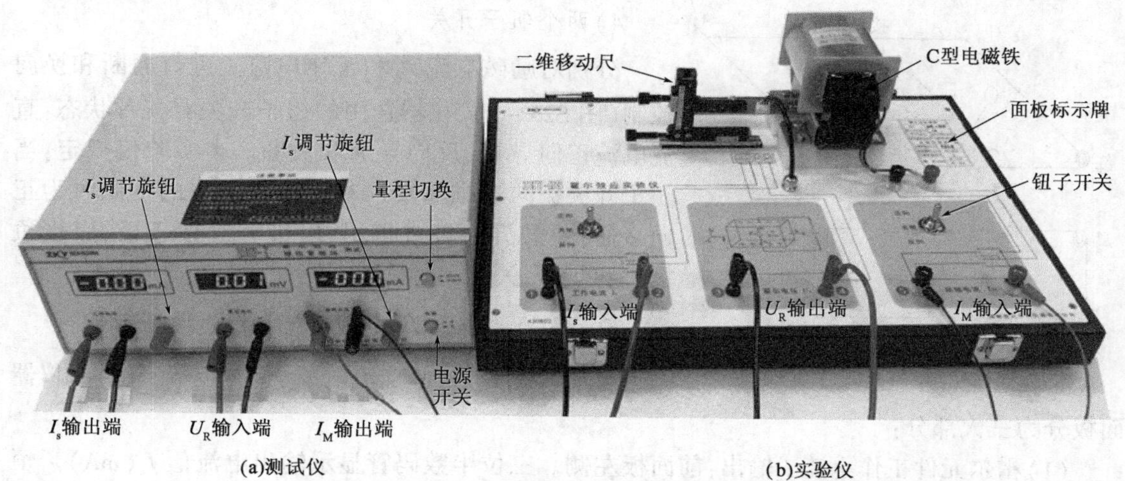

图 S25 - 6　霍尔效应实验仪与测试仪

1. ZKY－HS 霍尔效应实验仪

本实验仪由 C 形电磁铁、二维移动尺及霍尔元件、面板标示牌、两个钮子开关等组成。

1）C 形电磁铁

本实验中励磁电流 I_M 与电磁铁在气隙中产生的磁感应强度 B 成正比。导线绕向（或正向励磁电流 I_M 方向）已在线圈上用箭头标出，可通过“右手螺旋定则”以及磁力线基本沿着铁芯走的性质，确定电磁铁气隙中磁感应强度 B 的方向。

2）二维移动尺及霍尔元件

二维移动尺可调节霍尔元件水平、垂直移动，可移动范围：水平 0～50mm，垂直 0～30mm。霍尔元件相关参数见面板标示牌。

霍尔元件上有 4 只引脚（图 S25－7），其中 1、2 为工作电流（又称为控制电流）端，3、4 为霍尔电压端。同时将这 4 只引脚焊接在印制板上，四个引脚引线的定义见下排，然后引到仪器钮子开关对应的位置。

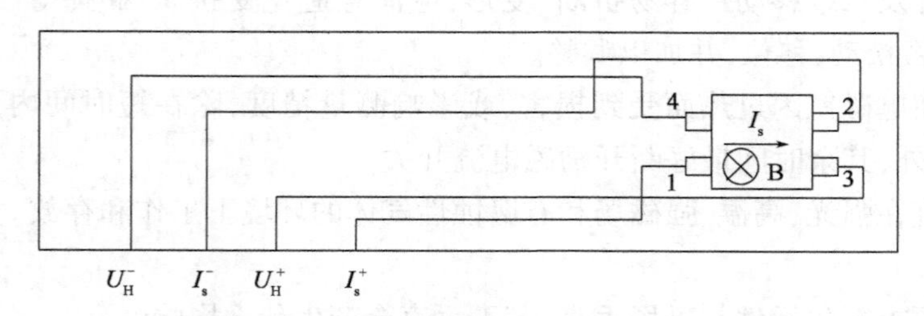

图 S25－7　霍尔元件的外形封装及管脚定义

3）面板标示牌

面板标示牌中填写的内容包括：霍尔元件参数（尺寸、导电类型及材料、最大工作电流）、电磁铁参数（线圈常数 C、气隙尺寸等）。由于本实验中励磁电流 I_M 与电磁铁在气隙中产生的磁感应强度 B 成正比，故用电磁铁的线圈常数 C 可替代测量磁感应强度的特斯拉计。电磁铁线圈常数 C 是指单位励磁电流作用下电磁铁在气隙中产生的磁感应强度，单位 mT/A。若已知励磁电流大小，便能根据公式 $B=CI_M$ 得到此时电磁铁气隙中的磁感应强度。

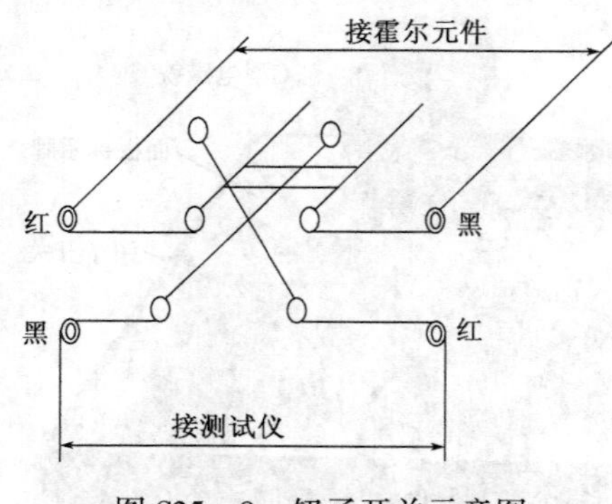

图 S25－8　钮子开关示意图

4）两个钮子开关

分别对励磁电流 I_M 和工作电流 I_s 进行通断和换向控制（图 S25－8），仪器选用的钮子开关有三种状态：直流电流正向导通、反向导通和关断。本实验仪规定：当钮子开关拨向二维移动尺和电磁铁所在的一侧时为正向接通，即电流为红进黑出，电极为红接正极、黑接负极；当钮子开关直立时关断电流。

2. ZKY－H/L 霍尔效应测试仪

仪器背部为 220V 交流电源插座（带熔断丝）。仪器面板分为三大部分：

（1）霍尔元件工作电流 I_s 输出：前面板左侧。三位半数码管显示输出电流值 I_s（mA）。恒流源可调范围：0～10.00mA（用调节旋钮调节）。

(2)霍尔电压 U_H 输入:前面板中部。四位数码管显示输入电压值 U_H(mV)。测量范围:0～19.99mV(量程20mV)或0～199.9mV(量程200mV),可通过按测试仪面板上的量程切换按钮进行切换。在本实验中,只用200mV量程,在实验前请先将霍尔电压测试仪的电压量程调至200mV。20mV量程应配套"螺线管磁场实验仪"使用。

(3)励磁电流 I_M 输出:前面板右侧。三位半数码管显示输出电流值 I_M(mA)。恒流源可调范围:0～1000mA(用调节旋钮调节)。

附 录

附表 1 国际单位制(SI)基本单位(Fundamental units)

物理量 Quantity	名称 Name	符号 Symbol	物理量 Quantity	名称 Name	符号 Symbol
长度 length	米 meter	m	冲量;动量 momentum	牛顿秒 newton second	N · s
质量 mass	千克 kilogram	kg	功;能量 power	焦耳 joule	J(N · m)
时间 time	秒 second	s	功率 power	瓦特 watt	W(J/s)
电流强度 electric current	安培 ampere	A	比热容(质量热容) specific heat	焦耳每千克开尔文 joule per kilogram kelvin	J/(kg · K)
热力学温度 thermodynamic temperature	开尔文 kelvin	K	黏度 viscosity	牛顿秒每平方米 newton second per square metre	N · s/m^2
物质的量 amount of substance	摩尔 mole	mol	电量 charge	库仑 coulomb	C(A · s)
发光强度 luminous intensity	坎德拉 candela	cd	电压;电动势 voltage	伏特 volt	V(W/A)
面积 area	平方米 square meter	m^2	电阻 resistor	欧姆 ohm	Ω(V/A)
体积 volume	立方米 cubic meter	m^3	压强 pressure	帕斯卡 pascal	Pa(N/m^2)
比容 specific volume	立方米每千克 cubic metre per kilogram	m^3/kg	力 force	牛顿 newton	N
频率 frequency	赫兹 hertz	Hz(1/s)	速度 velocity	米每秒 metre per second	m/s
密度 density	千克每立方米 kilogram per cubic metre	kg/m^3	角速度 angular acceleration	弧度每秒 radian per second	rad/s

注:引自国际标准化组织 ISO 发布的《国际单位制指南》。

附表 2　基本物理常数(Physical constants)

类别	物理量 Quantity	符号 Symbol	数值 Value	单位 Unit	相对不确定度 Uncertainty
通用常量	真空中的光速 vacuum speed light	c, c_0	299 792 458	$m \cdot s^{-1}$	精确(zero)
	真空中的磁导率 vacuum permeability	μ_0	$4\pi \times 10^{-7} = 12.566\ 370\ 614\cdots \times 10^{-7}$	$N \cdot A^{-2}$	精确(zero)
	真空中的介电常数 vacuum permittivity	ε_0	$1/(\mu_0 c^2) = 8.854\ 187\ 817\cdots \times 10^{-12}$	$F \cdot m^{-1}$	精确(zero)
	真空中特征阻抗 vacuum impedance	Z_0	$\mu_0 c = 376.730\ 313\ 461\cdots$	Ω	精确(zero)
	牛顿引力常数 gravitational constant	G	$6.673\ \ 84\ (80) \times 10^{-11}$	$m^3 \cdot kg^{-1}\ s^{-2}$	4.7×10^{-5}
	普朗克常数 Planck constant	h	$6.626\ 070\ 040(81) \times 10^{-34}$	$J \cdot s$	1.2×10^{-8}
			$4.135\ 667\ 662(25) \times 10^{-15}$	$eV \cdot s$	6.1×10^{-9}
	普朗克质量 Planck mass	m_P	$2.176\ 470(51) \times 10^{-8}$	kg	2.3×10^{-5}
	普朗克温度 Planck temperature	T_P	$1.416\ 808(33) \times 10^{32}$	K	2.3×10^{-5}
	普朗克长度 Planck length	l_P	$1.616\ 229(38) \times 10^{-35}$	m	2.3×10^{-5}
	普朗克时间 Planck time	t_P	$5.391\ 16(13) \times 10^{-44}$	s	2.3×10^{-5}
电磁常量	基本电荷 basic charge	e	$1.602\ 176\ 565(35) \times 10^{-19}$	C	6.1×10^{-9}
		e/h	$2.417\ 989\ 262(15) \times 10^{14}$	$A \cdot J^{-1}$	6.1×10^{-9}
	库伦常数 Coulomb constant	k_e	9.98774×10^{-19}	$N \cdot m^2 \cdot c^{-2}$	6.1×10^{-9}
	玻尔磁子 Bohr magneton	μ_B	$927.400\ 9994(57) \times 10^{-26}$	$J \cdot T^{-1}$	6.2×10^{-9}
			$5.788\ 381\ 8012(26) \times 10^{-5}$	$eV \cdot T^{-1}$	4.5×10^{-10}
	核磁子 nuclearmagneton	μ_N	$5.050\ 783\ 699(31) \times 10^{-27}$	$J \cdot T^{-1}$	6.2×10^{-9}
			$3.152\ 451\ 2550(15) \times 10^{-8}$	$eV \cdot T^{-1}$	4.6×10^{-10}
	电导量子 conductance quantum	G_0	$7.748\ 091\ 7310(18) \times 10^{-5}$	S	2.3×10^{-10}
原子和核常量	里德伯常数 Rydberg constant	R_∞	$10\ 973\ 731.568\ 508(65)$	m^{-1}	5.9×10^{-12}
	玻尔半径 Bohr radius	a_0	$0.529\ 177\ 210\ 67(12) \times 10^{-10}$	m	2.3×10^{-10}
	电子质量 electron mass	m_e	$9.109\ 383\ 56(11) \times 10^{-31}$	kg	1.2×10^{-8}

续表

类别	物理量 Quantity	符号 Symbol	数值 Value	单位 Unit	相对不确定度 Uncertainty
原子和核常量	经典电子半径 classical electron radius	r_e	$2.8179403227(19)\times10^{-15}$	m	6.8×10^{-10}
	康普顿电子波长 Compton wavelength of electron	λ_C	$2.4263102367(11)\times10^{-12}$	m	4.5×10^{-10}
	μ子质量 Mu mass	m_μ	$1.883531594(48)\times10^{-28}$	kg	2.5×10^{-8}
	τ子质量 Tau mass	m_τ	$3.16747(29)\times10^{-27}$	kg	9.0×10^{-5}
	α粒子质量 Alpha mass	m_α	$6.644657230(82)\times10^{-27}$	kg	1.2×10^{-8}
	质子质量 proton mass	m_p	$1.672621898(21)\times10^{-27}$	kg	1.2×10^{-8}
	中子质量 neutron mass	m_n	$1.674927471(21)\times10^{-27}$	kg	1.2×10^{-8}
物理化学常量	阿伏加德罗常数 Avogadro constant	N_A, L	$6.022140857(74)\times10^{23}$	mol^{-1}	1.2×10^{-8}
	原子质量常量 unified atomic mass constant	m_u	$1.660539040(20)\times10^{-27}$	kg	1.2×10^{-8}
	法拉第常数 Faraday constant	F	96485.33289(59)	$C\cdot mol^{-1}$	6.2×10^{-9}
	摩尔气体常数 mole gas constant	R	8.3144598(48)	$J\cdot mol^{-1}\cdot K^{-1}$	5.7×10^{-7}
	玻尔兹曼常数 Boltzmann constant	k	$1.38064852(79)\times10^{-23}$	$J\cdot K^{-1}$	5.7×10^{-7}
	理想气体的摩尔体积 (273.15K,100kPa) ideal gas mole volume	V_m	$22.710947(13)\times10^{-3}$	$m^3\cdot mol^{-1}$	5.7×10^{-7}
	斯特藩—玻尔兹曼常数 Stefan-Boltzmann constant	σ	$5.670367(13)\times10^{-8}$	$W\cdot m^{-2}\cdot K^{-4}$	2.3×10^{-6}

注:(1)括号内的数字是给定值最后两位数字中的一倍标准偏差的不确定度。

(2)国际数据委员会(CODATA)每四年对基本物理常数进行一次修正,本表节选自经由国家计量科学数据中心发布的《2018年CODATA基本物理常数推荐值全表》。

附表 3 希腊字母表(Greek alphabet)

字母 Alphabet	名称 Name	字母 Alphabet	名称 Name
Α α	Alpha	Ν ν	nu
Β β	beta	Ξ ξ	xi
Γ γ	gamma	Ο ο	omicron
Δ δ	delta	Π π	pi
Ε ε	epsilon	Ρ ρ	rho
Ζ ζ	zeta	Σ σ	sigma
Η η	eta	Τ τ	tau
Θ θ	theta	Υ υ	upsilon
Ι ι	iota	Φ φ	phi
Κ κ	kappa	Χ χ	chi
Λ λ	lambda	Ψ ψ	psi
Μ μ	mu	Ω ω	omega

附表 4 20℃时部分物质的密度 (Young's modulus of some metals at 20℃)

金属 Metals	弹性模量 (Young's modulus) 10^{11} N · m^{-2}	金属 Metals	弹性模量 (Young's modulus) 10^{11} N · m^{-2}
铝(Al)	0.68 ~ 0.70	镍(Ni)	2.14
钨(W)	4.15	铬(Cr)	2.35 ~ 2.45
铁(Fe)	1.9 ~ 2.1	合金钢 (alloy steel)	2.1 ~ 2.2
铜(Cu)	1.03 ~ 1.27	碳钢 (carbon steel)	0 ~ 2.1
金(Au)	0.81	康铜 (constantan)	1.60 ~ 1.66
银(Ag)	0.69 ~ 0.84	铸钢 (cast steel)	1.72
锌(Zn)	0.80	硬铝合金 (duralumin)	0.71

注:弹性模量值与材料的结构,化学成分及其加工方法关系密切。

附表5 20℃时部分物质的密度
(Densities of some materials at 20℃)

物质 Material	密度(Density) ρ/kg · m^{-3}	物质 Material	密度(Density) ρ/kg · m^{-3}
铝(Al)	2698.9	石英(quartz)	2500~2800
锌(Zn)	7140	冰0℃(ice)	917
铬(Cr)	7140	汽车用汽油(gasoline)	710~720
锡(白)(Sn)	7298	乙醚(diethyl ether)	714
铁(Fe)	7874	无水乙醇(ethyl alcohol)	789.4
钢(Steel)	7600~7900	丙酮(acetone)	791
镍(Ni)	8850	甲醇(methanol)	791.3
铜(Cu)	8960	煤油(kerosene)	800
银(Ag)	10492	变压器油(transformer oil)	840~890
铅(Pb)	11342	松节油(turpentine)	855~870
钨(W)	19300	蓖麻油15℃(castor oil)	969
金(Au)	19320	蓖麻油20℃(castor oil)	957
铂(Pt)	21450	石英玻璃(quartz glass)	2900~3000
硬铝(duralumin)	2790	苯(benzene)	879.0
不锈钢(stainless steel)	7910	钟表油(watch oil)	981
黄铜(brass)	8500~8700	纯水0℃(water)	999.84
青铜(bronze)	8780	纯水3.98℃(water)	1000.00
康铜(constantan)	8880	纯水4℃(water)	999.97
软木(soft wood)	220~260	海水(sea water)	1010~1050
纸(paper)	700~1000	牛乳(milk)	1030~1040
石蜡(paraffin)	870~940	无水甘油(glycerin)	1260
橡胶(rubber)	910~960	弗里昂-12(freon) (氟氯烷-12)(fCl alkanes)	1329
硬橡胶(hard rubber)	1100~1400		
有机玻璃(organic glass)	1200~1500	蜂蜜(honey)	1435
煤(coal)	1200~1700	硫磺(sulfur)	1840
食盐(salt)	2140	水银0℃(hg)	13595.5
冕牌玻璃(crown glass)	2200~2600	水银20(hg)	13546.2
普通玻璃(glass)	2400~2700	干燥空气0℃(air) (标准状态)	1.293
火石玻璃(flint glass)	2800~4500		
石英玻璃(quartz glass)	2900~3000	干燥空气20℃(air)	1.205

附表 6 部分物质中的声速(Velocities of some materials)

物质(Material)	声速(Velocity)/m·s^{-1}	物质(Material)	声速(Velocity)/m·s^{-1}
干燥空气(0℃)(air)	331.45	铅(Pb)	1210
氧气(0℃)(oxygen)	317.2	金(Au)	2030
氩气(0℃)(argon)	319	银(Ag)	2680
氮气(0℃)(nitrogen)	337	锡(Sn)	2730
氢气(0℃)(hydrogen)	1269.5	铂(Pt)	2800
一氧化碳(0℃)(CO)	337.1	铜(Cu)	3750
二氧化碳(0℃)(CO_2)	258.0	铝(Al)	5000
乙醚(20℃)(ether)	1006	不锈钢(stainless steel)	5000
乙醇(20℃)(alcohol)	1168	硼硅酸玻璃(borosilicate glass)	5170
水(20℃)(water)	1482.9	熔融石英(fused silica)	5760

注:干燥空气中的声速与温度的关系为 $331.45+0.54t$;固体中的声速为沿棒传播的纵波速度。

附表 7 常用光源对应的谱线波长(Spectrum of some lamps)

颜色 Color	主要谱线的波长(Wavelength)/×10^{-10}m						
	钠灯 Sodium	氢灯 Hydrogen	低压汞灯 Mercury(LV)	亮度 Luminance	高压汞灯 Mercury(HV)	亮度 Luminance	He—Ne 激光 Laser
红 red		6562.8			7081.9 6907.5 6716.5	极弱(very faint) 强(good) 弱(faint)	
橙 orange			6234.4	弱(faint)	6234.4 6123.3 6072.6		6328 强光(excellent)
黄 yellow	5895.9 5890.0		5790.7 5769.6	强(good) 强(good)	5889.6 5872.0 5859.4 5790.7 5769.6	弱(faint) 弱(faint) 弱(faint) 强(good) 强(good)	
绿 Green			5460.7	强(good)	5675.9 5460.7 5365.1 5354.1	弱(faint) 强(good) 弱(faint) 弱(faint)	

附表 8 部分物质的比热容(Specific heat of some materials)

物质(Material)	温度(Temperature)/℃	比热容(Specific heat)	
		kJ/(kg·K)	kcal[①]/(kg·℃)
铅(Pb)	25	0.128	0.0306
	20	0.128	0.0306
铂(Pt)	20	0.134	0.0326
银(Ag)	20	0.234	0.0566

续表

物质(Material)	温度(Temperature)/℃	比热容(Specific heat)	
		kJ/(kg·K)	kcal①/(kg·℃)
铜(Cu)	20	0.385	0.0920
锌(Zn)	20	0.389	0.0929
镍(Ni)	20	0.481	0.115
铁(Fe)	20	0.481	0.115
铝(Al)	20	0.895	0.214
黄铜(brass)	0	0.37	0.0883
	20	0.38	0.0917
康铜(constantan)	18	0.409	0.0977
钢(steel)	20	0.447	0.107
生铁(cast iron)	0~100	0.54	0.13
云母(mica)	20	0.42	0.1
玻璃(glass)	20	0.585~0.920	0.14~0.22
石墨(graphite)	25	0.707	0.169
石英玻璃(quartz glass)	20~100	0.787	0.188
石棉(asbestos)	0~100	0.795	0.19
橡胶(rubber)	15~100	1.13~2.00	0.27~0.48
石蜡(paraffin)	0~20	2.91	0.694
水银(mercury)	0	0.1395	0.03337
	20	0.139 0	0.03326
弗里昂—12(freon)	20	0.84	0.2
汽油(gasoline)	10	1.42	0.34
	50	2.09	0.5
变压器油(transformer oil)	0~100	1.88	0.45
乙醇(alcohol)	0	2.3	0.55
	20	2.47	0.59
乙醚(diethyl ether)	20	2.34	0.56
甘油(glycerin)	18	2.43	0.58
甲醇(methanol)	0	2.43	0.58
	20	2.47	0.59
冰(ice)	0	2.596	0.621
水(water)	0	4.219	1.0093
	20	4.175	0.9988
	100	4.204	1.0057
空气(定压)(air)	20	1	0.24
氢(定压)(hydrogen)	20	14.25	3.41

附表9　几种纯金属的“红限”波长及功函数

(Limited wavelength and work function of some pure metals)

金属(Metal)	λ_0/nm	W/eV	金属(Metal)	λ_0/nm	W/eV
钾(K)	550	2.2	汞(Hg)	273.5	4.5
钠(Na)	540	2.4	金(Au)	265	5.1
锂(Li)	500	2.4	铁(Fe)	262	4.5
铯(Cs)	460	1.8	银(Ag)	261	4

附表10　可见光波和频率与颜色对应关系

(The relationship between wave and frequency of visible light and color)

颜色 Color	中心频率 (Center frequency)/Hz	中心波长 (Center wavelength)/10^{-10} m	波长范围 (Wavelength range)/10^{-10} m
红(red)	4.5×1014	6600	7800~6200
橙(orange)	4.9×1014	6100	6220~5970
黄(yellow)	5.3×1014	5700	5970~5770
绿(green)	5.5×1014	5400	5770~4920
青(cyan)	6.5×1014	4800	4920~4700
蓝(blue)	7.0×1014	4300	4700~4550
紫(purple)	7.3×1014	4100	4550~3900

注:可见光波长范围:$3900\times10^{-10}\sim7800\times10^{-10}$ m;可见光频率范围:$7.7\times10^{14}\sim3.8\times10^{14}$ Hz

附表11　部分型号霍尔元件参数表

(Properties of some Hall components)

参数名称 Property	符号 Symbol	单位 Unit	HZ—1型 Ge(111)	HZ—2型 Ge(111)	HZ—4型 Ge(100)	HT—1型 InSb	HS—1型 InAs
电阻率 resistivity	ρ	Ω·cm	0.8—1.2	0.8—1.2	0.4—0.5	0.003—0.01	0.01
几何尺寸 size	$l\times b\times d$	mm×mm×mm	8×4×0.2	4×2×0.2	8×4×0.2	6×3×0.2	8×4×0.2
输入电阻 input resistance	R_1	Ω	110±20%	110±20%	45±20%	0.8±20%	1.2±20%
输出电阻 output resistance	R_0	Ω	100±20%	100±20%	40±20%	0.5±20%	1.0±20%
灵敏度 sensitivity	K_H	mV·mA^{-1}·kGs^{-1}	>1.2	>1.2	>0.4	0.18±20%	0.1±20%
不等位电阻 difference resistance	r_0	Ω	<0.07	<0.05	<0.02	<0.005	<0.003

续表

参数名称 Property	符号 Symbol	单位 Unit	HZ—1 型 Ge(111)	HZ—2 型 Ge(111)	HZ—4 型 Ge(100)	HT—1 型 InSb	HS—1 型 InAs
寄生直流电势 parasitic potential	U_0	μV	<150	<200	<100		
额定控制电流 limited current	l	mA	20	15	50	250	200
霍尔电势温度系数 Hall temperature coefficient	α	℃$^{-1}$	0.04%	0.04%	0.03%	-1.5%	
内阻温度系数 resistance temperature coefficient	β	℃$^{-1}$	0.5%	0.5%	0.3%	0.5%	
热阻 thermo-resistance	R_Q	℃ · m · W^{-1}	0.4	0.25	0.1		
工作温度 work temperature	T	℃	-40 -45	-40 -45	-40 -75	0 -40	-40 -60

附表 12　水在不同温度 t 下的饱和蒸气压 p_m (Saturated vapor pressure p_m of water at different temperatures t)

t/℃	p_m/mmHg	t/℃	p_m/mmHg	t/℃	p_m/mmHg	t/℃	p_m/mmHg
0	4.58	10	9.21	20	17.54	30	31.82
1	4.93	11	9.84	21	18.65	31	33.70
2	5.29	12	10.52	22	19.83	32	35.66
3	5.69	13	11.23	23	21.07	33	37.73
4	6.10	14	11.99	24	22.38	34	39.90
5	6.54	15	12.79	25	23.76	35	42.18
6	7.01	16	13.63	26	25.21	36	44.56
7	7.51	17	14.53	27	26.74	37	47.07
8	8.05	18	15.48	28	28.35	3	49.69
9	8.61	19	16.48	29	30.04	839	52.44
10	9.21	20	17.54	30	31.82	40	55.32

附表 13　不同温度下纯水的密度 ρ_w

(Density ρ_w of pure water at different temperatures)

ρ_w \ t/℃ t/℃	0	10	20	30
0	0.999867	0.999727	0.998229	0.995672
1	0.999926	0.999632	0.998017	0.995366
2	0.999968	0.999524	0.997795	0.995051
3	0.999992	0.999404	0.997563	0.994728
4	1.000000	0.999271	0.997321	0.994397
5	0.999992	0.999126	0.997069	0.994058
6	0.999968	0.998969	0.996808	0.993711
7	0.999929	0.998800	0.996538	0.993356
8	0.999876	0.998621	0.996258	0.992993
9	0.999808	0.998430	0.995969	0.992622
10	0.999727	0.998229	0.995672	0.992244

参考文献(References)

[1] 黄昆,韩汝琦. 固体物理学[M]. 北京:高等教育出版社,1988.

[2] 全国统计方法应用标准化技术委员会. 数值修约规则与极限数值的表示与判定:GB 8170—2008[S]. 北京:中国标准出版社,2009.

[3] 中国石油化工集团. 油产品运动粘度测定法和动力粘度计算法:GB/T 265—1988[S]. 北京:中国标准出版社,1988.

[4] Horowitz P, Hill W. The art of electronics[M]. 2nd ed. Cambridge:Cambridge University press, 1989.

[5] 林木欣. 近代物理实验教程[M]. 北京:科学出版社,2000.

[6] ISO/IEC Guide 98 - 3:2008 (en) Uncertainty of measurement — Part 3: Guide to the expression of uncertainty in measurement (GUM:1995).

[7] 陈惠钊. 粘度测量[M]. 修订版. 北京:中国计量出版社,2003

[8] 李保春,周海涛,董有尔. 大学物理实验[M]. 2 版. 合肥:中国科学技术大学出版社, 2016.

[9] 程守洙,江之永. 普通物理学[M]. 7 版. 北京:高等教育出版社,2016.

[10] 雷玉堂. 光电检测技术[M]. 2 版. 北京:中国计量出版社,2009.

[11] Bauer W, Westfal G D. 现代大学物理[M]. 英文版. 北京:机械工业出版社,2003.

[12] 李学慧. 大学物理实验[M]. 2 版. 北京:高等教育出版社,2012.

[13] 王银峰, 陶纯匡, 汪涛,等. 大学物理实验[M]. 北京:机械工业出版社, 2005.

[14] 杜义林. 实验物理学[M]. 合肥:中国科学技术大学出版社, 2006.

[15] 徐建强 ,韩广兵. 大学物理实验[M]. 3 版. 北京:科学出版社, 2020.

[16] 杨述武,孙迎春,沈国土. 普通物理实验(3)·光学部分[M]. 5 版. 北京:高等教育出版社, 2016

[17] 王宏亮. 大学物理实验[M]. 2 版. 北京: 机械工业出版社,2014.

[18] 刘跃,张志津. 大学物理实验[M]. 2 版. 北京: 北京大学出版社,2010.

[19] 赵凯华,陈熙谋. 电磁学[M]. 3 版. 北京:人民教育出版社,2011.

[20] 史金辉,邢健,张晓峻,等. 大学物理实验双语教程[M]. 哈尔滨:哈尔滨工程大学出版社,2014.

[21] 谢超然. 大学物理实验[M]. 北京:机械工业出版社,2015.

[22] 武颖丽,李平舟. Experimental college physics[M]. 西安:西安电子科技大学出版社,2016.

[23] 付浩,Balflour E A. Introductory physics experiments for undergraduates[M]. 北京: 科学出版社,2017.

[24] 李隆,昝会萍, 张琳丽,等. 大学物理实验[M]. 北京:高等教育出版社,2018.

[25] 李香莲,等. 大学物理实验[M]. 2 版. 北京:高等教育出版社, 2018.

[26] 王颖,戴俊. College physics experiment. [M]. 南京:南京大学出版社,2020.